T0073850

Science Wars

Science Wars

The Battle over Knowledge and Reality

Steven L. Goldman

Oxford University Press is a department of the University of Oxford. It furthers
the University's objective of excellence in research, scholarship, and education
by publishing worldwide. Oxford is a registered trade mark of Oxford University
Press in the UK and certain other countries.

Published in the United States of America by Oxford University Press
198 Madison Avenue, New York, NY 10016, United States of America.

Library of Congress Control Number: 2021942717

ISBN 978–0–19–751862–5

DOI: 10.1093/oso/9780197518625.001.0001

Contents

Acknowledgments   vii

Introduction: Why Science *Wars*?   1

1. Knowledge as a Problem   7

2. Is There a Scientific Method?   21

3. Was Galileo Right and the Catholic Church Wrong?   36

4. Newton and Knowledge of the Universe   50

5. Science Influences Philosophy   64

6. Science and Social Reform in the Age of Reason   80

7. What Is Science About?   101

8. The Knowledge Problem in Mature Science   115

9. Scientific Realism and the Romantic Reaction Against Reason   130

10. Early Twentieth-Century Philosophy of Science   145

11. Einstein Versus Bohr on Reality   161

12. In Quest of the Thinker of Science   178

13. A New Image for Science   194

14. The Opening Phase of the Science Wars   209

15. Taking Sides   226

16. The Science Wars Go Public   243

Notes   259
Bibliography   269
Index   275

Acknowledgments

This book is a much revised and expanded version of the audio/video course "Science Wars: What Scientists Know and How They Know It" that I created in 2006 for The Teaching Company, now called The Great Courses Company. That course, in turn, was based on the history and philosophy of science courses that I taught at Lehigh University during my thirty-nine years there as Andrew W. Mellon Distinguished Professor in the Humanities. I thank my wife Phoebe Weisbrot and my granddaughter Chagit Barash for encouraging me to write this book, which I dedicate to the memory of my late wife Risa Ebert Goldman.

Science Wars

Introduction: Why Science *Wars*?

Can human beings know anything, and if so, what and how? This question is really the most essentially philosophical of all questions. (Bertrand Russell)[1]

Doesn't all science live on this paradoxical slope to which it is doomed by the evanescence of its object in its very apprehension, and by the pitiless reversal that the dead object exerts on it? (Jean Baudrillard)[2]

What do scientists know and how do they know it? Straightforward questions, surely, and important ones, yet over the four hundred–year history of modern science, no answers have stood up to critical scrutiny.

No theory of how science works—no philosophy of science—has won universal acceptance. Scientists typically describe their work as producing objective knowledge of reality, knowledge that is, if not certainly true, then converging on it. Critics argue that what scientists call knowledge is actually pragmatically justified, probabilistic, subjective interpretations of experience. This may seem an abstract intellectual issue, but it is not.

Over the past two hundred years, society has become increasingly dependent on science-related technologies and the practice of science has become deeply embedded in social, political and economic institutions. Ambiguity about the nature of scientific knowledge thus has profound implications for the role of science in society, especially in the formulation of effective science-relevant public policies. The need for such policies is of particular moment today, relating, among other issues, to pandemics, global warming, energy alternatives, environmental degradation, and commercial applications of increasingly powerful biotechnologies, nanotechnologies, robotics, and artificial intelligence software.

For their part, scientists argue that policy decisions in these areas must be determined based on scientific knowledge because it is objective, hence value neutral. If this argument is contested, however, on the grounds that scientific knowledge is not objective, that it is itself political because it is a product of value judgments by members of an elite community, then science cannot be the neutral arbiter of controversial sociopolitical action agendas. Furthermore, if scientific theories are not true in the sense of corresponding to reality, that strengthens the demands of religious fundamentalists for the inclusion in science education of what the fundamentalists believe to be true about reality.

In this book, I argue that an understanding of how scientists produce knowledge has proven elusive because there is a logical inconsistency at the heart of modern science. Modern science is based on a conflation of deduction and induction, rationalism and empiricism, realism and conventionalism. The self-proclaimed goal of modern science from its beginning has been to use experience to transcend experience and reveal the mind-independent causes of experience. This cannot be done logically, but success is routinely proclaimed based on experimental confirmations of theory-based predictions, thereby proving the reality of scientific objects. I will argue that the contradiction at the heart of modern science is the result of equivocation over how the word "knowledge" is to be defined. This is pivotal because the definition of "knowledge" affects what "truth," "rationality," and "reality" mean.

Words matter. They matter because words often carry connotations, presuppositions and value judgments that shape how we respond to them. Consciously and unconsciously, we respond very differently to "knowledge" than we do to "belief" and "opinion." We use each of these words *intending* that they have different effects on listeners and readers. A scientific conviction typically is characterized as a matter of knowledge, while a religious conviction typically is characterized as a matter of belief and a political conviction as a matter of opinion. To claim to *know* something, especially to know something scientifically, is taken by just about everyone to be a fundamentally different kind of claim from stating opinions or beliefs about those same matters. "Knowledge" has more rhetorical force than the words "belief" and "opinion." Why is that?

Knowledge trumps belief and opinion because "knowledge" carries the connotation of a correlation with truth, and through truth with reality, with the way things "really" are *regardless* of beliefs and opinions. This elevates knowledge above belief and opinion, and it elevates people who possess knowledge, or can successfully claim to possess it, above those who "merely" express opinions and beliefs. Allowing the word "knowledge" this privileged usage implicitly assumes that knowledge really is essentially different from beliefs and opinions, but is this assumption justified? We can choose to *define* "knowledge" as statements that are necessarily true while opinions and beliefs may be false. But we could also choose to define "knowledge" as those opinions and beliefs for whose truth convincing but not compelling reasons can be given. In the latter case, "knowledge" would lose some of its rhetorical force because it, too, might be false and because what seems a convincing reason to one person might not be convincing to another.

From the time of the pre-Socratic Greek philosophers in the fifth century BCE to the present, the dominant definition of "knowledge" in Western culture makes knowledge fundamentally different from opinion and belief. On this view, there is only one truth of any matter and that is known only to the people who possess knowledge. I will call this definition of "knowledge" knowledge in the strong sense. A rival definition, also proposed in the pre-Socratic period, was that "knowledge" was a name for opinions and beliefs for whose truth more or less good reasons could be given, but never with certainty. I will call this definition of "knowledge" knowledge in the weak sense. Since 1600, natural philosophy, now called modern science, has promoted itself as the only source of knowledge for human beings. The historian Stephen Gaukroger began his history of the cultural context out of which modern science emerged as follows: "One of the most distinctive features of the emergence of a scientific culture in modern Europe is the gradual assimilation of all cognitive values to scientific ones."[3] Only scientists possess knowledge, only scientists possess truth, only scientific reasoning is rational, and only scientists possess an understanding of reality. This poses a profound challenge to all other responses to human experience, to art and literature, for example, no less than to religion and philosophy.

Given what is at stake, what *is* the correct definition of "knowledge"? *Is* knowledge fundamentally different from opinion and belief? Attempts to answer this question bump up against the fact that there is no Absolute Dictionary in which to look up the correct meaning of words. There is no uniquely correct meaning of a word because what words mean is a function of how people choose to use them. People routinely claim to have knowledge about matters that concern us, matters about which there are diverse opinions and beliefs. How can we be sure that their knowledge claims are valid, requiring that we give up our opinions and beliefs about those matters? How can we be sure that people are not just using the "knowledge" word in order to intimidate us into giving up our opinions and beliefs in favor of what are, in the end, only their opinions and beliefs? For that matter, how do we know that there actually *is* such a thing as knowledge in the strong sense, statements about the world, for example, that are necessarily true? Perhaps knowledge is not something universal and objective that transcends opinion and belief, but just those opinions and beliefs that are most strongly held in a given community at a particular time.

The question of whether *knowledge* of the world is possible cannot be answered without first settling what "knowledge" means in science. This will then determine what "objectivity," "truth," "reality," and "rationality" mean; it also affects what the object of scientific knowledge is. What is scientific knowledge, knowledge *of*? This relates to what scientists mean by "the world" and "nature." Do "world" and "nature" refer to that which is independent of human experience and reasoning, or to some conventionally conceptualized aspect of human experience? The definition we choose for "the world" makes a major difference to what we mean by knowledge of the world and to the criteria by which we justify claims of knowledge of the world, as opposed to having opinions or beliefs about the world. All of these questions place the word "knowledge" precisely at what the author Alex Rose called the "mysterious intersection of epistemology and ontology." At this intersection, our conceptions of knowledge and reality become mutually implicating. A commitment to knowledge as necessary truth, for example, implies a deterministic metaphysics in which time is an illusion: everything that happens in time is already implicit in the beginning. A commitment to knowledge as contingent truth, on the other hand, implies an evolutionary metaphysics in which time is real, a name for the emergence of unpredictable happenings. The intersection of epistemology and ontology is thus the intersection of our conceptions of knowing and of being.

Knowing is a process in the human mind, while being is the most universal feature of a reality conceived to be wholly independent of human experience. How can knowing say something about what is independent of our conceiving and our saying? There is an issue here of the most profound importance for any discussion of truth, rationality, and reality. Can the mind reason about the world in a way that transcends its biology and its socialized experience, allowing it to *know* a mind-independent reality? Is a knowable, mind-independent reality just an idea of ours?

The problem of defining a cluster of words cognate with the word "knowledge" is central to any understanding of science as knowledge of the world. Natural scientists routinely claim that their accounts of the world tell us what is real and what is not real, how the world works, and what it is *reasonable* to say about the world. Such accounts, if they are valid, surely deserve to be called knowledge, not opinions, but how can we know that they are valid? When scientists use the term "knowledge" do they mean it in the strong sense such that scientific knowledge claims are objective and accounts of reality, or in the weak sense, in which case they are inter-subjective and accounts of human experience only? The unhelpful answer given by scientists, as attested by the history and philosophy of modern science, is "Both"!

Since the mid-nineteenth century, growth in the *explanatory scope, predictive success*, and *practical applications* of scientific theories have been awe inspiring. This makes a strong pragmatic case that claims for the truth of such theories are far more than just the opinions and beliefs of the scientific communities that hold them. Through its association with technologies that have literally transformed the conditions of human existence, the idea of science has become deeply entrenched in contemporary social institutions and individual consciousness. This reinforces the prevailing identification of science with knowledge of the world as it really is. Social processes, including the way science is taught at all levels and how it is presented to the public, reinforce the claim that scientists possess objective knowledge of reality rather than empirically justified opinions and beliefs about experience. But how can this claim be verified, given that the human mind has no access to such a reality except through individual and collective experience, which science tells us is totally different from the way things "really" are?

Can scientists claim to know reality *because* their theories provide satisfying explanations of experience, make correct predictions, and give us a measure of (technological) control over experience? Logically, and historically, the answer is "No." Even all three of these accomplishments together do not provide proof that a theory is either about a mind-independent reality or that it is fundamentally different from opinion and belief. The history of science shows that theories that we now consider wrong were once considered true precisely because, for some period of time, they provided satisfying explanations, made correct predictions, and enabled new technologies. This suggests that current theories, as new experiences accumulate, also will be replaced by newer theories as all previous theories have been. And with their replacement, may not new realities be proclaimed (created?), as has been the case repeatedly since modern science emerged?

Any claim to universal and certain knowledge of reality is also belied by the nature of logical reasoning. Since the nineteenth century an increasingly strong consensus has grown among logicians, mathematicians, and philosophers that logical reasoning cannot prove the existence outside the mind of the conclusions of even the most rigorous of deductive arguments. All that such arguments can do is establish the logically necessary truth of their conclusions given the certain truth of their premises together with the truth of all the definitions, axioms, and postulates on which these premises ultimately depend. To avoid an infinite regress, however, the truth of these definitions, axioms and postulates must be stipulated or claimed to be self-evident, on nonlogical and non-empirical grounds.

All this suggests that knowledge in science cannot mean knowledge in the strong sense, but only knowledge in the weak sense: contextually justifiable opinions and beliefs. But how can successful explanation, prediction and control just be educated opinion, "miraculously" correlated only with human experience and not with the real underlying causes of experience? Isn't it reasonable to claim that our theories are at least converging on a definitive account of reality and thus in principle at least are knowledge in the strong sense?

The approach adopted in this book to answering the question of what scientists know and how they know it is historical. Its chapters trace changing conceptions of and controversies over what scientific knowledge is, what it is about, and how it is acquired, from the seventeenth century rise of modern science through the twentieth century. A historical approach allows us to watch unfold *within* the evolving practice of science a recognition that the nature of scientific knowledge is problematic by way of repeated attempts to clarify it. What we also see by adopting a historical approach is the persistence over centuries of opposing solutions to the problem of knowledge in general and of scientific knowledge in particular. This persistence itself calls into question any simplistic account of science as a progressive revelation of reality.

The first chapter locates modern science's conflicted conception of knowledge in a "perpetual battle" that has been raging for some twenty-five hundred years among Western philosophers over the meanings of "knowledge," "truth," "reality," and "rationality." Chapters 2–4 describe the very different conceptions of scientific knowledge promoted by prominent founders of modern science: Francis Bacon, Rene Descartes and Galileo Galilei, Christian Huygens, Isaac Newton, and Gottfried Wilhelm Leibniz.

Chapter 5 describes philosophical responses to the knowledge problem that were accommodated to the emergence of modern science, from Hobbes to Kant, while Chapter 6 describes the impact of modern science on eighteenth-century European intellectuals, leading to the creation of the social sciences, the proclamation of an Age of Reason, and proposals for social reform.

Chapters 7–9 describe nineteenth-century science, philosophy of science, and the Romantic reaction against both modern science and the science-based apotheosis of reason inherited from the eighteenth century. Chapters 10–16 range over twentieth-century defenses of, and attacks on, science as producing objective knowledge of the world, the openly proclaimed "Science Wars" of the closing decades of the century, and the implications of equivocation over the meaning of "knowledge" in science for society and for formulating public policies.

1
Knowledge as a Problem

The knowledge problem was born with Western philosophy, and like Jacob and Esau wrestling for primacy in their mother's womb, it was born into conflict.

Early in Greek philosophy, circa 475 BCE, Parmenides of Elea contrasted two irreconcilable approaches to knowledge. As laid down in his work *On Nature*, there was the way leading to logically necessary truth, his way of course, and then there was the way of mere opinion, the way taken by his misguided contemporary, Heraclitus. A century or so later Plato used this contrast to distinguish philosophers, lovers of wisdom who seek truth about a changeless reality, from sophists, who settle for opinions about an endlessly changing experience.

The Sophist is a Platonic dialogue dominated not by Socrates, but by an unnamed philosopher from Elea who is visiting Athens. To make the connection with Parmenides still clearer, Plato has the visitor declare that Parmenides is his [philosophical] "father." In the course of the dialogue, the visitor expounds Plato's version of Parmenides' deductive logic–driven philosophy and contrasts it with the teachings of sophists, pretend philosophers. At one point, the visitor describes the relationship between philosophers and sophists as "a sort of Battle of Gods and Giants because of the dispute they have with each other about [the nature of] beinghood [reality]."[1]

The Giants, with whom the sophists are allied,

> drag all things down out of the heavens and the invisible realm, literally grabbing rocks and trees with their hands. They grasp all such things and maintain strenuously that that alone *is* [has being and thus is real] which allows for some touching and embracing. For they mark off beinghood [reality] and body [materiality] as the same; and if anyone from the other side says that something is [real] that has no body, they despise him totally and don't want to listen to anything else.[2]

The Giants, in short, are empiricists, identifying reality with the material world of ordinary human experience.

The philosophers, allies of the Gods,

defend themselves [against the Giants] very cautiously out of some invisible place on high, forcing true beinghood to be certain thought-things and disembodied forms. But the bodies [the material objects] of their opponents and what these men call truth, they bust up into small pieces in their arguments and call it, instead of beinghood, some sort of swept-along becoming.[3]

The Gods are rationalists for whom reality is immaterial, is unchanging, and transcends human experience.

"*And between these two, Theaetetus, a tremendous sort of battle over these things has forever been joined*" [my italics],[4] a battle between two irreconcilable conceptions of reality, knowledge, and our ability to know truths about reality. Such truths lie at the intersection of ontology and epistemology, which is "mysterious" if it is claimed that the mind can know what is independent of the mind. How "real" is defined implicates the knowability of the real, and how "knowledge" is defined implicates what is real. For Plato, nothing less can be the object of philosophy than seeking universal, necessary, and certain knowledge of reality, defined as that which exists external to and independent of the human mind. Reality is analogous to mathematical objects—triangles, circles, spheres—which are eternal, immaterial, and changeless. Like knowledge of mathematics, knowledge of reality can be known only by a properly reasoning mind.

In the dialogue *Phaedrus,* Plato has Socrates say that "the reality with which true knowledge is concerned, [is] a reality . . . intangible but utterly real, apprehensible only by the intellect," without any involvement of the body and its empirical experience. Because the Gods in their "heavens" are detached from all materiality, they are able to see

absolute justice and discipline and knowledge, not the knowledge which is attached to things which come into being [and pass away], nor the knowledge which varies with the objects which we [based on our empirical experience] now call real, but the absolute knowledge which corresponds to what is absolutely real in the fullest sense.[5]

To the extent that the mind is able to rise above the body and its material experiences, it can "see" absolute knowledge of the real as the Gods do, and this is the goal of the philosopher. In the dialogue *Theaetetus,* which has as its subject the nature of knowledge, Socrates says that the senses give us empirical experience of a relentlessly changing natural world. Nature is a world of becoming and thus cannot be an object of certain knowledge, only of probable opinion and belief. But the mind, Socrates insists, perceives "through its own instrumentality,"[6] with no involvement of the body and its senses, that which makes unchanging knowledge possible because its object—echoing Parmenides—is the world of changeless Being.

The *Theaetetus* appears to end inconclusively, with Socrates agreeing that defining "knowledge" as "true opinions with a reason" comes closest to what knowledge is. However, what Plato means here by "a reason" is a deductive logical demonstration, and it is this that transforms true opinions into knowledge in the strong sense. In the *Meno*, Socrates says that

> true opinions are a fine thing and do all sorts of good so long as they stay in their
> place, but they will not stay [there] long. They run away from a man's mind, so they
> are not worth much until you tether them by working out the reason [why they are
> necessarily true] . . . once they are tied down they become knowledge.[7]

For Plato, truth is identified with knowledge in the strong sense: it is universal, necessary, and certain. Knowledge "does not reside in the sense impressions but in our reflection upon them."[8] This "reflection" is a distinctive way of reasoning that is the hallmark of the true philosopher. Plato calls it dialectical reasoning, and it alone is capable of grasping universal, necessary, and certain truths about reality.[9] The dialectical reasoner, "applying his pure unadulterated thought to the pure and unadulterated [wholly ideal] object" rises above hypotheses and assumptions and making "no use whatsoever of any objects of sense, but only of pure ideas" moves on "through ideas to ideas and ending with ideas."[10] The mind, from within itself, has the ability to comprehend reality, independently of the body and its experience of the material world to which the body, but not the mind, belongs.

Mathematics, for Plato, comes very close to exemplifying knowledge in the form of 'ideas about ideas ending in ideas.' Mathematics gives us universal, necessary, and certain knowledge because it uses deductive reasoning only, and mathematical objects are eternal and changeless. Where mathematics falls short is in its dependence on diagrams for geometric proofs and on hypotheses whose necessary truth is left as self-evident, not explained. For Plato, there is a slightly higher form of knowledge that the human mind can attain than mathematical knowledge, knowledge that is wholly ideal, the product of "pure" intellection.[11]

Because of the way that Plato defined "reality," and coordinately "knowledge," "truth," and "rationality," there can be no knowledge of nature because nature is constantly changing. To claim *knowledge* of something that is always becoming something else is self-contradictory. Knowledge can only have the changeless as its object, and it is the philosopher's ability to transcend sense-based experience and apprehend truths about changeless being that makes universal, necessary, and certain knowledge of reality possible. We can have more or less useful opinions and beliefs about the natural world, but not knowledge.

Aristotle was in full agreement with Plato that only what was necessarily true and "could not be otherwise" can be called knowledge. "What is understandable and understanding," he wrote in his *Posterior Analytics*, "differ from what is opinable and opinion because understanding is universal and through necessities, and what is necessary cannot be otherwise."[12] By contrast, opinion "is about what is true or false, but can also be otherwise."[13] In the *Nicomachean Ethics* he wrote: "We all suppose that what we know is not capable of being otherwise . . . Therefore the object of knowledge is of necessity . . . Therefore it is eternal . . . and things that are eternal are ungenerated and imperishable."[14]

Aristotle was in full *disagreement* with Plato, however, on changeless immaterial Being as the only possible object of knowledge. For Aristotle, reality was the ever-becoming material world that Plato dismissed as unreal and unknowable. "The most paradoxical thing of all," Aristotle claimed, "is the statement that there are certain things besides those in the material universe, and that these are the same as sensible things except that they are eternal while the latter are perishable."[15] Reality, for Aristotle, is the sum total of all of the particular material objects "out there" beyond the mind, and nothing else. Everything that the human mind knows about this reality begins with sense experience of individual material objects. Knowledge of nature in the strong sense is nevertheless possible because the human mind is able to acquire directly from its experience of individuals the universal truths that are required for reasoning deductively about the world.[16]

It is crucial for an understanding of Aristotle's position, and of modern scientific reasoning as well, to appreciate how different inductive and deductive reasoning are.

In a deductive argument, *if* the premises are true, then the conclusion *must* be true; the truth of the conclusion follows necessarily from the truth of the premises. But a deductive argument requires that at least one of its premises be a universal statement that is known to be true or is taken to be true hypothetically. If we accept as true that 'All men are mortal' and accept also that 'Socrates is a man,' then it *must* be true that 'Socrates is mortal.' If we accept that Euclid's definitions, axioms, and postulates are self-evidently true, then all of the hundreds of theorems of Euclid's geometry are necessarily true because they are deduced from true premises. Note that to reason deductively *about the world*, which Aristotle said was possible, we would have to know the truth of some universal statements about the world. But Aristotle held that all our knowledge of the world comes from sense experience, which is always of individuals, never of universals, so how can we know the truth of universal statements about the world?

By contrast with deduction, in an inductive argument, even if all the premises are true the conclusion can still be false. Even if every reported sighting of a swan by European observers for over thousands years was of a white swan, it is not necessarily true that all swans are white. Based on the evidence available for all that time, it was a valid inductive inference from experience that all swans are white, and a very *reasonable* generalization, but it turned out to be false when in 1697 the Dutch explorer Willem de Vlamingh discovered black swans in Australia. A crucial point to keep in mind throughout this book is that nothing can be *deduced* from particular observations, no matter how numerous. More or less probably true inductive generalizations can be drawn from observations, but no necessary truths, as required by knowledge in the strong sense.

We are free to *assume* that some universal statements about nature—the conservation of matter and energy, definitions of space and time—are true in order to *allow* reasoning about nature deductively. We might do this in order to draw deductive inferences from them and then compare those inferences to experiential facts, a strategy that became and remains central to modern science. In this way, the gulf between induction and deduction seems to be bridged, but it is hypothetical only, *contingent upon the assumed truth of universal statements based on generalizations from experience*. It is really inductive reasoning disguised as deductive reasoning because the truth of the assumptions is just that: assumed. How, then, could Aristotle claim that deductive knowledge of nature was possible if he also held that all knowledge of nature begins with sense experience of particulars?

The answer for Aristotle was that universals existed outside the mind, but only *in* individual material objects, not independent of material objects as they were for Plato. Individual objects somehow embodied the universals of which they were individual exemplifications, so universals were really "out there," not just conventional group names. That is the ontological part of Aristotle's solution to the problem of universals. The epistemological solution is his claim that the human mind possessed a special mental faculty that he called the active, or agent, intellect. This enabled our minds to recognize the universals that are *in* individuals, to "see" in an individual dog the universal dog, its variety, species, and genus, for example. Through the operation of the active intellect, universals "come to rest in the soul [mind]."[17] We know their truth directly without further need of logical demonstration.

The universal "first principles" required for deductive reasoning of any kind are somehow arrived at non-inferentially from our experience of the world of particular objects.[18] According to Aristotle, the mind also possesses an intuitive faculty that allows it to recognize the self-evident truth of first principles that allows deductive reasoning to begin.[19] These first principles were analogous to the purportedly self-evident truths on which Euclid based his deductive formulation of geometry. At least in principle, therefore, natural scientific knowledge in the strong sense was possible for the human mind, just as knowledge in mathematics was, because in both cases it comprised deductions about objects that were changeless and eternal (though mathematics played only a limited role in Aristotelian natural science).[20]

For Plato, for Aristotle, and for all the later philosophers who allied themselves with the Gods—all the rationalist and idealist philosophers who have dominated the Western philosophical tradition—the mind knows some universal truths and therefore can acquire, through deductive reasoning, knowledge that is universal, necessary, and certain. As the human mind cannot know the necessary truth of statements about reality from empirical experience—that would require an inductive generalization and result in probable knowledge only—the human mind must know some truths a priori or it could never begin to reason deductively. How does it know them?

Plato proposed, perhaps only half-seriously, that knowledge in the strong sense was possible because the mind could remember truths about reality that it had "seen" while it lived with the Gods, before it was born into a body. Aristotle proposed a mental faculty that grasped universals *in* the particulars of empirical experience together with an ability of the mind to intuit the truth of non-demonstrable first principles of reasoning. Augustine proposed that God gave humans selective access to His mind and thereby to the universals in it. Most medieval Muslim and Christian philosophers adopted some version of Aristotle's view; some adopted a version of Augustine's. Descartes proposed that the mind possessed innate ideas and an intuitive ability, a "natural light," to recognize truth. Kant created a philosophical system in which the mind itself generates necessary truths about experience.

The allies of the Giants, the sophists, that Plato and Aristotle took most seriously included Protagoras, Gorgias, and Antiphon. These were prominent Greek thinkers to whom Plato denied the title "philosopher," stigmatizing them as second-rate thinkers, fee-seeking teachers of rhetorical tricks who claimed to possess knowledge but didn't. It is a cliché that the winners get to write the history books, and that applies as much to intellectual history as to social, political, and military history. The teachings of Protagoras and Gorgias were derided by Plato, dismissed by Aristotle, and parodied by both. As few of the writings of the sophists have survived intact, much of what we know of their ideas is from quotation and discussion by other, often hostile, authors. Nevertheless, surviving texts and scholarly reconstruction show that Protagoras and Gorgias, and at least some of their fellow sophists, among them Antiphon, had carefully thought-out positions on knowledge, truth, reality, and rationality.[21]

Protagoras is most famous for a statement that could also be the motto for everyone on the side of the Giants, from antiquity to the present. It is typically translated as 'Man is the measure of all things.' Plato and Aristotle mocked this as entailing that knowledge, truth, and reality were all subjective. They claimed that Protagoras' statement meant that what appeared to be true for each individual *was* true, for that individual, and that what appeared to be real for each individual *was* real, for that individual. It followed, nonsensically to Plato and Aristotle, that there was no such thing as objective knowledge, no such thing as supra-individual truths, and no shared conception of reality.[22]

It is pretty clear that Protagoras did not mean by 'Man is the measure of all things' what Plato and Aristotle said he meant: that all claims to knowledge and truth are incorrigibly relative to each individual. Rather, Protagoras seems to have meant that humans are the *measurers* of all things. All of the measures by which we order our experience of the world are human inventions and conventions, including among these measures the concepts that we use to organize experience, as well as the quantitative metrics that we use to measure specific features of experience.

We have, for example, a concept of the volume of a closed space. There is no universal definition of "volume" or of "closed space," but if a group of people all agree on what "volume" and "closed space" mean in a particular instance, then we can measure the volume of that space. There is no universal metric for this volume either, but if the same group of people agree on one— say, inches, or feet, or yards, or meters—then the volume becomes the same for everyone in that group. They all come up with the same number. In the process, by way of shared metrics, the subjective becomes objective, in the sense of becoming intersubjective. This example scales up to the most complex scientific measurements, for example, measuring the size and age of the universe. If you accept the definitions of relevant terms used by astronomers and astrophysicists, and if you accept the truth of the concepts they employ in their reasoning about the universe, then given the same data and the same instruments, the calculated age of the universe will be the same for everyone.

Against this conventionalist view of metrics, Plato and Aristotle were essentialists, holding that there *were* natural metrics that could be applied to nature and to reality. Plato, for example, has Socrates say that all things "must be supposed to have their own proper and permanent essence, they are not in relation to us, or influenced by us, fluctuating according to our fancy, but they are independent, and maintain to their own essence the relation prescribed by nature."[23] The task of the philosopher is to discover what those essences are, in order to "carve" reality "at its [uniquely correct] joints" and not just any which way that humans please, "like a clumsy butcher," that is, like the sophists.[24]

If, however, Protagoras and the Giants are right that humans are the measure of all things, then there are no natural joints at which to carve nature and thus no one correct carving. The measures, conceptual and quantitative, that we collectively choose to apply to nature derive from, but are not uniquely deducible from, experience. Experience often changes unpredictably, provoking new measures and ever-newer "carvings." As an exemplary Giant, Protagoras does indeed "pull down from the heavens" what we mean by knowledge, truth, reality, and rationality. Knowledge becomes a species of opinions or beliefs about experience, which can be better or worse on pragmatic grounds, but not true or false in any absolute sense.

Borrowing ideas from the Pythagorean school of philosophers and from Heraclitus, Protagoras and Gorgias maintained that reality was not fully rational because every *thing*, in itself and in its interactions with other things, was composed of an ever-changing combination of opposing "forces" obeying their own laws of change (*logoi*). Nothing had an unchanging essence; there was only becoming and there was no changeless Parmenidean Being at all. For all allies of the Giants reason has a limited capacity for grasping a reality beyond experience. Antiphon argued that language was incapable of grasping the reality that existed external to the mind, and as reasoning is dependent on language, reasoning cannot grasp reality either. For Protagoras and Gorgias, poetry offered an aesthetic alternative to reason. It had the power to allow us to apprehend, non-discursively and non-rationally, something of the flux of the contending forces within all things, including within human beings. Through poetry, the mind felt something of the ultimate irrationality of reality and the ultimately tragic character of human existence.[25] Plato, by contrast, was distrustful of poetry and in the *Phaedrus* had Socrates say that there was a place beyond poetry that was accessible to pure reason, uncorrupted by the senses, and this alone made knowledge in the strong sense possible.[26] (These ideas of Antiphon, Protagoras, and Gorgias about language, reason, and the aesthetic will re-emerge in the eighteenth and nineteenth centuries.)

The battle lines between Gods and Giants have always been very clearly drawn. One or the other can be right, but not both. Knowledge claims can either be about a reality that is independent of our experience and our reasoning about it, or they are about our experience and reflect the ways in which we reason about it. Either knowledge claims transcend context and are essentially true, validated by their correspondence with an external-to-mind state of affairs, or they are contextual, a function of applying pragmatically validated reasoning to experience. To claim that knowledge is about reality puts you in the camp of the Gods and commits you to the strong sense of knowledge, with coordinate conceptions of truth and rationality. To claim that knowledge is about experience puts you in the camp of the Giants and commits you to very different conceptions of truth and rationality.

What it boils down to is this: Is there a *unique* truth that the mind can know of any matter pertaining to what is external to the mind, and if so, how does the mind know it? With respect to the study of nature—understood as that which is external to and independent of human experience and its cause—the creators of modern science claimed that there *was* a uniquely correct account of what nature was really like and of how it caused our subjective experience of it. At the same time, however, they claimed that they produced such accounts by virtue of a distinctive form of experience-based reasoning that they practiced, the "scientific method" as it later came to be called. The new, "modern" science of nature claimed to have reality as its object, taking the side of the Gods, but the reasoning by which knowledge of nature was produced was anchored in experience, taking the side of the Giants. How can induction from particular experiences produce uniquely correct knowledge of what is independent of experience?

The founders of modern science were acutely aware of the force of these questions. They claimed to have penetrated experience to discover its unexperienced causes by reasoning about that very experience. This should seem paradoxical to us, and it was paradoxical to them, yet the paradox, unresolved, remains integral to modern science. The rhetoric of science continues to reflect a strong realist commitment that science tells us the truth about what is real, yet it does that empirically, hence inductively, conducting experience-based experiments and reasoning about their results. Two recent examples of this realism of science are the "discovery" of the Higgs boson and of gravitational waves using extremely complex instruments: the Large Hadron Collider (LHC) and the Laser Interferometry Gravitational Observatory (LIGO), respectively. Already in the seventeenth century, new realities—that the Earth moved, that air has weight and exerts pressure, that there was a world of "animals" not visible to the unaided eye—were realist claims based on what newly invented instruments were said to have revealed. Then and now there was no way of confirming these claims using our senses. We have to take the scientists' word for it that their instruments were working correctly, that the outputs were not created by the instruments themselves, and that these outputs revealed, however indirectly, a reality independent of those instruments.

The use of increasingly sophisticated physical instruments as means to the end of exposing nature's hidden workings has been the hallmark of modern science from its beginning. In the seventeenth century there were the telescope and the microscope, the barometer, thermometer, air pump, and pendulum clock. There were as well "instruments" of the mind, including various versions of an impersonal methodology, symbolic algebra, the differential and integral calculuses, and early probability theory and statistical analysis. Today, there are terrestrial and space-based radio, X-ray, gamma ray, infrared, ultraviolet, neutrino, and gravity wave telescopes complementing giant computer-controlled optical telescopes; electron and atomic force microscopes; ultra-fast lasers to image chemical reactions in real time; particle accelerators; MRI and PET scan machines to image brain and mind activity; computer modeling tools, gene sequencing machines; and highly abstract forms of mathematics that are central to research in the natural and the social sciences.

To the extent that nonscientists and non-science studies specialists know anything at all about these instruments, they may find them wonderful or even awe-inspiring, but typically they accept what scientists tell us that their instruments *reveal* about nature, for example, "seeing" the Higgs boson and gravitational waves. What instruments actually "reveal," however, is always dependent to a greater or lesser degree on ideas in the minds of the inventors of an instrument and in the minds of its users. Features of the relativity and quantum theories, for example, are central to the design and operation of the LHC and LIGO and their respective detectors. These detectors generate as data tens to hundreds of millions of events that are automatically analyzed by specialized computer programs that extract from these data a handful of events that are announced as "sightings" of the Higgs boson and gravitational waves.

Instruments must be understood as extensions of our minds, not of our senses. Whether they are physical instruments or conceptual instruments they are embodiments of ideas and imagination. They are not at all neutral in the sense of being mind-independent "windows" onto reality because our experience-shaped minds are embedded in their design, construction, and operation, as well as in the interpretation of the data they generate. This is no less true of experiments themselves, however simple the instruments used in them may be. Experiments are intentionally designed for the purpose of generating an anticipated outcome or type of outcome. Furthermore, the data they generate cannot *explain* anything. Data are always open to multiple interpretations, so once again the mind enters into the practice of science at its most fundamental level: the intentional creation of data using instruments and tools of analysis in the context of some particular interpretation of their meaning.

Data by themselves do not explain, and instruments do not simply reveal, an already existing reality "out there" waiting to be uncovered.

Copernicus had effectively the same astronomical data available to him as Ptolemy did almost fourteen hundred years earlier, but Copernicus interpreted them in a dramatically different way than Ptolemy. And Tycho Brahe, using more precise versions of the same kinds of data as Ptolemy and Copernicus, interpreted his data differently than all of his predecessors, as supporting a new model of the universe that was widely accepted by astronomers until the last third of the seventeenth century. Robert Hooke knew of Newton's optical experiments passing sunlight through prisms but offered a different interpretation of what had happened from Newton's, one in which colors were not in the sunlight but were produced by the interaction between sunlight and the prism glass. Joseph Priestley performed the same experiments on combustion that Lavoisier did—in fact Lavoisier repeated Priestley's earlier experiments—but Priestley interpreted the outcome of his experiments as supporting the existing phlogiston theory of combustion rather than the new oxygen theory that Lavoisier proposed.

These examples are not the exception but the rule in the practice of modern science. It is important to recognize that the people who defended interpretations later judged incorrect were not judged to have been wrong because they reasoned incorrectly. They were judged to have been wrong because their reasoning was based on different assumptions from the ones that their successors later judged to be the right assumptions to make. On both sides, the interpretation of data and of experimental outcomes required the use of assumptions that were not themselves deducible from experience. It is assumptions that win scientific truth battles over time, not logical reasoning. It is rare in the extreme for a scientist's work to be challenged because he or she reasoned incorrectly. Challenges always target interpretation or analysis, both of which entail assumptions. The enterprise of producing scientific knowledge, then, is not a solely a matter of observation, experimentation, and data collection, but of assessing alternative interpretations of the outcomes of these activities.

Which leads us to what might be called the ultimate scientific instrument: reasoning itself. If we often take physical instruments as simply revealing what's really "out there," behind our experience, it is even truer that we take reasoning for granted as a neutral means to our understanding this reality. We reason about our empirical experience scientifically, and the result is said to reveal that the world "out there" in fact is very different from the way we experience it. We collect data from sensory and instrument-aided observations and from carefully planned experiments. We then analyze this data subject to assumptions that we make—including definitions, concepts, and conceptual tools such as mathematics and logical inferences—and announce that scientific reasoning has revealed what the world is really like. But is all this reasoning about the world really neutral? Is the fact that this reasoning goes on in a brain that was produced by an evolutionary process irrelevant to the conclusions it reaches? Can such a brain know what is independent of it and of what produced it?

All of our senses have limits that reflect our evolutionary history. What we see as "the world"—what we hear, smell, and taste—are all produced by the brain in ways that are distinctive of how the human nervous system works. Is what is true of our senses also true of our reasoning, or does reasoning somehow transcend our biology? Conceived as an instrument, as Plato and Aristotle did, can reasoning reveal a reality that our senses cannot? Or does reasoning also have an innately human character that limits our ability to know a reality that is independent of our reasoning about it?

The founders of modern science were highly self-conscious about these questions and of their implications for the knowledge claims they were making. Their self-consciousness must be understood against the backdrop of the rise of Protestantism and its rejection of the authority of the Catholic Church in matters of religious truth. This rejection, and the horrific violence it provoked, abating only with the Treaty of Westphalia ending the Thirty Years War in 1648, made claiming to know *anything* that was *certainly* true highly controversial. Protestant theologians used newly available skeptical arguments to undermine claims by the Catholic Church that it knew the truth about God and God's will as expressed in biblical texts. The Protestants enjoyed considerable success in this skeptical critique of the Catholic Church's claims, but then defenders of the Church used the same arguments against the various Protestant accounts of God's will as revealed by *their* reading of the Bible.

Early modern scientists—the word "scientist" is anachronistic applied prior to its adoption in the 1830s—like Copernicus, Vesalius, Tycho Brahe, Simon Stevin, William Gilbert, Johannes Kepler, Francis Bacon, Galileo Galilei, and Rene Descartes claimed that they were producing knowledge, true accounts of how nature "really" was and of how nature "really" worked, even though their reasoning (except for Descartes') began with and was validated by experience. This conflates the two mutually exclusive definitions of "knowledge" described earlier, along with their associated ontologies and epistemologies. By employing both deduction and induction selectively to produce what they claimed was the truth about reality, modern science in effect swallowed whole the battle between the Gods and Giants. That made the battle internal to modern science, a battle that many scientists were aware of yet that remains "joined" even today, four hundred years later.

For a long time the internalization of the Gods versus Giants battle was at most an abstract intellectual issue within science, one of little or no consequence to society. That began to change in the nineteenth century, when science, and science-based technologies, became the primary drivers of social change. But not *just* social change. Science became entangled with the physical consequences of the adoption of new science-related technologies. Science also became entangled with the creation of wealth and social power. It became deeply entrenched in governmental, commercial, and social institutions, including the educational system from kindergarten to college, and beyond. In *Objectivism and Relativism*, Richard Bernstein put it this way:

> The *agon* [Plato's Gods-Giants battle] between objectivism and relativism has been with us ever since the origins of Western philosophy ... but it is only in recent times that the complex issues that this debate raises have become almost obsessive and have spread to every area of human inquiry and life.[27]

The Quarrel Between Invariance and Flux by Joseph Margolis and Jacques Catudal also traces this *agon* in Western philosophy, while Thomas M. Lennon's *The Battle of the Gods and Giants* focuses on one moment in its history: the opposing theories of knowledge and reality of René Descartes and Pierre Gassendi. The argument in this book is that this *agon*/quarrel has been internal to modern science from its beginning, with consequences for the relations between science and society.

Beginning in the 1960s, even as it enjoyed unprecedented public support, science as a practice and as a profession, and scientific institutions and scientific knowledge, were subjected to bitter criticism from three directions: political, intellectual, and religious.

Politically, scientists and engineers were attacked as having struck a Faustian bargain in exchange for research grants, jobs, and fame. They produced new knowledge and new technologies, but in the process they served governments with military agendas, and corporations with a profits-at-all-costs ethos.

Intellectually, critics attacked scientific knowledge itself, arguing that it was not an objective account of reality at all, but a construct of the scientific community and as such not intrinsically superior to opinion and belief.

Concurrently, Fundamentalists attacked the validity of those scientific claims to truths about reality that conflicted with their religious truth claims, for example, the evolutionary theory of life versus creationism and intelligent design. If, as so many critics were now arguing, science produced knowledge in the weak, not the strong, sense, then religious teachings about nature should be taught alongside scientific teachings as parallel interpretations.

By the late 1980s, the attacks on science and counterattacks by scientists and their supporters had achieved sufficient visibility to a broad range of intellectuals to earn a name: "the Science Wars." Except for continuing battles with the religious right, the public Science Wars faded by the turn of the twenty-first century. The war internal to science, however, the "tremendous battle that has forever been joined" between the Gods and the Giants over knowledge, truth, reality, and rationality, continues to be waged. It is to the unfolding of that war in the course of the history of modern science that we now turn.

2
Is There a Scientific Method?

The emergence of modern science in the seventeenth century was more an evolution out of classical, medieval, and Renaissance sources than the result of a "scientific revolution." Among these sources were medieval nature philosophy, the medieval university, the humanist movement, and Renaissance natural philosophy and technology.

A foundational principle of modern science to this day is that nature is a closed system. Epistemologically, this requires that scientific explanations only include references to other natural phenomena. Ontologically, it entails that, for science, nature is all that there is.

Epistemological closure was articulated in the early twelfth century by Adelard of Bath in his widely read *Questiones Naturales* (*Questions About Nature*). Adelard and his nephew discuss seventy-six questions about natural phenomena. For the cause of the rainbow, the nephew cites the Bible text (Genesis 10:5) that God put the rainbow in the sky as a sign that He would never again cause a world-destroying flood. Adelard dismisses this out of hand. With all due respect to God, in natural philosophy natural phenomena can only be explained in terms of other natural phenomena; no supra-natural causes or explanations are allowed. Invoking God, he says, is a "miserable refuge," an evasion of a rational explanation. The cause of the rainbow must be sought by using reason to discover the relevant properties of light, air, and water and the way they interact to cause a rainbow.

By the end of the fourteenth century—following the invention of many kinds of water, and wind-powered, gear-driven machinery as well as the falling weight–driven mechanical clock—the heavens and nature in general were described by some philosophers, then Bishop of Paris Jean Buridan for one, as clockwork mechanisms. This made nature into a deterministic machine. Conceiving of nature as a mechanism led to the first suggestions of the conservation of motion in such a system, to preclude its running down over time, yet another idea that became fundamental to modern science.

The creation of the university in Bologna and Paris in the second half of the twelfth century, and its rapid spread over the next two hundred years, institutionalized the study of philosophy, and along with it natural philosophy.

For both of these subjects, university study centered on newly translated texts by ancient Greek and Roman philosophers, natural philosophers, and mathematicians, typically translated from Arabic manuscripts into Latin. The most influential and widely studied texts were the works of Aristotle—especially his books on physics and on logic, the latter collectively known as the *Organon*, or *Instrument* (for reasoning); Euclid's *Elements of Geometry*; and Ptolemy's *Almagest*, containing his Earth-centered mathematical model of the heavens. Alongside these were original works by Muslim and Christian philosophers.

Aristotle's philosophical writings dominated secular study in the medieval universities, in part because translations of Plato's works did not become available until the late fifteenth century. Aristotle was an excellent naturalist, but he limited the use of mathematics in physics, and his explanations of natural phenomena were based on passive observing and on his speculative metaphysics, not on experiment. Nevertheless, in spite of Aristotle's dominance, experiment—and mathematics-based *natural philosophy*—had emerged and was actively pursued at the universities of Oxford, Paris, Cologne, and Padua. These natural philosophers proposed explanations of phenomena, especially in optics and mechanics, in a way that is recognizably proto-modern. Some of the most prominent of them were Robert Grosseteste, Roger Bacon, Thomas Bradwardine, Nicole Oresme, John Pecham, John Buridan, Jordanus Nemorarius, and Campanus of Novara.

The problem of universals was a central issue for medieval philosophers. As they knew from their study of Greek philosophy, knowledge in the strong sense could only be produced by deductive reasoning, which required knowing the truth of some universal statements about the world. Absent this, only knowledge in the weak sense, produced by inductive reasoning, is possible and we cannot know anything for sure. There was a great deal at stake in this matter because of the tension between secular philosophy and theology in the new universities. Theology was threatened by the prospect of truths of reason conflicting with truths of faith, and it all hinged on the status of universals. The reality of universals, and thus the possibility of knowledge in the strong sense, was defended on one ground or another by the overwhelming majority of medieval philosophers. As it happened, the works of Aristotle were available when the university was created, and Aristotle's works on logical reasoning, which identified deductive reasoning as alone leading to truth, became required reading in all universities. Aristotle's position on universals—involving the active intellect and the intuition of first principles—dominated, both in its "pure" Aristotelian form and combined with Augustine's Christian-Platonic position that universals were ideas in the mind of God to which God gave humans limited access.

A minority of philosophers, allies of Plato's Giants, argued against the reality of universals. For them, universals were ideas in the mind only so that only knowledge in the weak sense was possible. Perhaps the most influential of these philosophers was the fourteenth-century Englishman William of Ockham (of "Ockham's razor" fame), who argued that, based on its sensory experience of the world of individual objects, people invented names for resemblances that they noted among these individuals. This allows clustering individuals by names such as "variety," "species," and "genus," but there is nothing outside the mind that corresponds to these names, only to the identified resemblances. For a conventionalist, there is no uniquely correct, natural ordering of individuals, no such thing as 'carving nature at its natural joints' as Plato required of a true philosopher of nature.

In the sixteenth century, some natural philosophers made important contributions to modern science and mathematics, including the work of Copernicus, Tycho Brahe, and Kepler in astronomy, Vesalius in anatomy, Tartaglia and Cardan in algebra and probability theory, and William Gilbert on the experimental study of magnetism. Nevertheless, Renaissance natural philosophy was very different from the medieval natural philosophy described earlier: it took a strongly "magical" turn, embracing alchemy and astrology along with a distinctive metaphysics. The magical natural philosophers attributed causal action in nature to immaterial, spirit-like forces, some within material objects and some outside them, forces that acted at a distance. These forces were immaterial but operated *within* the realm of nature and they acted lawfully, not whimsically or willfully. Humans could learn these laws and manipulate the forces. Practicing magical natural philosophy required observation, experimentation, and the use of mathematics, but now invisible, immaterial features of nature also needed to be taken into account, not just visible material things. In astrology, for example, the construction of horoscopes is a rigorously mathematical exercise based on the light-like, hence mathematically describable, "radiation" of influences distinctive of each heavenly body. The status of astrology as a science—accepted as such from classical antiquity until well into the seventeenth century—rested on its mathematical character, but is a theory true simply because it is rigorously mathematical? This question, fundamental to modern science and implicit in astrology, became explicit in Galileo's mathematical physics.

The magical natural philosophers also argued that reason by itself could not achieve knowledge of reality. For that, a "higher" mental faculty than reason was required, an intuitive sympathy for the real to which the imagination was central, but not what we normally think of as imagination. In creating the universe, God imprinted an image of the macrocosm in the human mind. We have access to this microcosmic image of reality first through intuition-imagination and only then by reasoning about it. Cultivating this supra-rational faculty required a natural philosopher to develop qualities of their person, especially spiritual qualities.

Immaterial forces acting at a distance, putting limitations on reason, and making knowledge of nature dependent on the person of the natural philosopher were all anathema to the founders of modern science. Descartes' mechanical philosophy of nature dismissed all immaterial forces acting at a distance, allowing only contact forces. Newton was attacked for suggesting that such forces were at work, for example, in gravitational attraction and chemical reactions. Explaining electricity, magnetism, heat, light, chemical reactions, and gravity using contact forces only proved impossible, so as modern science matured immaterial forces acting at a distance became incorporated into it. And the founders of modern science insisted that reason *was* able to know reality and that scientific knowledge was independent of personal qualities of the natural philosopher.

The idea that in the fourteenth century Europe entered a period of cultural rebirth was due to the humanists, a large network of scholars devoted to recapturing the intellectual culture of classical antiquity. To the founders of modern science, and to most historians of science until well into the twentieth century, the humanists were perceived as obstacles to the rise of modern science. They were lumped together with the magical natural philosophers for their obsessive study of texts and reliance on ancient authorities. It is true that the humanists revered classical culture and that with a few exceptions they had no interest in natural philosophy or mathematics. What they were interested in was recovering works by Greek and Roman authors and publishing accurate printed editions of them. In doing this, however, they made no distinction whatsoever between manuscripts of poetry, plays, history, and philosophy, and manuscripts of natural philosophy and mathematics.

Among the latter were texts by Archimedes on mathematical physics; Apollonius of Perga on the geometry of the conic sections (extremely influential in the seventeenth century); Ptolemy's *Geography*, from which the technique was derived, by Gerard Mercator among others, for representing the surface of the spherical Earth correctly on a two dimensional map; the works of Pythagorean mathematicians for whom mathematics was the essence of physical reality; and, at last, the complete works of Plato, acquired from scholars fleeing the fall of Constantinople to the Turks in 1453. They also translated and published in printed book form pioneering works on algebra by Muslim mathematicians, especially al-Khwarizmi and Abu Kamil, which stimulated the development of algebra by European mathematicians, a development crucial for the creation of modern mathematical physics. (Selections of al-Khwarizimi's work first became available in 1202 in Leonardo of Pisa's *Liber Abacus*.)

The humanists also influenced modern science by way of the methods they developed for reconstructing accurate versions of ancient texts from corrupt copies, and by way of their organization. In the fifteenth and sixteenth centuries, the humanists created an interactive, geographically distributed network of like-minded scholars who shared their achievements, submitted them to one another for comment and criticism, even shared their ongoing research, and built on the productions of fellow humanists. They formed the first "academies," informal intellectual organizations unaffiliated with the university or the Church. Members met regularly to discuss shared interests and disseminate new accomplishments. Mirroring this, in the seventeenth century, an interactive, geographically distributed network of natural philosophers emerged, with academies (or societies) whose members met regularly and published journals that circulated internationally. The most famous and long-lived of these are the Royal Society of London and its journal *Philosophical Transactions*, published continuously since 1665, and the slightly younger Royal Academy of the Sciences in Paris and its journal, published from 1666 as the *Memoires of the Academy of the Sciences*, but then renamed *Comptes Rendus* in 1835.

Through printed editions of books by classical-era skeptical philosophers, especially those by Cicero, Lucretius, and Sextus Empiricus, the humanists altered discussion of the problem of knowledge in the sixteenth century. Contemporary with the Protestant challenge to the authority of the Catholic Church in matters of religious truth, these books reinforced the voice of allies of the Giants on knowledge and reality, challenging the heretofore dominant voice of the allies of the Gods.

All of these intellectual influences affected early modern science together with Renaissance-era applications of mathematics to practical problems: painting (perspective drawing), mapmaking, navigation, music (tuning systems), machine drawing, and machine design and construction.[1] The routine application of mathematics to everyday life in these diverse areas was an important component of the rich legacy of ideas and intellectual tools inherited by the creators of modern science. Two of those creators were Francis Bacon and Rene Descartes, both of whom promised that the new science of nature would lead to further practical applications of knowledge that would be of great benefit to society. For both, the solution to the problem of knowledge lay in method, a distinctive methodical, impersonal approach to reasoning about nature that guaranteed knowledge of reality, not just probable opinions about experience. Unfortunately, the methods they proposed were diametrically opposed to one another!

Bacon was an English jurist and social reformer who was ennobled by King James and appointed Lord Chancellor of England, but he served only briefly before being dismissed for corruption. Bacon was a theorist of natural philosophy, not a practitioner, producing no new knowledge of nature. What he did produce was an experimental method to be used by others for acquiring knowledge of nature.

Bacon advocated a strictly inductive approach to discovering truths about nature, minimizing mathematics and deduction and maximizing the collection, production, and analysis of data. In a radical break with the Western philosophical tradition, Bacon limited the role of the mind in executing his method. He deeply distrusted the kind of abstract intellection, moving "from ideas through ideas to ideas," that Plato had argued alone made knowledge of reality possible. Bacon also distrusted deductive reasoning and Aristotle's claims of the intuition of true first principles from which knowledge of nature in the strong sense could be deduced.

For Bacon, the mind was the single greatest obstacle to knowledge of nature because it is inherently inclined to prejudices and falsehoods, and it is subject to flights of imagination that "being not tied to the laws of nature, may at pleasure join that which nature hath severed and sever that which nature hath joined, and so make unlawful matches and divorces of things."[2] Left to itself, the mind cannot be trusted to reason correctly about anything because, through "the daily intercourse and conversation of life it is occupied with unsound doctrines and beset on all sides by vain imagination."[3]

Bacon's most influential work was *The New Organon*, that is, the new *instrument* for reasoning about nature. Although subsequently a stand-alone book, it was planned to be the second of six parts projected by Bacon for an unfinished work, *The Great Instauration*, detailing a "total reconstruction of sciences, arts and all human knowledge raised upon proper [inductive-experimental] foundations."[4] The subtitle of part two of this work was *True Directions Concerning the Interpretation of Nature*. Its title, *The New Organon*, was a direct challenge to the then still dominant organon, Aristotle's books on logic. In *his* organon, Bacon called the prejudices and falsehoods that beset the mind "idols . . . which are now in possession of the human understanding and have taken deep root therein," preventing correct reasoning. He identified four idols: Idols of the Tribe, the Cave, the Marketplace, and the Theater, and only by his method of "true induction," he claimed, could they be avoided and "cleared away."[5]

Idols of the Tribe "have their foundation in human nature itself." They are generic to humanity, prejudices that are inescapable features of human reasoning affecting everyone. These include the subjective nature of sense perception, "supposing the existence of more order and regularity in the world than it finds," and valuing positive evidence in support of an opinion that one holds more than negative evidence against it.

Idols of the Cave are idiosyncratic, not generic. They "take their rise in the peculiar constitution, mental and bodily, of each individual," reflecting their personality and upbringing. Some people defer to tradition and some "seek novelty"; some incline to finding similarities in things and some to finding differences. Each of us makes assumptions that we employ uncritically in our reasoning, treating beliefs that we favor as truths.

Idols of the Marketplace, "the most troublesome of all," are prejudices and falsehoods that individuals internalize through their social interactions. The most trouble is caused by the influence of "words and names" on our reasoning. Language is intrinsically misleading because of the equivocal character of words and our predilection for giving names to abstractions, which seduces us into treating them as if they existed.

Finally, Idols of the Theater are the many prejudices and outright falsehoods that we are taught as truth in school, including in the university.

In order to acquire knowledge of nature, these inherent vulnerabilities of the mind must be strictly controlled, and nature must be allowed to reveal itself as it is. That is the task of Bacon's method, according to which the natural philosopher needed to follow a very particular form of reasoning, one that was nature driven at every step, not mind driven:

> In order to penetrate into the inner . . . recesses of nature, it is necessary that both notions and axioms be derived from things by a more sure and guarded way, and that a method of intellectual operation be introduced altogether better and more certain.[6]

The method begins with the collection of data on some problem or phenomenon unprejudiced by any assumptions whatsoever. These are studied for relationships or patterns among them, leading to a tentative, initial hypothesis of the connections among these data or their common cause. This initial hypothesis must then be subjected to experimental testing. If it is not verified, the hypothesis must be discarded or at least modified and tested again. If it is verified, then more data must be collected as suggested by the hypothesis, and further experimental testing targeting the hypothesis must be conducted. Often, multiple hypotheses are suggested by the available data, but what Bacon called "crucial" experiments, testing each hypothesis against the others one by one, eliminate the false ones and reveal the true one. In this way, experiment forces a convergence on the true cause of a phenomenon, ultimately revealing what he called universal "laws" of nature.

Bacon believed that his method, even though it begins with induction from particulars, would lead to universal truths about reality, not just the probable opinions about experience to which induction traditionally was limited: "God forbid that we should give out a dream of our imagination for a pattern of the world." His method, he writes, will lead to "a true vision of the footsteps of the creator imprinted on his creatures."[7] The "end of our foundation is the knowledge of causes and secret motions of things."[8] In other words, he believed that his experiment-controlled method of reasoning could bridge induction and deduction: reasoning inductively, but nevertheless reaching certainty about reality through confirming experiments. "Baconian science aimed for the certainty of mathematical demonstration via empirical data and scientific proper procedure," the latter being the method that Bacon revealed in his *New Organon*.[9]

The certain and demonstrable knowledge of "causes and the secret motions of things" was guaranteed by a rigorously depersonalized—we would say objective—and almost mechanical application of his method: repeated cycles of data collection, analysis, and experimental testing. The "entire work of the understanding [must] be commenced afresh, and the mind itself from the very outset be left not to take its own course, but guided at every step; *and the business itself [reasoning about nature] done as if by machinery*" [my emphasis].[10] Over time, Bacon's method became identified as "the" scientific method, even though it rejected mathematical modeling of phenomena and stigmatized intuition and deduction, all of which are fundamental features of the practice of modern science. And in spite of his reputation as the "father" of the experimental method of modern science, Bacon was quite wrong about the ability of experiment to bridge induction and deduction and reveal universal truths about reality.

In his *organon*, Aristotle had identified various fallacies of deductive reasoning. One of these is called the Fallacy of Affirming the Consequent (sometimes called the Fallacy of Affirming the Antecedent). Suppose that we know the truth of the sentence, "If X is true then Y is true." We can know this independently of knowing the truth of either X or Y if the relationship between X and Y is logical, not empirical. For example, let X stand for the general theory of relativity, and Y a logical consequence of that theory, say, the curved path of light rays in a gravitational field. We can know that the *sentence* "If the general theory of relativity is true, then light rays follow a curved path in a gravitational field" is true on logical grounds alone, but we don't yet know if the theory itself is true. We perform an experiment and discover that light rays passing close to our sun *do* follow a curved path as predicted by the theory. In other words, the prediction is confirmed, but this does not *prove* that the theory itself is true. The original sentence made a claim about the truth of Y given the truth of X. No claim was made about the truth of X given the truth of Y. As the history of science reveals, false theories can make true predictions!

It follows that the experimental method, insofar as it leads to claims about reality, is based on a deductive logical fallacy. Experiments can never prove the truth of a hypothesis or of a theory. What they can do is confirm a theory's consistency with the available data, but the theory is not unique in this respect, and new data can arise at any time, which implies that the theory is about our empirical experience, not about a reality independent of our experience. On the other hand, if a logical consequence of a theory is not confirmed experimentally (that is, if Y mentioned earlier is false), then logically the theory (X) is proven false, but the history of science is rich with theories being maintained as true even in the face of falsified predictions.

Copernicus' theory that the Earth orbits the sun rather than the reverse has as a logical consequence that the observed alignments of stars should change as the Earth moves. This change, called stellar parallax, is greatest at the two extremes of the Earth's orbit, and observing it was a direct test of Copernicus' theory. This parallax was not observed until the nineteenth century, yet everyone was a "Copernican" by the end of the seventeenth century. More recently, the persistent failure to detect the gravitational waves widely believed to be a logical consequence of the general theory of relativity was not taken as falsifying that theory. The Laser Interferometry Gravitational Observatory (LIGO), the largest and most expensive instrument built to detect such waves, operated for almost thirteen years without detecting a single event. Instead of taking that as a disproof of the theory, a very expensive upgrade was funded that in 2015 finally produced the first gravitational wave detection.

Bacon greatly overestimated the power of experimentation to overcome the inability of induction to reveal definitive truths about nature, as if by deduction. At the same time, he greatly underestimated the need for the mind to play an active role in the application of his method. It is not possible to collect data, or to analyze them, without making some assumptions in advance about relevance, assumptions about which data to collect and which to ignore, and without making some assumptions about the connections among the collected data. Some fifty years after Bacon's *New Organon* was published, at a time when the leading natural philosophers in England proclaimed themselves Baconians methodologically, Robert Hooke put it this way: "rude heaps of unpolish'd material [uninterpreted data] are worthless . . . They cannot lead to knowledge of nature."[11]

Data do not speak for themselves. To lead to knowledge, data must be interpreted, and interpretation is intrinsically pluralistic. That implies that hypotheses about data correlations and causes cannot be *deduced* from a (finite) data set, however large. Beyond simple generalizations, inductive reasoning involves imagination and intuition. Kepler did not simply *find* his first and third laws of planetary motion—that the planets move around the sun in elliptical orbits and that the square of a planet's distance from the sun is proportional to the cube of its orbital velocity—in the data made available to him by Tycho Brahe. Brahe for one did not see them there. Kepler intuited or guessed at those laws after studying the data and then went back to the data to test his guesses, ultimately pronouncing them truths about reality. (Three hundred years later, John Stuart Mill would argue that Kepler *did* see the ellipses in the data so that no interpretation was required.)

Soon after the publication of Bacon's *New Organon*, René Descartes, in two works, *Discourse on Method* (1637) and *Rules for the Direction of the Mind* (circulated earlier but first published, posthumously, in 1684), proposed a systematic *deduction*-based method as the only way to acquire knowledge of nature. Descartes was the son of a minor French nobleman. He was an intellectual prodigy who studied at the Jesuit college of La Fleche and already as a young man committed himself to refuting the skeptical philosophy that had become increasingly popular during the sixteenth century. He wound up becoming a founder not only of modern science, but also of modern philosophy—by making the problem of knowledge the primary concern of philosophy—and of modern mathematics, by his promotion of analytic geometry, contributing to algebra displacing geometry as the basis of European mathematics.[12]

For Descartes, empirical experience was the primary obstacle to acquiring knowledge of nature, and the mind was the means of overcoming that obstacle. This is exactly the reverse of Bacon's solution to the knowledge problem. So was Descartes' dismissal of inductive inference. Inferences based on experience "are frequently fallacious," he wrote, but deduction "cannot be erroneous when performed by an understanding that is in the least rational."[13] "Deduction is thus left to us as the only means of putting things together so as to be sure of their truth," and the deduction of "universal and necessary" truths cannot be based on particular or contingent facts.[14] Descartes did leave a role for what he called induction, but he used that word idiosyncratically, as a synonym for "enumeration."

While the highest form of proof is a simple intuition of the truth of clear and distinct ideas, most arguments in natural philosophy will have many steps. It is therefore necessary to enumerate every step in such a proof in order to guarantee that the truth of the conclusion follows deductively from the preceding chain of reasoning. If "we wish to make our science complete," we must take an "inventory . . . of all matters that have a bearing on the problem raised."[15] This enumeration of steps in a complex deductive proof and taking "inventory" of relevant data are what Descartes meant by induction.

Further differentiating his method from Bacon's, Descartes limited the role of experiment in acquiring knowledge of nature and maximized the roles of intuition and of mathematics. What he called "simple experiments," direct observation and manipulation of objects and phenomena, play an important role in his method as inputs to our reasoning, part of taking "an inventory." But "complex experiments," designed to test hypotheses as in Bacon's method, he considered deeply problematic. He recognized that the results of experimentation are always open to multiple interpretations, but because particular facts can often be deduced from universal principles in different ways, some experiments may be necessary in order to find out which way is the "true" way.[16]

Given these differences, and the still different methods of reasoning about nature employed by Galileo and Newton, the claim that modern science arose as a result of the invention of "the" scientific method is untenable. In the seventeenth century, and even today, there was no one such method. But there was a common denominator to all these seventeenth-century methods: a conviction that *some* method of reasoning was the solution to the problem of knowledge. "There is need of a method for finding out the truth," Descartes wrote, meaning by method "certain and simple rules such that if a man observe them accurately . . . [he will] arrive at a true understanding of all that does not surpass his powers."[17] His method, he claimed, "is a more powerful instrument [for acquiring] knowledge than any other that has been bequeathed to us by human agency," and it is the "source of all others [all other instruments that produce knowledge]."[18] It provides "a step by step" procedure, starting with the "intuitive apprehension" of simple truths about the world and ascending to universal knowledge from which the particular facts of natural phenomena can then be deduced.

Bacon had written that his method controlled reasoning "as if by machinery." Descartes wrote that his method "resembles indeed those devices employed by the mechanical crafts" that are "self-contained" and "themselves supply the directions for making their own instruments."[19] The difference between these two methods for controlling reasoning lies in the critical dependence of Descartes' deductive method on intuition and innate ideas. Descartes described intuition as a

conception of the pure and attentive mind which is so simple and distinct that we can have no further doubt as to what we understand, or, what amounts to the same thing, an indubitable conception of the unclouded and attentive mind which arises from the light of reason alone.[20]

And on innate ideas he wrote, "I am convinced that there are certain basic roots of truth [that are] implanted in the human mind by nature," and these, together with intuition, alone enable the human mind to employ deduction to acquire certain knowledge of the world.[21]

Descartes was implacably opposed to skepticism and thus rejected the definition of knowledge in the weak sense: "[W]e reject all such merely probable knowledge and make it a rule to trust only what is completely known *and incapable of doubt*" [my emphasis].[22] Mankind "has no other road toward certain knowledge open to it save those of self-evident intuition and necessary deduction."[23] His most popular philosophical work, *Meditations on First Philosophy*, was designed to disprove skepticism by proving that deductive knowledge of reality was possible. The mind, he argued, was able to grasp universal features of reality even from the corrupted particulars presented to it by the senses. From our experience of objects perceived by the senses as having shape, color, texture, and odor, the mind grasps that all of these "secondary" sensations are subjective, in the mind only. Philosophical reflection reveals that "out there," beyond the mind, matter is simply extension, without any of the properties that our senses tell us material objects possess.[24]

The basis of this reflection, he claimed, was the mind's self-evident ability to recognize truth thanks to a *lumiere naturelle*, a "natural light," that every human mind possessed.[25] (By self-evident, Descartes meant that the truth of this claim did not require a logical argument to prove it; its truth is "seen" directly by the mind, within the mind.) Descartes' "natural light" is clearly a secular version of the Divine Illumination theory of knowledge that St. Augustine had claimed allowed us to "see" the truth of universals in God's mind. For Descartes, it is this natural light that allows everyone to "see" the necessary truth of the proposition "I think, therefore, I am." It also lets us "see" the necessary truth of every idea whose necessary truth is "clear and distinct." He wrote, "I am convinced that certain primary germs of truth [are] implanted by nature in human minds," and the "human mind has in it something that we may call divine, wherein are scattered the first germs of useful modes of thought."[26] It is the intuited truth of simple ideas and of deductive logical connection, together with innate ideas, that makes knowledge of nature in the strong sense possible.

What is distinguishable in "clear and distinct thought" is also distinct and separable "out there," in reality, and can exist separately from thought.[27] This principle, linking epistemology and ontology, method and reality, is echoed in Spinoza's version of Cartesianism: "The order and connection of ideas is the same as the order and connection of things."[28] It is important to note that the truth of Descartes' "I think therefore I am" does not follow because "I am" is the conclusion of a deductive argument. Intuited truth is different from, and the basis of, deduction. The truth of the whole sentence "I think therefore I am" is "seen" immediately by the mind. This provides the mind with a criterion for all truth, in particular, the truth of the conclusion of a valid deductive argument and of each step in such an argument leading to the conclusion. Only a proposition whose truth is as clear and distinct as the truth of the sentence "I think therefore I am" can be trusted to be true, and the same applies to the intuited truth of all simple ideas.[29]

Like Plato, Descartes saw in mathematics an exemplification of knowledge in the strong sense. It exemplified the production of universal and necessary truth by deductive reasoning. Plato had ranked geometry slightly below the highest form of knowledge because of its use of diagrams. For Descartes, however, the use of figures in mathematics made a constructive contribution to deductive reasoning because imagination, properly employed, *increased* the power of deductive reasoning. In fact, Descartes called imagination, the senses, and the understanding (resting on intuition, innate ideas, and deductive reasoning) our only "instruments of thought." It is, however, understanding "alone [that is] capable of perceiving the truth," though "it ought to be aided by imagination, sense and memory, lest perchance we omit any expedient that lies within our power."[30]

As a young man, Descartes wrote in his *Discourse on Method* that he had thought mathematics "was *only* useful in the mechanical arts."[31] Under the influence of the Dutch natural philosopher Isaac Beeckman, however, with whom he collaborated for several years beginning in 1618, Descartes came to see mathematics as the privileged means of expressing truths about the natural world. (Beeckman also introduced the French priest-philosopher Pierre Gassendi to atomism and probabilism, both of which Descartes rejected. Beeckman's own natural philosophy was a fusion of scientific theory, applied mechanics/engineering, and mathematical atomistic mechanics.)[32]

The laws of mechanics [mathematical physics to us], Descartes wrote, are "identical with the laws of nature."[33] Ideally, mathematical physics begins by deducing universal truths about space, time, matter, and motion from intuited first principles and simple innate ideas. In practice, however, empirical experience and experimentation, which is also empirical of course, play a role. Experience can never on its own lead to knowledge. In fact, experience is profoundly wrong in its representation of the world in the mind, but it stimulates our thinking about that world. Together with carefully controlled experimentation, experience can act as a reality check on our deductive reasoning.

How would Descartes apply his philosophical method to understanding the cause of a natural phenomenon? He uses the magnet as an example. First of all, there "can be nothing to know in the magnet which does not consist of certain simple natures [whose truth is] evident in themselves," that is, seen in the mind to be clearly and distinctly true. We proceed to gain this knowledge by collecting "all the observations which experience can provide . . . of this stone." From these, we must "try to deduce the character of that intermixture of simple natures which is necessary to produce all these effects" associated with magnets: "This achieved, [the natural philosopher] can boldly assert that he has discovered the real nature of the magnet in so far as human intelligence and the given experimental observations can supply him with this knowledge."[34]

Note well what Descartes has done here. While highlighting deduction and necessary truth, hallmarks of knowledge in the strong sense, he has subtly incorporated a bridge between deduction and induction. The "deductions" employed in understanding the nature of the magnet are contingent upon experimental observations. New observations might lead to different "deductions," and there is no way of knowing in advance whether or not new observations of a relevant sort will or will not occur. As noted earlier, while the understanding "alone [is] capable of perceiving the truth . . . *yet it ought to be aided by imagination, sense and memory lest perchance we omit any expedients that lie within our power*" [my emphasis]. Also, recall that in defining "enumeration," Descartes had cautioned the natural philosopher to make an "inventory . . . [of] all matters that have a bearing on the problem raised."[35] But there is no logical way of proving in advance of experience that we have not omitted any "expedients," any facts that are relevant to the phenomenon being studied, nor that future experience will not reveal some new "matter" bearing on it. No less than Bacon, then, Descartes has conflated induction and deduction. Bacon promoted his method as purely inductive; Descartes promoted his method as purely deductive. Both were misrepresentations.

Ultimately, however, the goal was the same for both: universal knowledge of the physical world. While Bacon's method produced knowledge of nature from the bottom up, moving by induction from particulars to universal laws, Descartes' method produced knowledge of nature from the top down, moving from intuited universal truths about nature to the particulars of empirical experience. For Bacon, truths about the world will be revealed by the world itself, in empirical experience, but only if the mind accepts methodological constraints that allow the world to reveal itself. For Descartes, truths about the world are revealed in the mind, by the mind, *in spite* of how we experience the world. Descartes, like Bacon, also claimed that this knowledge would give humans power over nature. It would "render ourselves the masters and possessors of nature" and via new technologies "enable us to enjoy without any trouble the fruits of the earth and all the good things which are to be found there."[36]

Fully aware of the Fallacy of Affirming the Consequent, Descartes acknowledged that, on logical grounds, God could have created the world in too many different ways for humans to identify the one God actually used. But, he claimed, any way God chose had to have the same laws of nature as our world has. Once we know the latter, we effectively know how the world was created, because our knowledge of nature demonstrates [that is, deduces] "effects from the causes . . . showing from what beginnings and in what fashion Nature must produce them."[37]

Descartes' theory of knowledge became central to modern rationalist philosophy and won him many followers—among them Spinoza and Leibniz—and his mechanical philosophy of nature became central to seventeenth- and eighteenth-century modern science. His method, however, was not widely adopted. The use of deduction in justifying scientific theories, and of mathematics in formulating, them remain, at least in the physical sciences, standard features of science today, but both were already features of Galileo's method, as will be shown in the next chapter. The dependence of Descartes' method on innate ideas, on the touchstone of truth being something internal to the mind (the natural light), and on intuiting the necessary truth of first principles of reasoning about nature, and his distrust of experimentation, found less and less support as modern science matured. What happened in the course of that maturation was that practitioners of modern science selectively employed elements of the Baconian method and elements of the Cartesian method, conflating induction and deduction, the position of the Gods *and* the position of the Giants, in spite of the logical inconsistency this entailed.

3

Was Galileo Right and the Catholic Church Wrong?

Galileo Galilei is most famous for inventing the telescope and using it to study the heavens, and for his condemnation by the Catholic Church for defending the truth of the Copernican theory. In fact, he did not invent the telescope, nor the pendulum clock with which he is often also credited, nor was he first to use a telescope to study the heavens. And it is simplistic to portray him as a heroic defender of scientific truth crushed by authoritarian dogma. His condemnation by the Church for asserting the motion of the Earth was not a collision of rational science and irrational belief. It was a collision between conflicting definitions of "truth" and "knowledge" as applied to claims about nature.

Galileo (he is always called by his first name) was a highly accomplished musician and music theorist, as his father had been. He was a talented artist and a master of the perspective drawing technique, which for several years he taught to paying art students. He was a poet, a literary critic much admired for his insights and style, and a professional astrologer, as expected of professional mathematicians at that time. He was an inventor and a businessman, a gifted mathematician, and of course an extraordinary natural philosopher, the "father" of modern mathematical physics. He was, arguably, the first person one can call a scientist in the full modern sense because in his scientific work there is not a trace of Renaissance natural philosophy.

The previous chapter described the contrasting methods of Francis Bacon and René Descartes for producing knowledge of nature. In this chapter, I want to extend our exploration of how scientists reason by way of Galileo, moving from his distinctive conception of method (different from both Bacon and Descartes), to his application of it in physics, to his trial for heresy, which was related to his claims for knowledge based on his method.

Galileo's method is closest to what most people mean by "the scientific method." It is simultaneously experimental inductive and mathematical deductive. As he often says in his books, Galileo was strongly influenced in his method by Archimedes, the great Greek mathematician and physicist, several of whose works had been newly published by humanists in Latin translations in the mid-sixteenth century. Archimedes applied mathematics and deductive reasoning to problems in physics, most notably problems in hydraulics, hydrodynamics, and the action of the simple machines: the lever, pulley, and screw primarily, and secondarily the wedge and winch. He did not invent any of these machines, but he created a mathematical-deductive theory for each that explained how they worked and that led to improved versions of these machines being built. The goal of his method for reasoning about natural phenomena was a mathematical-deductive theory that described how the world "out there" worked. This is the method that Galileo adopted for his own studies of natural phenomena.

Galileo's career pivots neatly around the year 1610. He had been born in Pisa into a noble but poor Florentine family and attended the University of Pisa as a medical student without graduating. From 1592 to 1610 he was professor of mathematics at the University of Padua, then controlled by Venice, and a very active and highly respected member of a vibrant Northern Italian intellectual community. He invented and sold a "military compass" for calculating cannonball trajectories, belonged to three humanist academies, and after 1610 became a member of one of the first scientific versions of the humanist academies, the Academia dei Lincei (Academy of the Lynx-Eyed). During his eighteen years at the University of Padua Galileo filled notebooks with the results of experiments that he and his circle of natural philosophy predecessors, colleagues, and friends had performed, together with ideas of his own about applying mathematics to problems in physics, primarily to problems associated with motion and equilibrium. If he had died in 1609, he would have been mourned as having been a promising young man who had not lived up to that promise.

But Galileo did not die in 1609. He had by then taught astronomy for over twenty years without contributing anything original to it, though he had adopted Copernicus' theory as the true account of the planetary motions and was vocal about it. He mocked Tycho Brahe for his obsession with collecting ever more precise data about the changing positions of the planets and stars in the night sky. He also dismissed Kepler's 1609 *Astronomia Nova* (*New Astronomy*) in which Kepler announced that, using Brahe's data, he had determined that the planets moved around the sun in elliptical, not circular, orbits and at nonuniform speeds, contrary to all previous theories of the planets, including Copernicus'. Galileo's belief in circular orbits and uniform speeds, and in the natural, unforced character of circular motion generally, never wavered.

Then, late in 1609, Galileo learned that a Dutch spectacle maker had invented a device, a "spyglass" it was then called, that magnified the size of distant objects. Galileo's good friend and frequent collaborator Paolo Sarpi wrote to Galileo that he (Sarpi) already owned such a spyglass, and he described in his letter how it was constructed. Working very quickly, and pragmatically as he had no theory of optics to guide him—though Kepler had already published all the theory he would have needed to know[1]—, Galileo began making what he called "perspective tubes," the word "telescope" being coined very soon after (1612). He seized what he thought was his once-in-a-lifetime opportunity and presented one of his perspective tubes to the Venetian Senate as a gift. He suggested that in exchange it would be very nice if he were to receive a major increase in salary and a reduction in teaching responsibilities so that he could devote his time to discoveries that would enhance the glory of Venice. The Senate agreed.

The Dutch spyglass had been promoted as valuable for seeing distant objects more clearly, for example, ships at sea, but almost immediately Galileo used his to observe the heavens, starting with the moon. (Thomas Hariot in England had done the same thing slightly earlier, but Galileo's work wound up having much greater impact, in part because Hariot did not publish his scientific work.) Galileo discovered that, contrary to the accepted truth, the moon's surface was not perfectly smooth at all. It appeared to be highly irregular, with mountains and large dark areas that he interpreted as seas. Although it only had a magnification of twenty, his perspective tube revealed that the Milky Way, and some nebulae, rather than being "a denser part of the heavens" illuminated by sunlight or starlight, as previously thought, were "swarms of small stars placed exceedingly closely together." The tube revealed that there were many more stars in the sky than were visible to the naked eye, and that stars lost their twinkle when viewed through his tube. Most fatefully, by the end of 1609 it revealed four dots that moved across Jupiter in a periodic pattern. These were, he decided, moons orbiting Jupiter as our moon orbits Earth and *that* was front page–worthy news!

Galileo now saw what his *real* once-in-a-lifetime opportunity was. He quickly wrote up these astronomical discoveries in a small book with his own illustrations titled *Siderius Nuncius* (*Message,* or *Messenger, from the Stars*) and dedicated not just the book, but the four new "wandering stars" to the Grand Duke of Florence, Cosimo II Medici, naming them the "Medician stars." This was the first time in recorded history that a new heavenly body had been discovered, and the naming was a unique and unprecedented honor. It was also the first instance of heavenly bodies orbiting a center other than the Earth, which had implications for astronomical theory. Galileo's ambition was to move back to Florence, to be the highest paid university professor in the Italian city states, to be relieved of *all* teaching responsibilities, and to have the more prestigious title of professor of philosophy as well as professor of mathematics. Cosimo, who knew Galileo as a consultant, an astrologer, and a tutor to his children, agreed. Galileo, reneging on his deal with the Venetian Senate, relocated to Florence (leaving his common-law wife and family behind).

Galileo's observations, initially ridiculed by some, were soon confirmed as more and more people acquired telescopes. Galileo had promised Cosimo further dramatic telescopic discoveries once he was in Florence and he delivered on that promise. Within two years, he had calculated the periods of Jupiter's moons, which required just the kind of painstaking nightly observations for which he had earlier mocked Tycho. He observed that Venus showed phases, as our moon did, "proving" that Venus must orbit the sun, and by extension Mercury must do so as well although it was too close to the sun to see its phases. And he observed spots on telescopic images of the sun that he said were *on* the body of the sun. His observation of these spots revealed that they appeared, disappeared, and reappeared with a regular period of twenty-seven days. From this he inferred that the sun, like the Earth in the Copernican theory, rotated on its axis, though with a period of twenty-seven days not twenty-four hours. All of these were important discoveries, but they were not without controversy.

Jupiter's moons, Venus' phases, and the sun's axial rotation emboldened Galileo into openly promoting the truth of the Copernican theory, not just its usefulness as an astronomical hypothesis. But a *really* moving Earth and a *really* stationary sun contradicted biblical passages that described the Earth as stationary and the sun as moving. This triggered the first charges of heresy against him, charges that would dog him for the rest of his life and lead to his trial by the Inquisition in 1632. His claims about sunspots led him into a nasty dispute with the Jesuit mathematician and astronomer Christopher Scheiner, who had built a telescope based on Kepler's mathematical optics, not on Galileo's pragmatic design. The dispute was over priority: Who had discovered sunspots first, Scheiner or Galileo? (Scheiner had certainly published before Galileo, but again Hariot had preceded both without publishing.) The dispute was also over what the sunspots *were*. Scheiner argued that they were not on the sun, but solid bodies orbiting above the sun, like moons. Note that both men worked with the same data from the same kinds of instruments, and each claimed to *know* a reality that those data revealed, but the conclusions they reached were irreconcilable. In the course of this dispute, and others like it over the next decade, Galileo managed, by his vitriolic rhetoric, to make the Jesuits and the Dominicans implacable enemies.

What do Galileo's knowledge claims tell us about his response to the problem of knowledge? For Galileo, as for Bacon and Descartes, the goal of natural philosophy was knowledge of reality. When it suited his purposes, Galileo defended some knowledge claims as proven by observation, asserting that 'seeing is believing.' Just look through the telescope, he said, and see for yourself that the moon's surface is highly irregular, that Jupiter has four moons, that Venus has phases, that there are spots on the sun. "*The spyglass is very truthful*," he wrote in *Siderius*.[2] Without being able to explain how the device worked and knowing full well that his lenses suffered from distortion, Galileo insisted that just *looking* through the telescope *proved* these truths. He wrote that it was "certain," because he had seen it, "that . . . the moon has its surrounding vaporous orb," that is, an atmosphere. He was "certain" of this in 1610, but he quietly dropped this claim later, and with that the moon lost its atmosphere! *Is* there such a thing as simple seeing that counts as knowledge?

When it did not suit his purposes, as in his defense of the axially rotating and orbitally moving Earth, Galileo argued that reason overrides the plain evidence of our senses that the Earth is *not* moving. In *The Assayer* (1623), Galileo wrote that vision alone cannot differentiate the "real from the spurious."[3] And he added a lengthy argument distinguishing the "real" properties of matter, grasped by the mind as necessarily existing beyond the mind, from sensations that exist only *in* the mind, somehow caused by the essential properties of matter. Anticipating Descartes, he identified the former as shape, place, motion, size, and quantity, and the latter as color, taste, sound, and smell. And in his defense of the motion of the Earth against the evidence of our senses that it is motionless, he expressed the greatest admiration for Aristarchus of Samos and Copernicus, who "were able to make reason so conquer sense that in defiance of the latter, the former became mistress of their belief."[4]

What Galileo lacked was an objective criterion for determining in advance when the senses told us truths that reason had to accept and when the senses needed to be overruled by reason. The senses tell us that the Earth is stationary and the sun moves, but Galileo's reason tells him that the Earth moves and the sun is stationary. To the criticism that if the Earth orbits the sun, we should observe a stellar parallax, he could offer only Copernicus' ad hoc defense that the stars are so far away that the entire width of the Earth's orbit is no more than a point compared with that distance. Empirical evidence directly contradicts the axial rotation of the Earth, a motion of some one thousand miles per hour at the equator. We don't feel this motion and the flights of objects through the air—birds, arrows, cannonballs—are unaffected by it. Galileo could only offer another ad hoc defense: that the air close to the surface of the Earth must share in this motion.

Taking the offensive against deniers of the Earth's motion, Galileo argued, against readily available data, that the axial and orbital motions of the Earth were the causes of the ocean's tides. He was vehement in dismissing Kepler's claim (also Simon Stevin, Hariot, and others) that an attraction of the moon for the Earth was the cause. Support for this latter view came from a vast amount of tidal data that correlated well with the moon's orbital motion, but very poorly with the Earth's purported motions, yet Galileo never wavered in his insistence that the Earth's motion was the cause and not the moon. In an aside in his *Dialogue Concerning the Two Chief World Systems* (1632) Galileo chastised Kepler, who, despite "his open and acute mind," believed that the moon was responsible for the tides.[5]

As in his dispute with Scheiner about sunspots, the same data were used by both sides in the dispute over tides. In the case of sunspots, Galileo was right and Scheiner wrong; in the case of the tides it was Galileo who was wrong. He was wrong again in denying that comets were material bodies traveling through the solar system toward the sun. Galileo wrote, "Comets may be dissolved in a few days . . . their material is thinner and more tenuous than fog or smoke. In a word, a comet is more like a toy planet than the real thing."[6] Ten years later, in the *Dialogue*, he wrote that "comets are generated and dissipated in places above the lunar orbit" and doubted "whether comets are subject to parallax."[7] He was also wrong in insisting that hot bodies lose weight in giving off heat, persisting even in the face of careful experiments with his collaborator Paolo Sarpi that showed no detectable loss of weight whatsoever. His explanation of this negative experimental outcome was that the loss was too small "to be perceptible [by] any balance whatsoever."[8] He added that the matter given off by a hot body might be lighter than air, which would explain why some substances are heavier after burning than before.

Sometimes, reasoning led Galileo to conclusions subsequently deemed true and sometimes to conclusions subsequently deemed false, but as with sensory experience, Galileo had no independent criterion for determining when the conclusions of his reasoning were true and when they were not. And he seems not to have been aware that such a criterion was necessary in order to assess the validity of his knowledge claims. There is a common denominator to all of Galileo's disputes, one that covers all of his knowledge claims: a conviction that in appropriate contexts, data speak for themselves, at least to him, whether the data are derived from sensory experience, instruments, or experiments. He never saw his conclusions as reasonable *interpretations* of data, but as their *necessary* consequence.

This mindset is a corollary of his Archimedean method for solving problems in physics, using geometry to describe phenomena but also using geometric, hence deductive, reasoning in arriving at those solutions. In *The Assayer*, which contains many references to his methods, Galileo extended Archimedes' classic work on floating bodies. He wrote that to contradict geometry-style proofs in physics "is a bald denial of truth"[9]; it was irrational. Yet ostensibly quite rational people did deny his deductively arrived at conclusions. His response was to attribute their criticism to "a certain perverse urge to detract, steal or deprecate that modicum of merit I had earned."[10]

In was in *The Assayer* that Galileo laid down a fundamental principle of his natural philosophy:

> Philosophy is written in this grand book, the Universe, which stands continually open to our gaze. But the book cannot be understood unless one first learns to comprehend the language and read the letters in which it is composed. It is written in the language of mathematics, and its characters are triangles, circles, and other geometric figures without which it is humanly impossible to understand a single word of it; without these, one wanders about in a dark labyrinth.[11]

This is the basis of his use of geometric reasoning in mathematical physics, though of course it is not applicable to observational evidence, for example, to what his telescope revealed to be true about the moon, planets, and stars.

In his *Dialogue*, Galileo wrote that in the natural sciences "conclusions are true and necessary and have nothing to do with human will."[12] In these demonstrative sciences, knowledge of the world is reached by deductive reasoning. The natural philosopher begins "for the most part" with information from the senses, then comes careful observation, then experiment. The results of experiments become the conclusion of a deductive argument using as premises propositions that have been demonstrated previously or are asserted axiomatically: "You may be sure," however, that one must be certain of the conclusion *before* constructing the logical demonstration, not after![13]

Deductive reasoning is not the only route to revealing truths about nature. Galileo accepted William Gilbert's experimental studies of the magnet, summarized in his book *On the Magnet* (1600), as revealing its "true nature." Gilbert's "method of procedure in philosophizing" about nature, he acknowledged, bears "a certain likeness to my own," but because Gilbert was insufficiently mathematical, the reasons he gives in support of his conclusions are not rigorous.[14] Observations alone can sometimes produce knowledge of nature, as in his claim that it is "certain that Venus and Mercury must revolve around the sun," and sunspots were "conclusively proved [by the telescope] to be produced and dissolved and to be situated next to the body of the sun and to revolve with it or in relation to it."[15] The ideal route to truth, however, is logical demonstration.

In the *Dialogue*, Galileo's mouthpiece Salviati says that the human understanding can be taken in two ways, the intensive or the extensive. Extensively, that is, with regard to the multitude of intelligibles, which are infinite, the human understanding is as nothing even if it understands a thousand propositions . . . But taking man's understanding intensively . . . in so far as this term denotes understanding some proposition perfectly, I say the human intellect does understand some of them perfectly and thus in these has as much absolute certainty as Nature herself has. Of such are the mathematical sciences alone, that is, geometry and arithmetic.[16]

Galileo's most important scientific work by far was the *Discourses and Mathematical Demonstrations Concerning Two New Sciences, Pertaining to Mechanics and Motion.* (This was the title the publisher gave it; Galileo wanted it to be *Discourses on Motion*.) This was Galileo's last book, completed while he was under house arrest after his trial in 1632. The Church authorities would not allow him to publish this book, even though it contains nothing at all on astronomy or the motion of the Earth. Banned from publication in a Catholic country, two copies were smuggled out of Italy, one with the assistance of the French ambassador to Rome and one by an old friend and lifelong supporter, Fulgenzio Micanzio. Both copies wound up in the hands of a Protestant Dutch publisher, Elzevir. Most of the results announced in this book were in the notebooks Galileo had compiled during his eighteen years at the University of Padua, but here he has organized them into deductive proofs, in the manner of Archimedes. There are repeated references to Archimedes' mathematical physics as the model for Galileo's method. At one point, his mouthpiece Salviati says that he has "read and studied" Archimedes' books "with infinite astonishment."[17]

The first of the two new sciences Galileo announced pertains to strength of materials and can be considered the founding text of engineering based on mathematical physics. Galileo presents a solution, "demonstrated by geometry"[18] of the scaling problem: calculating the size that a beam must be to support progressively larger weights. Galileo praises Venetian artisans, whom he calls "first rank men," and who display expertise in applied mechanics and are "constantly" constructing new instruments and machines.[19] What Galileo adds to their practical expertise is a theory explaining why what they do pragmatically "works" and that allows deducing solutions to new practical problems, as Archimedes' theories had done.[20]

Discussing the vacuum and the force it can exert, Galileo wrote, "Experiment *leaves no doubt* [my emphasis] that the reason why two [smoothly machined] plates cannot be separated [after being pressed together] except with violent effort is that they are held together by the resistance of the vacuum."[21] Note the conflation here of certainty ("leaves no doubt") and empiricism. He takes as axiomatic that each effect "must have one true and sufficient cause," which is not at all self-evident and is in fact false. After discussing experiments aimed at measuring the speeds of sound and light, later of weighing air, Galileo describes in detail a thought experiment that proves conclusively that the speed of a falling body is not dependent on its weight. There is no need to climb the Tower of Pisa and drop weights off it.

Galileo proceeds to describe his discovery of the constancy of the period of a pendulum based solely on its length, regardless of the degree of displacement, and gives a geometric "equation" linking the two. This is another example of discovering nature's universal and necessary laws through the use of experiment-based reasoning. (Neither Galileo, nor his son, nor his assistants succeeded in building a clock based on the pendulum principle, though they all tried. Christiaan Huygens succeeded in 1656.) After discussing resonance phenomena and sympathetic vibrations, Galileo presents a lengthy discussion of the mathematics underlying stringed instruments. He is revisiting and formalizing work his father Vincenzo had done, some of it with Galileo's help, half a century before. Vincenzo was an important musical theorist who wrote two books on the mathematics of tuning systems and instrument design that were heavily dependent on experiment.

The account of the first of Galileo's new sciences concludes with a rigorously mathematical-deductive extension of Archimedes' theory of the balance equilibrium to "real-world" problems of design and construction, including the length and strength of levers and beams. Here Galileo gives a geometrical demonstration of his famous equation relating the distance covered by a freely falling body to the square of the time of fall. He gave a geometric proof of the relationship between distance and time ($S = kt^2$ algebraically), as he did for his equally famous claim that all bodies fall at the same rate. Note that these results rely on idealizations that are easily falsified by experiments, for example, by actually dropping weights from a tower. With respect to falling bodies and moving bodies generally, Galileo wrote that ignoring the resistance of the medium through which the body falls or moves will be "without sensible error," as confirmed by the period of the pendulum.[22] Galileo's use of thought experiments and of idealizations in solving problems in physics were major contributions to modern science. The universal character of theories and "laws of nature" in modern science almost always apply to idealizations and only work empirically in limited situations.[23]

The second new science, Galileo's new science of motion, begins on the third day of the dialogues. At one point Simplicio challenges Galileo's mouthpiece Salviati on an issue that speaks directly to the problem of knowledge as we have been discussing it. Simplico accepts that the mathematical proofs Galileo has produced are logically valid, but how, he asks, can we be sure that they apply to nature? Salviati replies as follows:

The request which you, as a man of science, make is a very reasonable one; for this is the custom, and properly so, in those sciences where mathematical demonstrations are applied to natural phenomena, as is seen in the case of perspective [drawing and painting], astronomy, mechanics [machine design and construction], music, and others where the principles, once established by well-chosen experiments become the foundation of the entire superstructure.[24]

Does this answer Simplicio's question, which goes to the heart of the claim that certain knowledge of nature is possible, or does it just restate Salviati's claim? *Does* the necessary truth of deductive reasoning transfer to what is external to the mind? Galileo seems to respond with a cleverly composed evasion of a question to which he could not give an honest answer. Galileo goes on to present important results from the application of his experimental-deductive method to bodies in motion. Among these results are the parabolic nature of ballistic trajectories, long known though not with the generality of Galileo's proof, and that a body can have multiple simultaneous motions as a result of multiple forces acting on it. This is a result of great significance and became a cornerstone of modern physics.

Galileo's method emphasizes the centrality to acquiring knowledge of nature of evidence derived from the senses, of empirical observations, and of experiments, all of which sounds Baconian. But Galileo's method is keyed to deduction, not induction, to mathematics as the "language" of nature itself, to reason overriding empirical evidence, and to experiments, including thought experiments and idealizations, designed to confirm conclusions already reached that will then be reverse engineered, so to speak, into deductive arguments from which the necessary truth of the conclusion can be deduced. Galileo shares with Descartes a commitment to deduction in order to arrive at certain knowledge of nature as it "really" is, as well as to an essential role for mathematics and an active role for the mind in producing knowledge of nature. Unlike Descartes, however, experiments are a primary feature of Galileo's method, and he rejects Descartes' innate ideas and intuited first principles.

As did Bacon and Descartes, Galileo invented a method of reasoning about nature that conflates induction and deduction, empiricism and rationalism, and the view of the Giants and the view of the Gods. It purports to use experience and reasoning to give us certain and not just probable knowledge of a reality that transcends experience and that cannot itself be experienced. But there is a deep problem with any attempt to get behind experience: how can we know that logically valid proofs apply to nature? Galileo recognized the validity of this question and allowed Simplicio to ask it, but he had no answer other than pragmatic success. (Recall, however, the Fallacy of Affirming the Consequent.) If experience is untrustworthy, what is the standard against which we correct it? Everyone agrees that we have no access to the reality behind experience except *in* experience, so what does it mean to correct experience? Relative to what? The Baconian, Cartesian, and Galilean methods claimed to provide a correction mechanism and standard, but do they?

Galileo's most famous work, the *Dialogue Concerning the Two Chief World Systems*, demonstrates the depth of Galileo's commitment to truths about reality being the goal of natural philosophy. It also reveals something significant about Galileo personally and about the limitations of his method for producing truths about reality.

By 1616, Galileo was fully aware of how dangerous it was to hold that the Copernican theory was physically true. In that year he was officially warned by the Church in the person of Cardinal Roberto Bellarmine that while he was free to teach Copernicus' theory as a mathematical theory and as a physical hypothesis, he must not teach it as the truth. The Church was not in the astronomy business, but Scripture, particularly in *Joshua* (10:12), states that the Earth is stationary and the sun moves. As everyone knew, there were many passages in the Hebrew Bible whose literal meaning the Church had set aside in favor of an alternative, Christian, meaning. Galileo argued that the Church should do the same with the astronomically problematic passages, taking into account the evidence of our senses and logically demonstrated truths.

But Galileo did not take into account the seriousness of the situation the Church was in. It was under attack by the Protestant revolution against the authority of the Catholic Church, an attack that in 1618 erupted into a brutal thirty-year-long war in which millions died. A core principle of Protestantism was that the Catholic Church was not the final authority on the meaning of Scripture. In 1565, the Catholic Church had convened a Council at Trent to coordinate a counterattack against the Protestants. A "primary decree" of that Council read as follows:

> In matters of faith and morals pertaining to the edification of Christian doctrine, no one, relying on his own judgment and distorting the Sacred Scriptures according to his own conceptions, shall dare to interpret them contrary to the sense which Holy

Mother Church, to whom it belongs to judge their true sense and meaning, has held and does hold, or even contrary to the unanimous agreement of the fathers [of the Catholic Church].[25]

Galileo knew all this, of course, but he and his circle of intellectuals talked and acted, sometimes discreetly and sometimes not, as if it were obvious that Church teaching needed to accommodate itself to truths of reason. Galileo professed to be a good son of the Church, yet he seemed not to appreciate that the Church's primary mission was the salvation of souls, not to mention the survival of the Church itself. It was not the Church's mission to protect the free exercise of abstract reasoning by a minute fraction of the population. Bellarmine only forbade Galileo to teach the Copernican theory as a physical truth established by reason, because that threatened the Church's authority with respect to the meaning of Scripture. Bellarmine argued, and Galileo agreed, that all motion was relative so that Galileo could not *know* with certainty whether the sun or the Earth moved.

In 1623, Cardinal Maffeo Barberini, who had a long-standing friendly relationship with and respect for Galileo, was elected Pope, taking the name Urban VIII. Knowing full well the position Galileo was in with respect to Copernicus' theory, he invited Galileo to write a book defending that theory but without calling it true. Galileo responded by producing the *Dialogue Concerning the Two Chief World Systems* and it was doubly deceptive: it deceived the Pope and it deceived its nonexpert readers. Galileo wrote the book in Italian rather than in Latin in order to reach the maximum number of readers in his home country. (It would soon be translated into a number of European languages.) And he constructed the book in such a way that only a fool could doubt the physical truth of the Copernican theory. This not only deceived Urban VIII; it also put the Pope in an extremely embarrassing position given Galileo's many enemies in the Church hierarchy, especially in the Jesuit and Dominican orders. Urban yielded to the pressure from Galileo's enemies and approved an investigation by the Roman Inquisition into Galileo's possible heresy.

The outcome of this investigation and the subsequent trial are well known. Galileo was convicted and forced to publicly and definitively proclaim the falsity of the Copernican theory and the truth of the motionlessness of the Earth, which he did (under threat of torture). His sentence was house arrest for life and monitoring of his activities to make sure he had no opportunity to discuss astronomical theory with anyone.

Galileo was wrong in arguing that the truth of the Copernican theory was *proven* by his telescopic observations. It was made more probable by them, but in no sense was it proven. Galileo's telescope did not, indeed could not, provide compelling evidence that the Earth moved. Moreover, there was an alternative model with a motionless Earth that was consistent with all the data available to Galileo, and Galileo knew it, as we will discuss shortly. Furthermore, Galileo was wrong in defending the truth of the Copernican theory because it is *not* true. Copernicus held that the planets moved in circular orbits around the sun at uniform speeds, that the sun was located an arbitrary distance away from the center of the Earth's orbit, that no force was needed to keep the planets in their orbits, and that all the stars were roughly the same distance away from the sun. In 1609 Kepler had claimed, based on precise observational data, that the planets moved in elliptical orbits at nonuniform speeds, but Galileo dismissed these claims as erroneous.

The *Dialogue* deceived its readers by presenting a face-off between two "chief" world systems, the Ptolemaic and the Copernican. But there was another "chief" world system in 1632, one that was accepted by most European astronomers and that Galileo knew well. He did not believe it, however, and he suppressed any mention of it in the *Dialogue*. This was the theory of Tycho Brahe. Long before Galileo saw the phases of Venus through his telescope, Brahe had proposed a kind of halfway house between Ptolemy and Copernicus, whose theory he had rejected after careful study. In Brahe's model of the heavens, Venus and Mercury orbit the sun, and the sun, together with Venus and Mercury, orbits the Earth, which is stationary at the center of the universe. The outer planets then orbit the Earth-centered and the sun-Venus-Mercury subsystem. This may strike us as bizarre today, but Brahe's world system is consistent with all the evidence available to Galileo, including all telescopic evidence. Galileo knew this, yet there is not a single reference to Brahe's system in the *Dialogue*. There are scattered, and mostly negative, references to Brahe as an observational astronomer, but not a single reference to Brahe's system. It is impossible not to conclude that Galileo deliberately omitted Brahe's system *because* it was empirically indistinguishable from Copernicus'. To have included it would have made it impossible to portray Copernicus' system as the system proven true by reason.

Galileo passionately *believed* that the Copernican theory was the one correct, true account of astronomical reality, but he did not *know* it. In may seem to us that any intelligent, open-minded, rational person in the early seventeenth century should have adopted the Copernican system. The problem, however, is unlearning twenty-first-century astronomy and thinking like a person in the early seventeenth century. Most astronomers in Galileo's lifetime sided with Brahe, not Copernicus, on the grounds that Brahe's was the more scientific choice![26] It was supported by all the available evidence, made the same predictions of the future positions of the planets in the night sky as

Copernicus' theory did, and it did not require the kind of ad hoc explanations that the Copernican theory required in order to explain away the absence of any evidence of the Earth's motion.

Which astronomical theory to choose? Which theory of the tides, of sunspots, of comets, of heat, of the constitution of matter to adopt? One possible response is 'why choose?': use the theory that works best for you. But that's not an option if a nonnegotiable criterion of truth is correspondence with reality. If that is the case, then the fact that a theory works empirically does not mean that it is necessarily true. And it is quite possible, indeed probable, that more than one theory works equally well empirically, as with the Copernicus and Tycho theories. This highlights the dependence of scientific reasoning on assumptions that must be adopted in order to reason scientifically but cannot be deduced from empirical data, and so cannot be known with certainty to be true. And the history of science shows that assumptions change over time, even those that were once considered unquestionable.

4
Newton and Knowledge of the Universe

In his lifetime, "Newton" had become as synonymous with scientific genius as "Einstein" is today. His death elicited Alexander Pope's adulatory epitaph "Nature and Nature's laws lay hid in night, God said 'Let Newton be' and all was light."

Galileo was prickly and self-aggrandizing and so was Newton, but Newton was also vindictive, deceitful, manipulative, and petty to an extraordinary degree. To what extent, if any, do such personal characteristics affect our understanding of scientific knowledge?

Galileo was an inspiring teacher, but he was also arrogant, combative, thin-skinned, self-destructive, self-promoting, and reluctant to share credit. These are personal characteristics that seem to conflict with the popular conception of science as an impersonal pursuit of truth. For his part, Newton, and Galileo as well, could not tolerate disagreement with what he had concluded was true and could not acknowledge that contemporaries contributed in any way to his achievements, or that objections to his conclusions had any merit at all. Newton suppressed the fact that in 1679, eight years before he published his theory of universal gravitation in *The Mathematical Principles of Natural Philosophy* (the *Principia*), Robert Hooke had written to Newton asking if he thought that the planetary orbits might be explained by an attractive force from the sun that obeyed an inverse square law. Newton dismissed out of hand Hooke's earlier criticism of his theory of colors based on Newton's famous prism experiments. In his capacity as President of the Royal Society, Newton anonymously "rigged" in his own favor a determination of whether he or Leibniz first invented the calculus. Equally disturbing was his betrayal of the trust of John Flamsteed, the Astronomer Royal of Great Britain, by publishing observational data accumulated by Flamsteed after promising Flamsteed not to do so.

And then there are Newton's extensive studies of alchemy, religion, ancient history, and biblical chronology, amounting to at least two-thirds of the words he left behind. Newton was profoundly religious and he did not compartmentalize his life: here his physics, there his faith, here his biblical studies, there his mathematics.[1] It was only in the mid-twentieth century that all of

Newton's notebooks and manuscript pages were fully studied by scholars.[2] At that point, it became clear that Newton's contributions to physical science and mathematics were only part of an integrated pursuit of the whole truth about the universe. This truth began with the creation of the universe by the all-powerful God of the Hebrew-Christian Bible and included the Creator's continuing involvement as sustainer of the natural world, including the many hidden forces at work producing natural phenomena, and a "scientific" account of how the universe behaves at the level of empirical human experience. Studying alchemy, the Bible, ancient history, and biblical chronology complemented physics and mathematics, which, alone, did not tell the whole truth about the world. Furthermore, the partial truth they told revealed that they were part of a greater whole: "It is in Newton's belief in the unity of Truth, guaranteed by the unity and majesty of God, that one may find a way to reunite his many brilliant facets, which however well-polished, now remain incomplete fragments."[3]

Newton had no interest in practical alchemy, but he needed to study alchemy in order to understand aspects of nature not dealt with in physics and mathematics. Newton studied the Bible intensively because he believed that it contained clues to the nature of reality as well as to the nature of God and of God's relation to, and role in maintaining, the world. In the second edition of his *Principia*, Newton added a General Scholium to the last proposition of his theory of gravity. It begins by refuting Descartes' vortex-based theory of gravity and then discusses at some length God's nature and role in the universe. "It is agreed," he wrote,

> that the supreme God necessarily exists and by the same necessity he is always and everywhere . . . All the diversity of created things, each in its place and time, could only have arisen from the ideas and the will of a necessarily existing being.

Newton ends by saying that "to treat of God from phenomena is certainly a part of natural philosophy."[4]

In the *Opticks*, Newton posed a series of "Queries" in natural philosophy whose answers would reveal that God was "behind" nature. What, if anything, he asked, is there in space "almost empty of matter"; how does gravity act on bodies without dense Matter between them; why does nature do nothing in vain, and what is the source of its order and beauty; why do the planets all move the same way as they orbit the sun while comets move every which way; what keeps the stars "from falling upon one another"; how did the organs of animals come to be so well "contrived" and matched to one another; how does the will move the body and what is the nature of instinct? Once these questions are answered, Newton confidently predicted, the phenomena of nature would reveal the existence of a Being "incorporeal, living, intelligent, omnipresent," for whom infinite space is the "Sensorium" by which it is aware of everything that goes on in the universe it created.[5]

Newton believed that, in addition to gravity, electricity, and magnetism, there were other hidden forces at work in nature, forces that acted at a distance to produce the world as we experience it. What we call chemical reactions, in which atoms selectively combine to form molecules and molecules combine to form compounds, were the result of forces of selective attraction and repulsion between atoms, causing them to combine in certain ways and not in others. Alchemy was the theoretical and experimental study of those forces and, as with his biblical studies, Newton accepted that understanding these forces required uncovering clues encoded in esoteric alchemical manuscripts.

In his magisterial biography of Newton, Robert Westfall described Newton's passion for knowing the Truth about Reality this way:

> For eight years [Newton] had locked himself in a remorseless struggle with Truth. Eight years of uneaten meals and sleepless nights, eight years of continued ecstasy as he faced Truth directly on grounds hitherto unknown to the human spirit . . . By 1672, Newton had lived with his theory [of colors] for six years and it now seemed obvious to him. For everyone else, however, it still embodied a denial of common sense that made it difficult to accept. Their inability to recognize the force of his demonstrations drove Newton to distraction. He was unprepared for anything except immediate acceptance of his theory.[6]

Westfall captures the depth of Newton's lifelong and life-absorbing ontological commitment to uncovering the truth about the universe. He believed that reason could uncover this truth and that in his natural philosophy, but also in his biblical studies, he had succeeded in uncovering some parts of it. He rejected Descartes' method of reasoning about nature because it produced merely hypothetical accounts of how natural phenomena *might* be caused and of how reality *might* be. Newton's goal was to uncover the "true causes" of natural phenomena. In his theory of colors, Newton wrote that,

> A naturalist would scarce expect to see the science [of colors] become mathematical, and yet I dare affirm that there is as much certainty in it as in any other part of Opticks. For what I shall tell concerning [colors] is not an Hypothesis but most rigid consequences, not conjectured by barely inferring 'tis thus because not otherwise

or because it satisfies all the phenomena . . . but evinced by the mediation of experiment concluding directly and without any suspicion of doubt.[7]

Newton claimed that his prism experiments decisively revealed what sunlight really was. Sunlight was not "homogeneal," as it appears to our senses, but heterogeneous, a mixture of colors ranging from red to violet. These experiments also revealed that there was a mathematical basis to the decomposition of sunlight into these colors. As Galileo did with respect to his telescopic observations, Newton insisted that his experimental data spoke for themselves, that experimental outcomes, without any interpretation or argument, reveal truths about natural phenomena. He was frustrated by the resistance of others to the conclusions he drew from his experiments. Everyone *had* to draw the same conclusions from his experimental data that he did *because* the data spoke for themselves.

Newton's theory of colors required dismissing the perceived homogeneity of sunlight as an illusory sense experience. This experience was "corrected," and the "true" nature of sunlight revealed, by his experiment. He called this a "crucial experiment," adopting Francis Bacon's language, because it decided between two possible alternatives: the homogeneity of sunlight or its heterogeneity. No logical argument was necessary in support of his claim of heterogeneity because just *looking* at the results of his experiment revealed the *certain* truth of heterogeneity, as clearly as Galileo's telescope revealed that Venus had phases. Note, though, how problematic this *looking* is.

Sunlight *looks* homogeneous but this is false, while *looking at* the spread of colors produced by passing sunlight through a prism of a particular composition and shape under particular experimental conditions specified by Newton reveals, veridically, that sunlight is heterogeneous. We *see* that this is so and we *know* it, according to Newton; we know it universally, necessarily, and certainly. It follows that experiment alone can produce certain knowledge out of misleading empirical experiences. But what is the criterion by which we reject certain empirical experiences as false, for example, seeing the homogeneity of sunlight, and others as true, not just seeing the spread of refracted colors emerging from Newton's prisms but seeing the colors as *proving* the heterogeneity of sunlight?

Newton's greatest achievements in physics, his universal theory of gravity and his theory of the motion of material objects (mechanics), were not deduced from experience, though they were tested against empirical observations. His mechanics, which became as dominant in modern physics as Euclidean geometry was in mathematics, is based on his three laws of motion. These laws, in turn, are based on his definitions of space, time, matter, and motion. It is important to note that these definitions were not deduced from empirical experience and could not have been proven experimentally. They were assumptions, not at all self-evidently true.

Here is what Newton had to say in his *Principia* about the foundations of his mechanics:

> Although time, space, place, and motion are very familiar to everyone, it must be noted that these quantities are popularly conceived solely with reference to the objects of sense perception. And this is the source of certain [erroneous] preconceptions; to eliminate them it is useful to distinguish these quantities into absolute and relative, true and apparent, mathematical and common.

Sense perception must give way to "truths" that only the mind can produce. Consider Newton on motion: "Absolute motion is the change of a body from one absolute place to another; relative motion is change of position from one relative place to another . . ." It is certainly very difficult to find out the true motions of individual bodies and actually to differentiate them from apparent motions, because the parts of that immovable space in which the bodies truly move make no impression on the senses. Difficult but not hopeless, as it is possible to draw evidence partly from apparent motions and partly from the forces that are "the causes and the effects of the true motions."[8]

Newton *defines* it to be that case that space and time are "things" that are independent of and unaffected by what is in space or what happens in time. Space and time would exist with the specific properties Newton assigned to them even if the universe were empty of material objects and nothing happened. These definitions allow Newton to distinguish absolute and relative motion, but they were contested by Christiaan Huygens, and after him by Gottfried Leibniz, both of whom argued that space and time were relationships, not things, which implied that there was no such thing as absolute motion. Newton was also aware of Pierre Gassendi's mathematical mechanics based on the claim that space, time, and motion, like matter, were atomic, based on indivisible, discrete units. In these rival versions of mechanics, all motion was relative as in Einstein's relativity theories. Furthermore, relativity theory not only treats space and time as relationships, it also adds that the properties of both are variable, a function of what occupies space and happens in time. Space and time are local for Einstein, not universal as for Newton, their properties dependent on the local density of matter and energy. In addition, all measurements of time, spatial position, and motion are relative to some particular, freely chosen frame of reference. According to relativity theory, Newtonian space and time do not exist.

Newton laid down other definitions that his mechanics required in order to support his equations of motion, among them, definitions of matter and "measures" [recall Protagoras] of motion, inertia, and force. These definitions, too, were challenged, by Leibniz for one, who maintained that there were no such things as Newtonian material objects. Instead, matter and the properties we assign to it are "really" the expression of forces whose action can be described by mathematical laws, the very same equations that Huygens and Newton had discovered, but without requiring Newton's ontology of material objects exerting those forces. (Leibniz' mechanics was developed more fully in the eighteenth century by Roger Boscovich, who replaced solid matter with mutually impenetrable point force centers.)

Newton's famous three laws of motion also are not deductions from experimental data. Newton called them "Axioms, or the Laws of Motion," because they are presumed, not "deduced" from phenomena. The point at issue here is the move from particular experience, in the form of experiment-generated data, to universal truths about natural phenomena. Like Galileo, Bacon, and Descartes before him, and effectively all of his contemporaries, Newton was fully aware how fateful this move was for any claim to knowledge of reality.[9] And like Galileo, Bacon, and Descartes, inter alia, he was aware that only a systematic method of reasoning could validate it.

In a letter to Oldenburg written in 1672 regarding his newly announced theory of colors, Newton had written (in his first scientific publication),

> For the best and safest method of philosophizing [about nature] seems to be, first
> diligently to investigate the properties of things and establish them by experiment,
> and then to seek hypotheses to explain them. For hypotheses ought to be fitted
> merely to explain the properties of things and not attempt to predetermine them
> except in so far as they can be an aid to experiments. If any one offers conjectures
> about the truth of things from the mere possibility of hypotheses, I do not see how
> anything certain can be determined in any science; for it is always possible [to for
> mulate] hypotheses, one after another, which are found rich in new tribulations.[10]

In the second edition of his masterwork, the *Principia*, however, Newton added a General Scholium in which he wrote,

> I have explained the phenomena of the heavens and of our sea [the periodicity
> of the tides] by the force of gravity, but I have not yet assigned a cause to gravity.
> Indeed, this force arises from some cause that penetrates as far as the centers of

the sun and planets without any diminution of its power to act, and that acts not in proportion to the *surface* of the particles on which it acts (as mechanical causes are wont to do) but in proportion to the quantity of *solid* matter, and whose action is extended everywhere to immense distances, always decreasing as the squares of the distances . . . I have not as yet been able to *deduce* from phenomena the reason for these properties of gravity, and I do not feign hypotheses. For whatever is not *deduced* from the phenomena must be called a hypothesis; and hypotheses, whether metaphysical or physical, or based on occult qualities, or mechanical, have no place in experimental philosophy. In this experimental philosophy, propositions are *deduced* from the phenomena and are made general by induction. The impenetrability, mobility, and impetus of bodies, and the laws of motion and the law of gravity have been found by this method. And it is enough that gravity *really exists* and acts according to the laws that we have set forth and is sufficient to explain all the motions of the heavenly bodies and of our sea [my emphases].[11]

In his *Opticks* he wrote:

the main business of natural philosophy is to argue from Phaenomena without feigning Hypotheses, and to deduce Causes from Effects, till we come to the very first Cause, which certainly is not mechanical; and not only to unfold the Mechanism of the World, but chiefly to resolve these and suchlike questions.[12]

What *was* Newton's attitude toward hypotheses, and what, more specifically, was his method of reasoning? As regards hypotheses, it is important to recognize the equivocal character of Newton's use of the term: disdainfully and dismissively when describing Descartes' hypotheses, which serve explanatory ends for Descartes; constructively when describing his own use of them as disposable means in his own, very different method of reasoning. For Descartes, a hypothesis based on "clear and distinct" universal truths from which phenomena could be deduced was the truth and the end of the matter for natural philosophy. That was the goal of natural philosophical inquiry. To be sure, Descartes acknowledged that an omnipotent God could have accomplished the same effects using other causes, but why would He (other than to mock us)?

Newton dismissed the use to which Descartes put hypotheses in his method of producing knowledge. Newton's own method revealed the "true causes" operative in nature, not some logically possible, rationalist fantasy. Newton had no qualms about using hypotheses in his reasoning, but only as an intermediate means to non-hypothetical ends. He was even comfortable openly speculating, using queries as a means of announcing what he believed the truth to be but was as yet unable to mount a compelling, evidence-based argument for.

For example—and perhaps surprisingly, given what he had written in his attack on Descartes' use of hypotheses in the second edition of the *Principia*, cited earlier—in the third edition, Newton added to the General Scholium a concluding paragraph in which he refers to "a certain subtle spirit pervading gross bodies and lying hidden in them." The "force and actions" of this spirit are, he says, responsible for a wide range of phenomena, from chemical reactions to sensation and the willful behavior of animals, presumably including humans, "by the vibrations of this spirit being propagated through the solid fibers of the nerves from the external organs of the sense to the brain and from the brain into the muscles." All this, he writes, is too complicated to explain "in a few words," and in addition "there is not a sufficient number of experiments to determine and demonstrate accurately the laws governing the actions of this spirit."[13]

In the *Opticks*, Newton offered numerous hypothetical speculations, in the form of Queries, about light and matter that he could not yet prove to be true, but that he clearly believed to be true. Many of these are remarkably prescient, among them: that matter and light "are convertible into one another"[14] and "act mutually on one another"[15]; that material bodies "act upon light at a distance and bend its rays,"[16] the light bending "before arriving at bodies"[17]; that "Rays of light are very small Bodies emitted from shining Substances."[18] In the midst of these speculations, Newton inserted an extended description of the method of reasoning that he used to produce truths about nature:

As in Mathematicks, so in Natural Philosophy, the investigation of difficult Things by the Method of Analysis [Resolution], ought ever to precede the Method of Composition [Synthesis]. This Analysis consists in making Experiments and Observations, and in drawing General Conclusions from them by Induction, and admitting of no Objections against the Conclusions, but such as are taken from Experiments or other certain Truths. For Hypotheses are not to be regarded in experimental Philosophy. And although the arguing from Experiments and observations by Induction be no demonstration of general Conclusions; yet it is the best way of arguing which the Nature of Things admits of, and may be looked upon as so much the stronger, by how much the Induction is more general. And if no exception occur from Phaenomena, the Conclusion may be pronounced generally. But if at any times afterwards any exceptions shall occur from experiments, it may then begin to be pronounced with such Exceptions as occur. [Descartes, recall, had made the identical *caveat*.] By this way of Analysis, we may proceed from Compounds to Ingredients, and from Motions to their Causes, and from particular Causes to more general ones, till the Argument end in the most general. This is the Method of Analysis.[19]

Newton goes on to describe his use of this method in arriving at his theories of light and colors. The method of Synthesis, also called the method of Composition, "consists in assuming the Causes discover'd, and established as Principles, and by them explaining the Phaenomena proceeding from them, and proving the Explanations."[20] That is, "proving the Explanations" by deducing particular phenomena from the general principles arrived at through the method of Analysis or Resolution, a process Newton sometimes calls induction and sometimes "deducing [general conclusions] from the Phaenomena" themselves. This is as perfect an illustration as one could wish for of being *both* a God *and* a Giant by conflating induction and deduction. This was, alas, the only way that one could claim to base universal, necessary, and certain knowledge of reality on empirical experience. Newton knew very well that, logically, particulars cannot prove the truth of generalizations based on them. Logically, facts are only "rude heaps of unpolish'd material" that need to be interpreted nonlogically. Though Newton repeatedly calls the inferences he drew from experimental data "deductions"—Sherlock Holmes repeatedly does the same thing!—he knew that induction and deduction cannot be bridged logically, but he believed, as Bacon, Galileo, and Descartes before him, that they could be bridged by methodical reasoning, each proclaiming their method the right one for doing so.

In this conflation of induction and deduction, Newton was typical of many the best natural philosophers of his generation. Huygens, his older contemporary and nearly his equal as a founder of modern mathematical physics, wrote, in the introduction to his *Treatise on Light* (1692),

> There will be seen [in this work] demonstrations of those kinds which do not produce as great a certitude as those of Geometry, and which even differ therefrom, since whereas the Geometers prove their Propositions by fixed and incontestable Principles, Here [contrary to the Fallacy of Affirming the Consequent] the Principles [universal theories] are verified by the [experimental] conclusions to be drawn from them; the nature of these things not allowing of this being done otherwise. It is always possible to attain thereby to a degree of probability which very often is scarcely less than complete proof. To wit, when things which have been demonstrated by the Principles that have been assumed to correspond perfectly to the phenomena which experiment has brought under observation, especially when there are a great number of [observations/data], and, further, principally, when one can imagine and foresee new phenomena which ought to follow from the hypotheses which one employs, and one finds that the facts correspond to our prevision . . . It must be ill if the facts are not pretty much as I represent them here.[21]

What Huygens "represents" in this work, for those who "love to know the Causes of things," is his spherical wave theory of light. This was in direct opposition to Newton's corpuscular theory of light, a theory whose truth Newton claimed to have "deduced" from *his* experimental observations. Both claimed that their theories corresponded to reality as "proven" by experimental observations and predictions, and both theories differed from Descartes' theory of light, which he also had claimed corresponded to reality. Huygens was sensitive to how problematic claims of knowledge of reality were. In his *Cosmotheoros*, published posthumously in 1698, he had written, "we do not here advance anything with complete conviction," as Galileo, Descartes, and Newton had done. For his part, Huygens says that he is "content with conjectures concerning whose likelihood each is free to judge for himself." If someone were to ask, why propose

conjectures about things which we ourselves admit cannot be comprehended with certainty, I would answer that the entire study of physics, to the extent that it is concerned with finding the causes of phenomena would be subjected to disapprobation for the same reason, the greatest glory being to have found likely theories . . . But there are many degrees of likelihood of which some are nearer the truth than others; it is above all in the evaluation of these degrees that one must show good sense.[22]

This is consistent with what Huygens had written earlier, that in physics there are no strictly logical demonstrations/proofs. Identification of causes requires making generalized "suppositions," or assumptions, about the experimental data available on the effects being studied. These suppositions must then be tested against other similar effects to see if those effects too can be deduced from the same suppositions, recognizing that even this does not *prove* the truth of the suppositions. But lack of such demonstrative proof "must not cause us to conclude that everything is equally uncertain."[23] The more experiments people do that are consistent with the truth of the suppositions, the more probable their truth and the truth of the causes identified.

Of course, one must ask how Huygens can know that one set of suppositions is closer to the truth than another in the absence of any theory-neutral access to the truth. Huygens, who began his career as a natural philosopher as a Cartesian, soon recognized that Descartes' criterion of "clear and distinct" ideas cannot give us certainty. He concluded that nearly everything we claim to be true, "perhaps even the demonstrations of the mathematicians," is only probable. But this Giant-like humility is offset by the confidence that we can establish that one claim to truth is closer to it than another. It is clear, for example, that he believed that his relational definitions of space and time, and his spherical wave theory of light, were closer to the truth than Newton's.

Does *anyone* know the truth about natural phenomena, about what's really "out there" causing our experience, or can we only have empirically supported conjectures as to how nature works, while the ultimate causes remain a logically impenetrable mystery? Newton was inconsistent on this, as he was on hypotheses. The weight of the evidence strongly supports the conclusion that Newton's goal was truths about reality, not just empirically validated accounts of experience in the form of equations that "worked." In the General Scholium cited earlier, Newton forcefully asserted that he did not know and made no claims about the *causes* of gravity. His claim is only that it is a real force obeying an inverse square law that really accounts for the motions of the planets, the Earth's moon, the tides, and falling bodies. Newton here backed off his earlier position—in the first edition of the *Principia*, later in the *Opticks*, and in various letters—that gravity was a force inherent in matter, a force that in the case of the planets and the moon, at least, acted at a distance across space that was empty of matter.

In the wake of their lengthy and bitter dispute over priority for the invention of the calculus, Leibniz attacked Newton's entire natural philosophy. In 1715 and 1716, Leibniz and Dr. Samuel Clarke, who acted as a proxy for Newton, exchanged five letters. (Leibniz died before responding to the fifth.) In these letters, Leibniz articulated his criticisms of Newton, all having to do with claims about reality, and Clarke/Newton responded. Leibniz began with Newton's theory of gravity, dismissing Newton's depiction of gravity as a force that acted at a distance across empty space. To claim that such a force was real was, Leibniz warned, not only wrong but a dangerous reversion to the occult qualities of Renaissance magical nature philosophy. It was the antithesis of "modern" natural philosophy.[24] This was a very serious charge indeed, though by 1715 it had been made often since the *Principia* appeared in 1687, especially by continental natural philosophers who were largely Cartesians. And it cut deeply enough that Newton had added the General Scholium to the second edition of the *Principia*, backing down from claiming to know *how* gravity affected bodies. Leibniz clearly thought this was posturing on Newton's part and that Newton really believed that forces did act at a distance, hiding behind the Queries of his *Opticks*.

Leibniz went on to attack Newton's attribution of an active role for God in maintaining the natural world. Newton, he said, claimed that God used the force of gravity to prevent the finite natural world from dispersing into infinite space and to prevent the stars from collapsing onto one another *because* of gravity. He mocked Newton's claim that it was God who conserved the amount of motion in the world, without which the world would run down because of the loss of motion in collisions among material bodies. For Leibniz, such a role for God was profoundly erroneous, theologically, metaphysically, and philosophically. Theologically, it implied that God was not powerful enough or intelligent enough to create a world that was self-sustaining. Metaphysically, it misrepresented what was real "out there," which for Leibniz was totally different from Newton's ontology of atoms moving in absolute space and time. Philosophically, attributing an active role to God reflected Newton's misunderstanding of two foundational principles of philosophizing: the Principle of Sufficient Reason—that every thing/event has a unique cause from which it follows necessarily—and the Principle of the Identity of Indiscernibles—that two or more things that have no uniquely distinguishing features are just one thing—from which it follows that Newton's absolute space and time cannot be real.

Leibniz cited Newton himself on God as the cause of the conservation of motion in nature. In his *Opticks*, Newton had written,

For there is a necessity of conserving and recruiting [motion] by active principles such as are the cause of gravity . . . and . . . fermentation . . . For we meet with very little motion in the world besides what is owing to these active principles or to the dictates of a will [to wit, God].

This is the text in the first English edition (1704) of the *Opticks* and in the first Latin translation (1706), the one Leibniz had read and reacted to in his letters to Clarke. In the second English edition of the *Opticks*, however, published in 1716 in the midst of the Leibniz–Clarke exchanges, Newton tellingly deleted the phrase "or to the dictates of a will."[25]

The underlying issue in these exchanges is who to believe and what to believe when scientists make competing knowledge claims, and on what grounds. Note that in none of the controversies in early modern science, not even the vitriolic ones involving Galileo, did one party accuse the other of illogical reasoning. Logical reasoning, therefore, cannot be the criterion for deciding between competing knowledge claims. Logical consistency alone is equivocal with respect to the physical truth of a theory. The criterion for accepting a knowledge claim as true certainly cannot be authority or reputation, which is antithetic to the spirit of modern science. Galileo may have been the greatest natural philosopher of his generation and he was right about sunspots, but he was wrong about comets, the tides, and objects losing weight as they heated up, yet the style of his reasoning was the same in all these cases, a point raised by Thomas Kuhn in his *The Structure of Scientific Revolutions*

(1962). Deciding to side with Galileo because he is Galileo, or with Einstein because he is Einstein, is not an option in science. It is, however, a common-place of the practice of science that scientists can be passionately committed to the truth of a theory, but on what grounds?

Can the answer to the question of who to believe and what to believe in science lie in focusing on data? Data come from the world after all. Perhaps, as Bacon argued, we need to make data the grounds of any inferences about the world that may legitimately be drawn from them. All we need do is let the data "speak to us." But as we have seen in the cases of competing astronomical models and competing theories of color, raw data are explanatorily equivocal. Ideally, correspondence with reality is a decisive criterion for determining the truth of a scientific knowledge claim, but we have no direct, non-experiential access to reality to confirm such a correspondence. We are forced, therefore, to employ indirect criteria, among them, logical consistency, consistency with the available data, and prediction of experimental outcomes. At the end of the nineteenth century, the physicist Heinrich Hertz wrote that, "As a matter of fact, we do not know, nor have we any means of knowing, whether our conceptions of things are in conformity with them [correspond to the way they really are] in any other than this one fundamental respect, anticipation of future events."[26] But successful "anticipation of future events"—successful predictions deduced from a theory and confirmed by experiment—cannot give us certain knowledge of the truth of a theory. This is blocked logically by the Fallacy of Affirming the Consequent. It is also blocked by science itself. Over the four hundred years of its practice, modern science has generated a continually growing "scrap heap" of discarded theories. In almost all cases, the discarded theories were successful for a time in making experimentally con-firmed predictions, in addition to being logically consistent and accounting for what was then considered all the relevant data.[27]

Simplicio's challenge to Galileo on the third day of the *Two New Sciences* di-alogue cited earlier, echoes through the centuries: 'How do you know, Salviati, that your logically correct and experimentally confirmed equations corre-spond to physical reality?' Can scientists *know* this or are all such claims con-jectural? This question is worth exploring in some detail at this point because it was already an issue for Newton, Huygens, and Leibniz given their rival theory-based, empirically supported ontologies.

How can anyone *know* that the terms in a mathematical description of a natural phenomenon correspond to actual features of that phenomenon? Or, more accurately, how can we know *which* of the many terms that occur in for-mulating and using a mathematical theory correspond to physical realities and which are just terms produced by manipulating its mathematical expressions?

Newton said that the term *mv* in his equations of motion corresponded to a feature of reality—momentum—while the term mv^2 did not. Leibniz said mv^2 corresponded to a feature of physical reality that he called *vis viva* (kinetic energy), and *mv* did not. As it happens, both were partially right and partially wrong. The conservation of *mv* remains a cornerstone of physics and so does the conservation of *vis viva*, within the broader principle of the conservation of energy.

To summarize, at the end of the first century of modern science two exclusive views of conceptualizing the world had emerged, echoing Plato's "battle" between the Gods and the Giants. One of these views may be called archeological, the other, interpretational. According to the former, what scientists do is analogous to what archeologists do when they uncover buried objects. Such objects are what they are regardless of what individual archeologists might want them to be. A fossil bone, a gold coin with the profile of a face and an inscription, a decorated bronze shield, a jar with or without handles and with or without a picture painted on it all are simply "there" in the ground, as they are, and uniquely so because there is only one correct description of an uncovered artifact. Analogously, scientists can be said to uncover objects and processes that are "buried" in the sense of being hidden from ordinary experience. As with archaeological artifacts, there can be only one correct description of these objects and processes. On this view, arriving at a uniquely correct description of reality is the goal of science.

Alternatively, scientists produce interpretations of experience based on available evidence and contingent assumptions. As opposed to the archaeological view with its goal of one correct account of reality, interpretations are by their nature pluralistic, very much like induction. There is no such thing as a uniquely true interpretation of anything, just as there is no single true inference to be drawn from the premises of an inductive argument. All interpretations are relative to specific assumptions, and as experience evolves, assumptions evolve as well. With new assumptions come new interpretations that scientists *call* "reality," identifying universal "laws" of nature that are considered to be true only for as long as they prove fruitful. The nineteenth-century French biologist Claude Bernard wrote that, for him, theories were like scalpels: for as long as they are sharp, use them, but once they become dull, discard them and use new ones.

So, do scientists uncover reality or interpret experience? The Gods say, "uncover reality." The Giants say, "interpret experience." Working scientists say "yes" and go about their work as if these two alternatives were mutually consistent, somehow justified by "the scientific method," while knowing full well that logically this is not possible.

5
Science Influences Philosophy

The pursuit of certainty, of necessarily true knowledge, has been the hall-mark of mainstream Western philosophy since Parmenides. From antiquity, skeptical philosophy has been defined by its rejection of the possibility of certain knowledge. In the sixteenth century, Erasmus mocked the obsession of philosophers and theologians with the pursuit of certainty. In the nineteenth century Kierkegaard called it 'the comedy of the higher lunacy.'[1] In the twentieth, John Dewey described the quest for certainty as religion disguised as philosophy.[2] Nevertheless, the battle between the Gods and the Giants remains joined, even though the allies of the Gods have yet to achieve necessarily true knowledge of anything. The attraction of defining "knowledge" in the strong sense and making reality its object is clearly very powerful.

Even as early as the mid-seventeenth century, it seemed that while mainstream philosophers were still *seeking* certain knowledge, the new natural philosophers were actually *producing* it, in the form of universal truths about nature. And they were doing this by the systematic application of observation, experiment, and methodical reasoning to human experience. Given the challenge to philosophy this represented, it is not at all surprising that modern philosophy is identified with epistemology becoming the central concern of Western philosophers and that Descartes is identified as the "father" of modern philosophy. As a philosopher on the side of the Gods, Descartes had to justify the claims to certain knowledge of reality that he made in his natural philosophy. More broadly, the new natural philosophy affected *all* claims to knowledge, in any branch of philosophy. Thomas Hobbes is one example of a mid-seventeenth-century philosopher whose ideas were affected by the methodological and ontological claims of early modern science.

Hobbes is best known as a social and political philosopher who argued in his book *Leviathan* that government was based on a social contract and that an absolute monarch was required in order to protect people from one another, given that life is a 'war of all against all.' Nevertheless, while he was even less of a natural philosopher than Bacon, Hobbes' thinking was affected by the new empirical-experimental science although he was hostile to it. He was aggressively critical of its methodology, of its claims to truth, and of its social organization. At the same time, he was careful to anchor his own claims in social and political philosophy in a theory of knowledge and of how the mind acquires it.

Hobbes can seem to be an empiricist, as he held that all the information that the mind has available to it for reasoning about the world comes through the senses. With respect to the world, at least, the mind is a tabula rasa, so Hobbes can sound very much like John Locke (and Aristotle). But unlike Locke and the later British empiricists, Hobbes, like Descartes, was committed to a definition of knowledge in the strong sense. Only what was arrived at deductively, as in Euclidean geometry, deserved to be called knowledge. Experiment-based theories were at best probable accounts of phenomena, never rising to the level of certain knowledge of reality. To the extent that the new natural philosophy was inductive and experimental, its claims to knowledge were inflated and logically unjustifiable. Hobbes had a protracted dispute about this with Robert Boyle, a leading exponent of experimental natural philosophy, challenging the discoveries about air that Boyle, working with Robert Hooke, claimed to have made using an air pump of their own design. These experimental discoveries included a mathematical "law," later called Boyle's law, relating the pressure and the volume of a gas, and the role of air in combustion and in sustaining life. This last claim implied that air, traditionally considered one of the four basic elements, was "really" a composite: only a specific part of it was responsible for sustaining life and combustion.

I referred earlier to the role that newly invented instruments played in the rise of modern science and to the problematic character of observations using these instruments. The slide rule and the mechanical calculator are not problematic because their results can be checked independently, using traditional arithmetic. Observations made using the telescope (in astronomy) and the microscope, however, cannot be checked independently, nor can the outputs of the barometer or the thermometer or the air pump. In all these cases, the question arises of whether these instruments are revealing truths of nature, or producing artifacts by their own operation, or some combination of the two.

Hobbes dismissed the discoveries announced by Boyle, and he rejected all claims to *knowledge* of nature based on experiment because, he argued, as Descartes had, that experiments were equivocal. They were open to multiple interpretations. We cannot *know* which of these is the correct one based on the data alone. Furthermore, we cannot be certain that data collected experimentally have been affected, or even produced, by the instruments themselves and/or the experimental setup. We cannot be sure that the instruments are operating "correctly." As a result, we cannot know for sure that the results of the Boyle–Hooke air pump experiments revealed truths about air or about the behavior of air interacting with the air pump.[3] (The Boyle–Hooke air pump was in fact a very finicky machine, using greased leather seals to create a partial vacuum in a bell jar, and it was difficult to keep it working "right" for very long. Contemporaries who tried to replicate their results, of whom Huygens was one, found it difficult to do so.)

However reactionary this may sound to some, Hobbes had a valid point, one that became increasingly relevant as instruments became increasingly complicated. As microscopes became more powerful, and especially after specimen-staining techniques were developed, it was not at all obvious what the microscope was revealing. People had to be trained to "see" what was claimed to be there. The same was true with the development in the twentieth century of increasingly powerful X-ray, CT, and MRI technologies. What was important and what was not, in the output of these machines? What was normal and what was abnormal? Again, only people trained to see what they were told was there could "make sense" of what they were "seeing." The problem is compounded for instruments whose design and operation is theory and computer dependent, for example, digital output telescopes; electron, scanning tunneling, and atomic force microscopes; particle accelerators; and gene sequencers. There is no *seeing* at all of the phenomena said to be revealed by these instruments, only an output constructed by a computer program, that is, by a set of instructions for how the data are to be interpreted.

Although primitive to us, in its time the air pump was a novelty and vulnerable to Hobbes' critique: How can we *know* what the machine is telling us about nature, or that the machine is telling us *anything* about nature? At the same time, there is a certain irony in this Hobbes–Boyle/Hooke dispute. Recall Hooke's rejection of Newton's claim to have discovered via his prism experiments that sunlight, traditionally considered homogenous, was in fact "really" heterogeneous, composed of multiple colors. Against this claim, Hooke argued that the colors exiting the prism were produced by the prism and were not in the sunlight. With respect to his own air pump experiments, on the other hand, Hooke defended his and Boyle's claim that the air pump revealed truths about air, including its composite nature, while Hobbes argued that their results were produced by the air pump itself! On what grounds did Hooke's experiments with Boyle's reveal truths about air, while Newton's experiments did not reveal truths about light? Hobbes' point was a good one and still is. All experimental outcomes are equivocal and require interpretation; they never reveal just one truth.

Hobbes rejected the claim that the mind could produce knowledge of nature based solely on reasoning about sensory input to the mind, but he argued that the mind *could* produce demonstrative knowledge. Such knowledge came from reasoning about ideas in the mind. Ideas always begin with sensory experience, but they are transformed by the action of imagination, memory, and reasoning. Reasoning about ideas could, in principle, produce certain knowledge of nature—though Hobbes personally was uninterested in this—and such reasoning could produce knowledge relevant to action in the human lifeworld, as in moral, social, and political philosophy, which Hobbes *was* interested in. Hobbes characterized reasoning as a form of calculating, by analogy with the operations of arithmetic. In this he was consistent with his commitment to a rigorously materialist-mechanist conception of nature, arguing that even the mind was material, a radical position for his day and one that went beyond what Descartes was prepared to commit to, publicly at least.

Hobbes' view that reasoning was calculating may reflect an influence of contemporary technology on philosophical ideas. Bacon and Descartes, as noted earlier, for all their differences on knowledge, mind, and method, had both claimed that correct reasoning should be done 'as if by machine.' Galileo, too, compared reasoning about nature to the systematized know-how of artisans. Where did this idea come from? It seems relevant that in 1622 the slide rule had been invented by William Oughtred, at roughly the same time as the first mechanical calculators by Blaise Pascal and Wilhelm Schickard. Hobbes extended these mechanizations of arithmetic calculations to reasoning generically. By the end of the century, others, among them Leibniz, proposed that reasoning could be reduced to the rule-bound manipulation of unambiguous symbols in a "rational" language. This idea disappeared for a while but was revived in the nineteenth century with the invention of symbolic logic and surfaced again in the 1960s in artificial intelligence research.

Hobbes is one example of an influential seventeenth-century philosopher who felt that in order to do social-political philosophy, he had to make it consistent with contemporary natural philosophy. Spinoza is another, far more influential in philosophy than Hobbes, whose work explicitly incorporated a natural philosophy of the Cartesian type. He wrote a *Treatise on the Emendation of the Intellect* that echoes Descartes' *Rules for the Direction of the Mind, Principles of Cartesian Philosophy*, whose title speaks for itself, and, most famously, he cast his *Ethics* in a geometric form such that all its claims are deductive consequences of a set of putatively self-evidently true definitions, axioms, and postulates with which he begins the book. Spinoza devotes Part II of his *Ethics* to necessarily true conclusions about mind and knowledge, including knowledge of nature, taking an even stronger position on the latter than Descartes.

Like Descartes, Spinoza dismissed the senses as a source of knowledge. None of the ideas in the mind that trace back to the senses are reliable. But the mind is capable of forming "adequate" ideas, ideas that can grasp the essences of the things that comprise reality and thus acquire universal, necessary, and certain knowledge of it. In his *Ethics*, Spinoza "proved" that the "order and connection of [adequate] ideas is the same as the order and connection of things [in nature]." In an explanatory Scholium he wrote, "And so, whether we conceive Nature under the attribute of Extension or under the attribute of Thought or under any other attribute, we find one and the same order, or one and the same connection of causes, that is, the same things following one another."[4] Reasoning, properly conducted, ipso facto is reasoning about reality.

It is important to remember that, for those pursuing certain knowledge of nature, "cause" must be defined as a necessary and sufficient condition of an effect: without the cause, the effect will not happen, and with the cause, the effect necessarily happens. The ability to know causal connections among natural phenomena is itself a necessary condition of certain knowledge of nature. Deny that causal connections are real outside the mind, or deny the knowability of causal connections as facts "out there," and certain knowledge of nature is impossible. A qualification must be added with respect to Leibniz.

Like Descartes, Leibniz was both a philosopher and an important natural philosopher, mathematician, and logician. His theory of mind, of knowledge, and of knowledge of nature follows from his metaphysics, a core precept of which is that what we call things "out there" are each and every one, from mud to human beings to God, the nexus of an apprehension of the entire universe characteristic of that type of thing and unique to each individual thing within its type. As applied to humans, this means that each mind uniquely apprehends *in* the mind everything "out there," to a greater or lesser degree, more or less consciously, including the necessary and sufficient connections among all these things. There is more than a hint here of the Renaissance magical nature philosophy notion of the universe-macrocosm being imaged in the human mind-microcosm. For Leibniz, however, no two images of the world in human minds were the same, while in Renaissance magical nature philosophy the images are the same for everyone.

Echoing Spinoza on the necessity of all things and the impossibility of will as a "free cause,"[5] Leibniz identified what he called the Principle of Sufficient Reason (PSR) as the cornerstone of *all* rational philosophy and thus for the reasoning of all allies of the Gods: nothing happens without a necessary and sufficient reason, or cause, for its happening. It follows that every moment of reality is "pregnant" with all of the consequences that must follow from it, as that moment itself was a necessary consequence of previous moments going back to the beginning of time. Reality is rigorously deterministic. It unfolds in a uniquely necessary way, exactly the way that the theorems in Euclid's geometry unfold from its definitions, axioms, and postulates. Note that this is a logical "unfolding," an unfolding that is intrinsically timeless. To our limited minds, it takes time to discover theorems, but in fact they are already "in" the definitions and premises, which are thus "pregnant" with all possible theorems. (This is precisely the view of science and time defended by Pierre Laplace in the early nineteenth century and by Einstein among many others in the twentieth.)

We are again at the intersection of ontology and epistemology, of the ontological knowledge claims of early modern "scientists" and the pursuit of epistemological certainty by philosophers. The allies of the Gods may have been dominant in the Western philosophical canon, but there were always allies of the Giants at work at that intersection as well. A community of skeptical philosophers had emerged in the sixteenth century, among them Francisco Sanchez and Michelle de Montaigne. The emergence of skepticism was contemporary with the Protestant rejection of the Catholic Church's claim to possessing certain knowledge of God and the one true meaning of Scripture. It was contemporary as well with the publication of newly printed works by classical skeptical authors, among them Cicero, Lucretius, and Sextus Empiricus. The skeptical philosophers dismissed the pursuit of certainty as quixotic. In the spirit of Protagoras, there was nothing outside the mind that one could know in the strong sense of "know." Pierre Gassendi, for example, was a contemporary of Descartes who was very well known and respected in his lifetime, though little known today. He was a philosopher and a theologian who was simultaneously deeply affected by modern science and deeply engaged with it.

Gassendi was one of the philosophers—Hobbes was another—who Descartes asked to critique the manuscript of his *Meditations on First Philosophy*, which was then published with those critiques and Descartes' responses to them. Gassendi was 180 degrees from Descartes. In spite of being a Catholic priest and theologian, Gassendi was a skeptic, an empiricist with a probabilistic conception of knowledge, and an atomist, holding that space, time, and therefore motion, were atomic. Descartes was a rationalist, a pursuer of certain knowledge who rejected any atomic theory of matter or space, though he may have held an atomic theory of time. Gassendi was up to date on the mathematical physics of his time, including Galileo's, but his natural philosophy was integrated into a broader philosophical and theological vision. As an empiricist, Gassendi held that all of our ideas about the external world come from sensory experience, that all of our sensations, primary and secondary, are real, and that we could only reason about the world inductively. It followed that all of our knowledge of the world was only probable, that mathematical accounts of nature were fictions, non-unique interpretations of data. For Gassendi, as for the sophists, probable knowledge *was* knowledge, and it was the only kind of knowledge there was.[6]

Given the place of the British empiricists in the history of Western philosophy, it is easy to assume that they rejected defining knowledge in the strong sense. The truth is more nuanced than this, however. John Locke, like Hobbes and Spinoza, was a philosopher for whom pursuing philosophy required taking the new natural philosophy into account. Again like Hobbes, Locke was primarily a political, social, and moral philosopher. He was explicit, in his highly influential *Essay Concerning the Human Understanding* (1690), that a theory of mind and knowledge consistent with the new *natural* philosophy was required in order to do philosophy. Locke was a long-time friend and correspondent of Robert Boyle's, so he was kept current on developments in experimental philosophy. He began a lasting friendship with Newton, also a friend of Boyle's, shortly after the *Principia* was published. (Boyle, Newton, and Locke shared an interest in alchemy, among other things.)

In the *Essay*'s preface, Locke wrote,

> The commonwealth of learning is not without master-builders, whose mighty designs, in advancing the sciences, will leave lasting monuments to the admiration of posterity: But everyone must not hope to be a Boyle, or a Sydenham; and in an age that produces such masters, as the great [Huygens] and the incomparable Mr. Newton, with some others of that strain; 'tis ambition enough to be employed as an under-labourer in clearing the ground a little, and removing the rubbish that lies in the way to knowledge.[7]

Locke was unable to follow the mathematical arguments in the *Principia*, but he inquired of those who could—Huygens was one—and was told that the mathematics was sound and that the *Principia* was a major breakthrough in our knowledge of nature. Although the *Essay* was effectively done by the time Locke read the *Principia*, it was crucial for him that the ideas about mind and knowledge in his *Essay* be consistent with the *Principia*. His inclusion of "the incomparable Mr. Newton" in the Epistle to the Reader, alongside Boyle, Sydenham, and Huygens, as "master-builders" of newly won knowledge of the world, reveals his confidence that this was indeed the case.

Locke's theory of knowledge was that knowledge was a matter of relationships among ideas within the mind—which sounds like Hobbes—but all ideas trace back to sensations, so that reasoning about ideas *can* be reasoning about the world. Like Galileo and Descartes, among others, Locke distinguished primary sensations (shape, mass, motion), which were properties of the objects "out there" stimulating the senses, and secondary sensations (color, sound, texture, taste, odor, feelings), which were subjective, produced by the mind in response to external stimuli. As all sensations are particular, the familiar problem arises: How can we have universal, necessary, and certain knowledge of the external world beginning with sensations? The obvious answer is that we cannot, but Locke could not ignore the claims to knowledge of nature by his contemporary "master-builders" of knowledge.

Locke argued that from our sense data–based ideas we could not have knowledge of the essences, the essential natures, of the objects that comprised the world external to the mind. But from primary sensations the mind could know for sure that there *was* an external world, and it could know the "powers" that these objects have to act on our sense organs and cause our secondary sensations as their effects. By reasoning about these ideas, the mind forms particular and general ideas about these objects and their mutual relationships. All our knowledge claims about the external world are actually claims about the relationships among the ideas of the external world that the mind forms based on its sensory experiences. But what sort of knowledge is this? Here, as with Hobbes, we see how unsatisfactory categories can be. Locke was an empiricist, but he accepted the rationalist definition of "knowledge," that knowledge was universal, necessary, and certain.

Locke is very clear that his empiricism implies that we cannot have universal, necessary, and certain knowledge of anything outside the mind (not even the existence of God). But Locke was not a skeptic: he acknowledged that the natural philosophers were producing knowledge of nature. Locke could have chosen the path of Gassendi, that knowledge of nature was possible only if you defined "knowledge" as probable. Instead, he accepted the strong definition of "knowledge" as the correct one and adopted a new definition of "certainty," one that had emerged out of a circle of intellectuals and theologians to which Locke belonged: "moral certainty."

Certain knowledge of what is external to the mind is impossible for a consistent empiricist like Locke. But the new experimental philosophy showed that the mind could reason to knowledge claims about the world (and God) that are so highly probable that for all practical human purposes they are not merely probable, but *effectively* certain, or "morally certain." Locke wrote the *Essay* shortly before Huygens wrote his *Treatise on Light* (1692) in which Huygens would acknowledge that, logically, empirical-experimental philosophy could never give demonstrative knowledge of nature, but "It is always possible to attain [by experiment-backed theorizing] to a degree of probability which very often is scarcely less than complete proof." When experiments confirm a theory's predictions, it "must be ill if the facts are not pretty much as I represent them here."[8]

Huygens equivocated: we can't be absolutely certain that our theories are true, but they can come so close to certainty that they can be taken as if they were certain. Locke gave the name "moral certainty" to such near-but-never-attainable full certainty. Inconsistently, however, because of Newton's theories, Locke kept the possibility of certain knowledge of nature open:

> Though the systems of physics that I have met with afford little encouragement to look for certainty, or science [knowledge in the strong sense] . . . yet the incomparable Mr. Newton has shown, how far mathematics, applied to some parts of nature, may, upon principles that matter of fact can justify, carry us in the [certain] knowledge of some . . . particular province of the incomprehensible universe.

If Newton's method became the standard for others, "we might in time hope to be furnished with more true and certain knowledge than hitherto we could have expected."[9]

Bishop George Berkeley was, with Locke and David Hume, the second of the three canonical British empiricists, but his response to Newton, and to experimental natural philosophy generally, was unique. Berkeley developed an empirical theory of knowledge in support of an anti-materialist theory of reality. He, too, was very familiar with the new science, and its algebraic mathematics—he exposed a flaw in Newton's calculus that was not repaired until the nineteenth century—and he foresaw that the new science would inevitably lead to an atheistic materialistic determinism. To forestall that, Berkeley argued that the new natural philosophy contained an internal contradiction. It claimed to begin with and to end in empirical experience, eschewing the abstract metaphysics characteristic of traditional philosophy, but in fact it was itself built on a metaphysical concept: matter. As an empiricist, Berkeley asked, where is our experience of matter in and of itself, as opposed to our experience of the *properties* of matter? As a matter of empirical fact, we have no experience of matter per se, only of its sense-based properties. We take it for granted that there is something "out there" that underlies and is responsible for the more or less stable clusters of properties that make up the objects of our empirical experience. Matter is not a fact but a metaphysical concept that satisfies our desire for stability amid change and continuity.[10]

Berkeley rejected the distinction between primary and secondary sensations, arguing that all sensations are in the mind. Why, then, attribute the cause of our sensations to inert matter? In Berkeley's philosophy, the real world is a projection of ideas in God's mind and of our experience of that projection. This implies that the being of things apparently "out there" beyond the mind, existing on their own, is "really" dependent on their being perceived by us, but in context, Berkeley had an important point. Why postulate a metaphysical, unexperienced, hence non-empirical, inert substance as the source of empirical experience, and why make it the foundation of purportedly empirical natural philosophy? Everything empirical and all scientific accounts of the empirical stay the same if you postulate a metaphysical, unexperienced, and unexperienceable God as the source of the empirical. What logical or empirical criterion is there for making reality material rather than ideal/spiritual? Why choose one metaphysical ground over another? And the consequences of this choice are profound.

Materialistic determinism entails a fundamentally inhuman universe, intrinsically valueless and meaningless. If, on the other hand, you choose God as the ground of empirical experience, the universe becomes intrinsically value rich, meaningful, and human, with no loss of scientific explanatory power. Berkeley believed that he had pulled the rug out from under claims by the new science to knowledge of *reality*, but note that this leaves untouched scientific reasoning to theories as accounts of *experience*. Berkeley's critique of matter was complemented by David Hume's critique of a concept equally fundamental to scientific knowledge: causality.

Hume was Locke pushed to the limit, a limit in which skepticism is acknowledged as the inescapable consequence of empiricism; and reason, especially logical reasoning, is revealed to have sharp limits. Hume begins as all empiricists do, with the claim that all our ideas begin with sensory experience and there is nothing in the mind that is independent of sensory experience. Along the way from the senses to the conscious mind, memory, imagination and reason act on sensations and create ideas on various levels of abstraction, generality, and speculation. Hume argued that our belief in an external world was based only on an internal perception of the markedly greater "liveliness" of sensations immediately experienced compared with sensations remembered or reasoned about. Logical reasoning, he argued, has not proven, and cannot prove, that there *is* a world "out there" beyond the mind. That's something all humans *know*, in the sense of believing with complete certainty, but its truth is beyond reason.[11]

We cannot have knowledge in the strong sense of anything that is external to the mind because all we have to reason about are individual "matters of fact" presented to the mind through the senses. The same holds for proving the existence of other minds, again something that we *know* with certainty but non-cognitively. For Hume, this reveals the limits of reason and the mistake of defining knowledge as universal, necessary, and certain. In fact, there is nothing at all inside the mind, with the exception of simple arithmetic, that qualifies as knowledge in the strong sense, so that defining "knowledge" in the strong sense must be dismissed out of hand. The pursuit of certainty, he might have said, stops here.

Hume was appreciative of modern science, albeit from a distance, like Locke, and he accepted its accomplishments as true in a sense he needed to clarify. The language of his *Treatise of Human Nature* echoes that of Newtonian mechanics as if he were seeking a physics of the mind, a theory of the mind that would account for its operations in a law-like way, though neither mathematically nor deductively, as Newtonian physics is, and of course without the Newtonian ontology. He announced three laws (rules is more accurate) governing the association of ideas in the flow of consciousness: resemblance, contiguity in space or time, and causal connection.[12] Hume argued that causal connection, understood as a necessary and sufficient condition for something to happen, was an idea only. It could not be traced back to a sensation, hence it was not a fact about the world but a fact about the mind. It is an idea that gives a name to patterns we have noted, and remembered, in our empirical experience. Certain events are remembered as "constantly conjoined," repeatedly occurring in specific before-and-after sequences, and the mind explains this by inventing a relationship between events that it calls causal connection. Sometimes we are mistaken in the relevant before-and-after events, and sometimes a sequence is broken in new experiences, but believing in causal connections is generally validated by experience. We can only reason about the world inductively, but there is an intrinsic flaw in inductive reasoning. We generalize inductively *assuming* that the future will be like the past, but we have neither logical nor empirical grounds for such an assumption. We have a feeling that past relationships will continue in the future, we believe it, but we cannot know it.[13]

Causal connection is metaphysical, only an idea in the mind. Inductive reasoning about nature is not logically valid. If left standing, these conclusions would be devastating not only for rationalist philosophy, but also for empiricists who wanted to avoid skepticism. Causal connection as a fact about the world is a cornerstone of knowledge of nature in the strong sense and central to modern scientific explanation. It is a corollary of Leibniz' Principle of Sufficient Reason. Make causality an idea in the mind only, just a way that the mind organizes its experiences with no guarantee that any particular causal relation will be preserved in the future, and knowledge as revelation of reality is impossible. For a skeptic, making a holy grail out of knowledge in the strong sense is completely mistaken, so these conclusions are reinforcing for the weak sense of "knowledge." Modern science is indeed knowledge of the external world, but only as conjectural, probable, and always corrigible by future experience. That is satisfying to the Giants, but not to the pursuers of certainty who came before Hume, and not to Immanuel Kant after Hume.

The first edition of Kant's *Critique of Pure Reason* appeared in 1782. Its unfolding in the series of works that followed made Kant one of the most important and influential of modern philosophers. In 1770, after a prolific earlier period writing on an eclectic mix of topics in natural philosophy, mathematics, and logic as well as traditional philosophy, Kant was appointed full professor of philosophy. In 1771 he published the inaugural lecture he gave on that occasion, and then for the next eleven years he published nothing. What happened?

The subject of Kant's inaugural lecture was the relationship between the senses and the intellect, an important philosophical subject obviously, but of particular importance to Kant because he thought Newtonian physics was true of the world. He was keenly aware of the then current version of the knowledge problem: How can science produce necessarily true, certain, knowledge of the world beginning with the senses? Kant sent copies of his lecture to four philosopher friends for criticism and in letters said that while his lecture did not fully solve the problem, it brought him very close to a solution, which it should take him perhaps a year to complete. One of these philosopher friends was J. G. Hamann, who suggested that Kant reread Hume's *Treatise*. Kant did and was suddenly struck by the force of Hume's critique of causality in Book 1. He was already familiar with Hume's writings, having included them, and even a sympathetic treatment of skepticism, in his earlier teaching. But now, reading Hume's critique as he was attempting a theory of his own that would link the senses to the intellect in a way that would validate modern science's claims to produce knowledge of nature, Kant was shaken. Hume was right. Causality is an idea in the mind only, but causality is pivotal to modern scientific explanation that claims to reveal truths about reality. How is knowledge in the strong sense possible if causality is only an idea?

In 1783, Kant looked back on the preceding twelve years, which had changed his life and the direction of Western philosophy:

[T]he reminder [in 1771] of David Hume was the very thing which many years ago first interrupted my dogmatic slumber and gave my investigations in the field of speculative philosophy a quite new direction. I was far from following him in the conclusions at which he arrived [skepticism] by regarding, not the whole of the problem, but a part, which by itself can give us no information. If we start from a well-founded but undeveloped thought which another [Hume] has bequeathed us, we may well hope by continued reflection to advance farther than the acute man to whom we owe the first spark of light.[14]

Kant's *Critique of Pure Reason* (*CPR*) is precisely a "continued reflection" on Hume's thinking, putting it in the context of a broader philosophical inquiry than Hume did, while retaining the germ of Hume's thinking. Hume forced Kant to accept that the connection between cause and effect was wholly in the mind. For Hume, this meant that empirical science was dependent on a non-empirical principle and was more the mind's interpretation of experience than knowledge of the world as it is. Hume's argument that causality was only an idea in the mind was compelling, but Hume's skeptical conclusion was unacceptable to Kant. His "continued reflection" on the "first spark of light" struck by Hume occupied the rest of Kant's life.

First, Kant thought through a comprehensive system of philosophy that was, as he said in the introduction to *CPR*, a "revolution" in philosophy analogous to Copernicus' revolution in astronomy. Kant's revolution pivoted about the possibility of what he called "synthetic *a priori* propositions," propositions with empirical content but known to be necessarily true a priori, wholly independent of experience. It will not come as a surprise that one of these synthetic a priori truths was that causes were necessary and sufficient conditions of effects. For two thousand years before Kant philosophers had asked how knowledge in the mind conforms to objects independent of the mind, and for two thousand years this question (our knowledge problem) did not receive a satisfactory answer. Kant now reversed the question: What follows if we suppose that the objects of empirical experience conform to our knowledge of them?[15] This was his Copernican revolution in philosophy, and Kant's philosophical system is an answer to that question, one that deductively proves that mathematics and physics give us universal, necessary, and certain knowledge. The object of this knowledge, however, is not a reality independent of the mind. Reality as it is in itself is completely unknowable. The object of our universal, necessary, and certain knowledge is experience itself, but experience is now defined to be a necessary construct of the human understanding, not a window onto the external world.

Kant opens the first edition of *CPR* as follows:

There can be no doubt that all our knowledge begins with experience . . . But though all our knowledge begins with experience, it does not follow that it all arises out of experience. For it may well be that even our empirical knowledge is made up of what we receive through our [sense] impressions and of what our own faculty of knowledge (sensible impressions serving merely as the occasion) supplies from itself.

If that is the case, and Kant offers deductive arguments that it is, then only with "long practice of attention" will we be able to distinguish the contribution to what we call empirical knowledge of sensory input and the contribution of the mind. *CPR* is the product of Kant's "long practice of attention" to this problem.[16]

The introduction to the second edition begins as follows:

Experience is beyond all doubt the first product to which our understanding gives rise, in working up the raw material of sensible impressions . . . . Experience tells us, indeed, what is, but not that it must necessarily be so, and not otherwise. It gives us no true universality; and reason, which is so insistent on this kind of knowledge is therefore more stimulated by [experience] than satisfied. Such universal modes of knowledge [as Newtonian physics], which at the same time possess the character of inner necessity, must in themselves, independently of experience, be clear and certain.[17]

Kant's entire philosophical enterprise from *CPR* on was motivated by the need to create a theory of knowledge and mind from which it would follow that Newtonian physics was both empirical *and* universal, necessary, and certain. If he succeeded, the Gods win and the battle between the Gods and the Giants at long last is over. If he loses, the battle goes on. "What we need," Kant wrote, "is a criterion by which to distinguish with certainty between pure [knowledge] and empirical knowledge. Experience teaches us that a thing is so and so, but not that it cannot be otherwise."[18] Experience cannot, so to speak, know itself, know why it is the way it is. That is the task of reason, which for Kant is capable of disengaging from experience, transcending experience, and discovering, from pure reason alone, independently of all experience, why experience is the way it is and must be so.

In Kant's philosophical system, what reason discovers is that experience is the way it is because it is constructed by the understanding responding with a strict innate logic to stimuli from an external world. Space and time, for example, are discovered by Kant's reasoning to be sensory "intuitions," first responders, as it were, to external stimulation. Space and time are of the mind, which has no way of getting behind experience to know if space and time exist at all in the world outside the mind, let alone exist in the way that we experience them. Furthermore, space as we experience it is necessarily the space of Euclidean geometry, which is the only possible geometry for us because it is the geometry of our spatial intuition. (A century later, Kantians would have to deal with non-Euclidean geometries.)

Coordinately, our experience of time is necessarily of Newtonian time, not because it is the time of the universe in itself, but because that's the way our temporal intuition works. We experience the world spatially and temporally, but that's a fact about the way our understanding works, not about the world. Analogously, the understanding employs twelve concept types, one of them being causality, that determine how we can reason about experience. For example, just as Euclidean geometry is the only way we can reason spatially about the world as we experience it, Newtonian physics turns out to be the only way we can reason about the objects that we discover in experience as organized by our spatial and temporal intuitions. Newtonian physics is universally, necessarily, and certainly true because the mind makes it so. (In the twentieth century, Kantians would have to deal with the relativity and quantum theories.)

The system presented in *CPR* was the product of Kant's "continued reflection" on Hume's "well-founded but undeveloped thought." It led, however, to a philosophy that stood in stark contrast to the one Hume erected on that thought.

Hume had concluded that we could not have demonstrative knowledge of the external world because knowledge claims about the world beyond the mind were dependent on ideas in the mind—in particular the idea of causal connection—that could not be known to have a correlate in that world. There was, therefore, no such thing in natural philosophy as knowledge in the strong sense, only knowledge in the sense of more or less probable opinions and beliefs about the world. A corollary of Hume's argument is that reason has been greatly overestimated in the Western philosophical tradition. As a matter of existential fact, human beings simply "know," non-cognitively, things that reason cannot prove (logically), among them, the existence of a world external to their minds and the existence of other people with minds like their own. Furthermore, for Hume, reason is incapable of determining action in the world—he called reason the "slave of the passions,"[19] in agreement with Aristotle, who held that desire, hence will, determines action, not reason.[20]

Kant also concluded that the world outside the mind was totally unknowable, but he claimed that in *CPR* he had proven that there *was* a world outside the mind and that this world provoked the self-activity of the understanding into the necessary determination of the form of our experience of the external world, into constructing experience. Hume had attributed our accounts of experience to the self-activity of the mind, for example, his three "rules" for the association of ideas, but Kant's project from the beginning was guided by his conviction that universal, necessary, and certain knowledge was possible in science and mathematics. Hume forced Kant to make experience and not the world itself the object of that knowledge. In the process, Kant restored reason to preeminence in human affairs, and he emerged as a lead actor in Kierkegaard's "comedy of the higher lunacy," the obsessive pursuit of certainty.

6
Science and Social Reform in the Age of Reason

In the nineteenth century, modern science had a profound *direct* effect on society by way of a flood tide of technological innovations. In the eighteenth century it had an equally profound *indirect* effect by way of three interconnected ideas: the idea of progress, the idea that science exemplified reason-based progress (in knowledge of nature), and the idea that a reason-based science of society was possible. These ideas reinforced a realist interpretation of scientific knowledge claims, whether of nature or of society, in spite of open acknowledgment that such claims were logically problematic.

The idea of progress emerged in the Renaissance, introduced by the humanists. It was reinforced by the introduction of new technologies that had a major impact on European society. The humanists are typically characterized as obsessed with classical literary culture. That is true, but Petrarch—"father" of the humanist movement—compared what he and his fellow humanists were starting to do in the mid-fourteenth century to the work of bees. Bees take pollen and convert it into honey, which is better than pollen. The humanists studied the works of classical antiquity in order to produce even better works. Classical works established a baseline, so to speak, setting intellectual and stylistic standards for better works, hence progress.

The humanist idea of progress affected only a small intellectual elite. New technologies, on the other hand, affected society as a whole: gunpowder weapons; printing using movable metal type; "realistic" painting enabled by the geometry-based technique of central vanishing point perspective drawing; the application of realistic drawing to the depiction of machines, including cutaway and exploded views, leading to the design and construction of more complex and more capable machines; new forms of music; advances in mathematics with the introduction of algebra and the application of mathematics to mapmaking; new ship designs that, together with cannons, the new maps, the magnetic compass, and new navigational tools, allowed Europeans to circumnavigate the globe, projecting their power at will.

By the middle of the sixteenth century, the reality of progress was widely proclaimed. Renaissance Europeans were better than their classical-era ancestors. They had advanced way beyond what the ancients knew and what they could do. The reality of material and intellectual progress during the Renaissance seems obvious to us, but at the time this was a controversial claim. It was in conflict with the entrenched belief that, just as everything in our experience deteriorates with age, so do the planet and the humans who inhabit it. The idea of progress seemed also to be in conflict with the Christian teaching that mankind had been steadily deteriorating since Adam and Eve rebelled against God by eating the fruit of the tree of knowledge of good and evil.

Into the seventeenth century, the evidence for claiming the reality of progress came from advances in know-how. Suddenly, evidence for progress became identified with advances in knowledge, specifically, the knowledge being produced by the new science of nature through its methodical application of reason to experience. This displacement of know-how by knowledge reflects a continuing prejudice among Western intellectuals that was already manifest in ancient Greek philosophy. It ranked the abstract above the concrete, the universal above the particular, knowledge (*episteme*) above know-how (*techne*), theory above practice, the necessary above the contingent, and reason above will and feeling. With rapid advances in astronomy and mathematical physics from Kepler through Galileo and Huygens to Newton, and with discoveries enabled by the telescope, barometer, air pump, and microscope, the new science became the standard bearer of progress, relegating know-how to a subordinate position. One lesson of the identification of progress with modern science for many intellectuals—among them the French thinkers called the *philosophes*, but also their contemporaries in England, Scotland, Germany, Holland, and Italy—was that reason could be the engine of progress in human affairs as it already was in knowledge of nature, if only society allowed it to play that role.

The reality of progress was not universally accepted. For deniers, all of the advances cited earlier, knowledge or know-how alike, were superficial only. From a Christian perspective, as people, we were not only no better than people were when Jesus lived but worse, because we were further removed from God's word. From a secular perspective, opponents of progress focused on wisdom, literature, and philosophy, claiming superiority for the ancients. Controversy over the reality of progress found expression in a popular form of sixteenth-century entertainment: verbal or staged "battles" between defenders of the ancients, denying progress, and defenders of the moderns, asserting progress. With the rise of modern science in addition to the introduction of new technologies, the moderns would seem to have won a decisive victory: war over. But while explicit "battles" as public or private spectacles slowly disappeared, the war most definitely was not over. It continued to simmer into the early eighteenth century before erupting into open conflict with the rise of the Romantic movement.

Jonathan Swift's *A Tale of a Tub*, published in 1704, quickly went through five editions. The introduction, titled "Battle of the Ancients and the Moderns," is an allegorical "battle" that takes place in the King's library between anthropomorphized books with ancient or modern authors (think of the animated Pixar movie *Toy Story*). The books first argue among themselves about superiority, then take up weapons and fight it out. The ancients win, consistent with Swift's mocking assessment of the value of modern science in *Gulliver's Travels*. In that extremely popular novel, Gulliver's third voyage carries him to the flying island of Laputa, governed by Baconian natural philosophers, who, using their putative growing knowledge of nature, continually devise new, supposedly improved, devices and ways of doing things that they claim will make life better. Gulliver finds that the opposite is true: the projects of the "Projectors" are patently ridiculous and always make things worse than when life was based on practical know-how rather than on theoretical knowledge.

Swift's view of science is the exact inverse of what Bacon had promised in his fictional work *New Atlantis*. This was an account of an imagined island governed by wise Baconian natural philosophers whose applications of knowledge *do* make life better. But Swift had a point. Bacon's promise of new know-how that would flow from true knowledge of nature was, in the early eighteenth century, still just that, a promise, with nothing concrete to show for it. Fifty years after *Gulliver's Travels*, even as pro-science a *philosophe* as Denis Diderot—creator and editor-in-chief of one of the greatest intellectual accomplishments of the eighteenth century, the French *Encyclopédie*—had lost hope that the new physics would ever have practical applications. The increasingly abstract mathematics in which physics had become couched was, he decided, too complex for theoretical knowledge to lead to new know-how. His early co-editor, the mathematician and mathematical physicist Jean d'Alembert, acknowledged in his introduction to the *Encyclopédie* the "meagre application and usage" of newly discovered mathematical and physical truths.[1] As a result, Diderot concluded that advances in useful know-how would come not from theoretical science but from within know-how itself, and he saw to it that the *Encyclopédie* provided as comprehensive an account of contemporary know-how as it did of contemporary knowledge.

Quite apart from practical applications, the claim that reason was the key to progress in human affairs did not go unchallenged in the eighteenth century.

In 1750, Jean-Jacques Rousseau wrote a prize-winning essay, *Discourse on the Arts and Sciences*—"arts" here referring to the mechanical, not the fine arts—in which he argued that progress in both of these was specious; their advance was actually corrupting mankind, not improving it. He developed this argument further in his 1755 essay *Discourse on the Origin of Inequality Among Men*, linking the arts and sciences to the political and economic subjection of the masses. In *On the Social Contract, or Principles of Political Right* (1761), *The New Heloise* (1762), and *Emile, or On Education* (1762) Rousseau made will, feeling (*sentiment*, in French), and religious spirituality, not reason, and certainly not modern science, the basis for improving human well-being. True progress required returning to a primordial state of nature by dismantling social institutions altogether, not by constructing new ones based on reason.

The dissemination of Rousseau's works was a major factor in the emergence in the late eighteenth and nineteenth centuries of a Romantic movement in literature, poetry, drama, and philosophy that opposed will, feeling, and intuition to the claimed preeminence of reason as the means to improving the human condition. The Romantics typically rejected claims of the potentially unlimited improvement of humankind, seeing human being in the world very much as Protagoras and Gorgias did, as ultimately tragic.

The Romantics notwithstanding, however, and in spite of the paucity of practical benefits, modern science became for most educated people the strongest evidence for the reality of progress. What was clear to all was that the new science truly was progressive, in the sense that the knowledge it produced advanced cumulatively. We take this for granted today, but it was startling in the eighteenth century. Historically, the growth of *know-how* has been cumulative, primarily by way of incremental improvements but with episodic "revolutionary" innovations as well, as cited earlier. Philosophical (and theological) knowledge production, on the other hand, manifestly was not cumulative. Each thinker—Plato and Aristotle, Augustine and Aquinas, Descartes, Spinoza, Leibniz, Locke, and Kant—proposed their own solutions to the same philosophical (or theological) problems. There was nothing in philosophy or theology like the cumulative advances achieved in seventeenth-century mechanics, astronomy, mathematics, and optics, for example, and nothing like Francis Bacon's promise that, using his experimental method, truths about nature would steadily accumulate over time as successive practitioners of the new science built on the work of their predecessors.

There could be no doubt that cumulative progress in knowledge of nature and in mathematics had been made in the course of the seventeenth century and continued to be made in the eighteenth century. The basis of this progress was Newtonianism, a fusion of Newton's experimental-mathematical physics, Descartes' mechanical philosophy of nature, and a dash of Leibniz (internal forces and the Principle of Sufficient Reason). Although Newton would not have approved, this produced exactly the atheistic materialistic determinism that Bishop Berkeley had anticipated and feared. Mathematical physics displayed dramatic advances, as did the calculus, with important advances in probability theory. In addition, there was clear cumulative growth in astronomical knowledge, in knowledge of electricity (with the invention of numerous instruments for producing and storing static electricity), in the subject that at the turn of the nineteenth century was given a new name, "biology," and in chemistry, which morphed from practical know-how to theoretical knowledge in the course of the eighteenth century.

At the heart of this cumulative growth was the methodical application of reason to the study of natural phenomena. The next step to take was to transpose the application of reason to the study of human affairs. This required assuming that human affairs, both at the individual and social levels, were rational so that there could be a *social* science, as Baron Turgot, another of the *philosophes*, named it. Reason was proclaimed by the *philosophes* to be the basis not only for *understanding* human affairs, but also for *improving* human well-being, again at the individual and social levels. This proclamation of the human progress that could be expected, given the hegemony of reason in human affairs, is why the mid-to-late eighteenth century was called in its own time an Era of Enlightenment, an Age of Reason, a Century of Light. (Note that unlike natural science, social science was born with explicit action agendas.)

It is important to keep in mind the example set by modern science of the reality of progress in knowledge accomplished through the hegemony of reason. The example was of an enterprise that was rigorously secular and naturalistic, one that rejected tradition and authority, whether political or religious, in favor of reason-substantiated truths. The reality of progress in scientific knowledge, and the belief that it provided knowledge of reality, was a conscious example for the people who created the new social sciences as complements to natural science. Voltaire was perhaps the most famous of the *philosophes*. He was not a natural philosopher, but in 1738 he published *Elements of the Philosophy of Newton*, making Newton rather than Descartes central to French thinking about knowledge, reason, and progress. The physics and mathematics in this work, barely comprehensible to Voltaire, was provided by his lover Gabrielle de Breteuil, the Marquise du Chatelet. (This brilliant woman went on to produce, together with the mathematician Alexis Clairaut, the first, and for the next 250 years the only, French translation of Newton's *Principia*. In addition, she also wrote *Institutions of Physics*, which introduced Leibniz' natural philosophy to French thinkers.) Voltaire published the *Elements* for the *philosophes*, as a manifesto of the power of reason, not for the benefit of French natural philosophers and mathematicians, who could read Newton in Latin.

Diderot's *Encyclopédie* was a majestic expression of the *philosophe* mindset and the role played in it by a realist interpretation of modern science: that modern science provided knowledge of reality, not merely an interpretation of human experience. The *Encyclopédie* was inspired by the publication in England in 1728 of Ephraim Chambers' two-volume *Cyclopaedia, or an Universal Dictionary of the Arts and Sciences*. This was an immediate success, and a French book dealer, André le Breton, proposed that Diderot translate the *Cyclopaedia* into French. Instead, Breton wound up agreeing to Diderot's counterproposal for a far more ambitious work to be edited by Diderot and the mathematician Jean d'Alembert. This became the *Encyclopédie, or Reasoned Dictionary of the Arts, Sciences and Trades*. (Note well the addition of "Trades," that is, of know-how, to the title.)

The first volume appeared in 1751 and was controversial from the outset. D'Alembert resigned as co-editor in 1758, though he remained actively involved as a contributor, writing over one thousand entries, but Diderot saw the project through to its completion in 1772. By that time, the *Encyclopédie* comprised seventeen volumes of text with 71,818 entries by scores of authors, and eleven volumes of illustrations, including 1800 engraved plates, primarily of workshops depicting the how-to of know-how. The work was perceived by the Church and the government as subversive of religion, social hierarchy, and traditional political authority, which is why d'Alembert resigned his editorship. Diderot was jailed several times for short periods, but because of powerful supporters in Louis XVI's court, he was always released and allowed to continue his work, in part by way of the transparent fiction that the volumes were being published by a Swiss publisher and thus were outside French jurisdiction.

For the entry "Encyclopédie," Diderot wrote that the aim of the work was to change the way people think, to allow people to inform themselves about current knowledge but also about current know-how, bypassing official channels of information and education, as well as control of information about know-how by the secretive craft guilds. D'Alembert wrote that he and Diderot were "impelled to go directly to the [most capable] workers," initially in Paris but then across France, in order to collect information about everything made in France and how it was made. Sometimes, he and Diderot had to do the work themselves the better to understand it and describe it correctly.[2] Diderot's motivation for this extraordinary effort by an intellectual to understand know-how was his belief that people should have access to know-how that was applicable to daily life, not just abstract knowledge. By allowing readers to inform themselves about both contemporary knowledge and contemporary know-how, readers could make more rational decisions about how their lives should be conducted and what to do with them. No surprise, therefore, that the French establishment saw the *Encyclopédie* as subversive, and indeed it did play a role in instigating the Revolution of 1789.

D'Alembert wrote a lengthy introduction to the *Encyclopédie* titled "A Preliminary Discourse," which became influential in its own right as a theory of knowledge and as a manifesto for reason as an engine of human progress. In this essay, he described the *Encyclopédie's* two goals: to reveal the order and connection of all of the diverse parts of human knowledge, and to provide a "Reasoned Dictionary" of the sciences, arts, and trades. The order of the 71,000+ entries in the *Encyclopédie* is alphabetical, but this is purely for convenience. The content of the entries, ranging over all the kinds of knowledge that humans possess or pursue, takes the form of a metaphorical "tree" of knowledge, one whose trunk, limbs, branches, and leaves display the *real* order and connection of the content of the entries.[3] As d'Alembert explains, this "tree" of knowledge reflects the influence of Francis Bacon, who made correlating the many parts of knowledge and their organization by type and topic a necessary preliminary for the systematic production of new knowledge. D'Alembert's "tree" also calls to mind the many attempts by seventeenth- and eighteenth-century naturalists to formulate a classification scheme that would show how all living things were ordered and connected in reality. Recall Plato's injunction that the true philosopher "carves nature at its joints," not any which way he or she pleases. To do that, for knowledge as well as for nature, requires showing how all the parts of nature and knowledge are "joined."

The second goal of the *Encyclopédie*, d'Alembert explained in his introduction, was to describe the general principles and the most essential facts of each of the sciences and each of the arts, whether the liberal and fine arts, or the mechanical arts, such as printing, dyeing, or baking. This goal as well as the "tree" of knowledge are, so to speak, static, providing a comprehensive account of the mid-eighteenth-century status quo of knowledge and of know-how. As with Petrarch and the Humanists, however, the further goal of the *Encyclopédie* was to provide a foundation for the *advancement* of knowledge and of know-how. The *Encyclopédie* might seem only an ambitious compendium of contemporary knowledge, but it incorporated an action agenda for progressing beyond the status quo. To that end, d'Alembert offered a theory—selectively borrowing from Bacon, Descartes, Locke, Hume, and Condillac—of how knowledge was produced, what it was knowledge of, and the role it could play in human affairs. (Note that modern natural science then and now *lacks* an action agenda. Its claim to objectivity entails that it be value free, hence neutral as to what can be *done* with the knowledge it produces. Science may inform action, but action cannot be derived from science.)

Knowledge, according to d'Alembert and the *philosophes* generally, begins with sensations. "Nothing is more indisputable than sensations," he wrote, and these teach us the existence of objects external to the mind that cause them. We cannot know this logically, however. Reason cannot prove that sense objects exist external to the mind or that there is a world external to and independent of the mind. Nevertheless, we "know" this, according to d'Alembert, but we "know" it non-rationally. Echoing Hume, who for years lived in Paris among the *philosophes*, d'Alembert wrote that we know this by an "irresistible impulse" that connects sensations and the world in a way that reason cannot. In this way sensations provide us with "direct," though particular, knowledge of the world. The mind then processes this direct knowledge, via memory, reason, and imagination, producing history, philosophy, and the fine arts, respectively.[4]

D'Alembert emphatically rejected Cartesian innate ideas, but he adopted Descartes' view that only deduction counts as knowledge, because only deduction gives us the kind of certainty that allows us to know truths independently of authority, whether secular or religious. We see in this the same conflation of necessity and contingency, of realism and relativism, that we saw in Bacon, Galileo, Descartes, Newton, Huygens, and Leibniz. The goal of science must be a uniquely correct account of how things really are "out there." This requires deductive reasoning, leading to necessary truths, but we can only begin our reasoning with particular experiences, whether sensations or experiments, together with a belief that these are correlated with the world beyond the mind. Thus, according to d'Alembert, the "physico-mathematical sciences, by applying mathematical calculations to experiments, sometimes [can] deduce from a single and unique observation a large number of inferences that remain close to geometrical truths by virtue of their certitude."[5] This sounds very much like Huygens' apology, cited in the previous chapter, for taking as deductively certain what logically is only inductively probable. It skips over d'Alembert's recognition that reason cannot connect sensations about which we reason to the external world of objects causing them. Science has a rational structure as produced by the mind, yet it has as its object an external world whose existence reason cannot establish because, by his own account, reason is a form of "reflexive knowledge," thus produced by the mind.

Logical inconsistency notwithstanding, d'Alembert, like all of his fellow *philosophes*, claimed that reason alone must be "sovereign" in human affairs, that reason alone, not prayer and not petitioning the political authorities, can improve the human condition. The *Encyclopédie* was subversive because it proclaimed that liberty of action and thought alone can lead to progress in human affairs. At the same time, it was widely recognized by the *philosophes*, including d'Alembert, that there were clear logical limits to claiming that the conclusions of reasoning applied to reality. These limits were explored by skeptical philosophers such as Pierre Bayle, whose *Historical and Critical Dictionary* was very popular in France throughout the eighteenth century, and by David Hume, but also by Kant in his idiosyncratic, certainty-driven philosophical system, and of course by the Romantics. These limits were recognized by the *philosophes*, as by eighteenth-century natural philosophers generally, but they glossed over them, dazzled as it were by the power of reason to reveal reality.

The *philosophes* proclaimed that reason could reveal the inner workings of human affairs as it was already revealing the inner workings of nature, because in principle human affairs were themselves reasonable. That this was the case was obscured by religion, tradition, and authoritarian social institutions. Nevertheless, reason *could* be applied to moral, social, and political philosophy in the same way that it was successfully being applied to natural philosophy. In addition to being a radical and a subversive idea, this was also a liberating idea. It liberated thinkers from the claim that human affairs were the ineluctable products of tradition, authority, non-rational "free" choices, destiny, or a mystical soul, whether of a person or a people or a land. In place of the inherited institutionalization of ineluctable factors to which people needed to adapt, modern *social* science was born with an agenda of social reform, namely, the creation of rational social, political, and economic institutions that would better serve human needs than existing institutions.

The relevance of all this to the thesis of this book is that, as manifested by their action agenda, the social sciences were inherently realistic. They proposed deliberate action to alter social realities on the basis of theoretical knowledge claims that, epistemologically, suffered from the same problem that knowledge claims in the natural sciences suffered from: conflating the strong and weak senses of "knowledge." The newly created social sciences were also inherently paradoxical. They saw as their exemplar of the application of reason to the study of social reality the new, modern science. The latter depicted reality as rigorously deterministic, yet social "scientists" proposed deliberate action to alter the social status quo in order to improve the future of human affairs, to make that future better than it would have been without deliberate human intervention. The Principle of Sufficient Reason and deliberate intervention to change the future are incommensurable!

The new social science quickly became the social *sciences* of social physics (renamed "sociology" by Auguste Comte in the nineteenth century), cultural anthropology, political science, psychology, and economics. (Linguistics and history emerged as social sciences in the nineteenth century, while psychology was then turning itself into a natural science.) Three books by the Baron de Montesquieu were, arguably, the basis for what the Baron Turgot later recognized as the beginning of social science generally and of "social physics" in particular. These books were *The Persian Letters* (1721), *Considerations on the Greatness of the Roman Empire* (1734), and *The Spirit of the Laws* (1748), the last placed by the Catholic Church on the *Index of Forbidden Books* in 1751. In *The Persian Letters*, Montesquieu describes French society and culture from the perspective of two visitors from Persia. The book aims to provoke critical reflection on social institutions and values that we ordinarily take for granted by "watching" as the Persian visitors to France become self-conscious of their own institutions and values. The book also promotes a form of cultural relativism: that different peoples live differently and there is no one set of institutions and values that is right for everyone. In *Considerations* and even more strongly in *Spirit of the Laws*, Montesquieu argued that political institutions and prevailing social and personal values are the effects of underlying and determining "moral and physical causes," not free choice, not accident, and not the will of God. The physical environment, including geography and climate, also plays a determining role.

Earlier than Montesquieu the claim had been made that reason, incorporating empirical observations, and the use of mathematics could lead to improved, because rational, economic policies. In the late seventeenth century, William Petty, John Graunt, and Edmund Halley had collected data on mortality and morbidity (Graunt and Halley), money flows, and trade (Petty). They analyzed these data mathematically, creating longevity and illness profiles, calculating annuities and insurance premiums, and identifying the sources and distribution of England's national wealth. The objective was to base government policies and commercial decisions on facts rather than conjectures. In the eighteenth century, John Law, a Scotsman, argued against Petty that national wealth was driven by trade, not by money flow. It was increased by exports and diminished by imports, which implied that the government needed to adopt policies that encouraged exports and discouraged imports. Law became Controller General of France in 1716. He created the first national central bank; argued for the elimination of the much-hated private tax "farming"; and promoted a form of proactive government entrepreneurship, the use of paper money in place of gold, and stimulation of trade by increasing the money supply. Prescient in many respects, he was dismissed in disgrace after the collapse of a speculative "bubble" he had led the government to initiate.

In the fourth book of his very influential *The Art of Conjecture* (1705) the Swiss mathematician Jacob Bernoulli suggested that probability theory could lead to improved governmental policies. It was the dramatic growth in the sophistication of probability theory in the eighteenth century that led to Turgot's embrace of this suggestion, effectively uniting collection of statistics, proto-economic theory as in John Law, and probability theory. In 1774, Turgot became Minister of Finance to King Louis XVI at a time when the French government was effectively bankrupt. Turgot warned Louis of how close the population was to revolution and counseled, among other things, that the king should not raise taxes and should eliminate the corvée—the involuntary, unpaid labor required of every citizen, except of course the aristocracy and clergy—and also the exemption from taxation of the aristocracy and the clergy, by far the wealthiest elements of the population. Turgot introduced the use of probability theory and statistics in order to formulate reason-based government policies. He was dismissed in 1776, after less than two years as Minister of Finance, and he died in 1781, but the Marquis de Condorcet, his protégé in the application of probability theory and statistics to politics, lived on to play a significant role in the French Revolution.

Condorcet, like Turgot an aristocrat, enthusiastically embraced the French Revolution and gave up all his titles and privileges. But he identified with the rationalist wing of the revolution, that is, with those who saw the revolution as an opportunity to implement governmental policies and standards—for example, national weights and measures—based on reason. The then-recent American Revolution was for him an example of a successful overthrow of government by authority in favor of creating a government and associated social institutions based on reason. Condorcet was charged with writing a constitution for the new French Republic. He included a mathematics-based system for voting that would reflect the will of the majority of voters, but by the time he completed it, the revolution was in the control of decidedly irrational elements, led by Danton and Robespierre, busily slaughtering some seventeen thousand "enemies" of the revolution.

With his aristocratic birth and his incorporation of scientists and support for science into the proposed new constitution, Condorcet was denounced by the Committee for Public Safety and ordered arrested, effectively a death sentence at the time. He evaded capture for over a year, but in 1794 he was recognized and jailed. In jail, shortly before he was found dead in his cell, Condorcet wrote a remarkable document, *A Sketch of a Historical Picture and Progress of the Human Mind*. In this, he traced the progress of the human mind from primitive beginnings through subjection to arbitrary authority and religion to the era of modern science. Now, he argued, through the application of reason to human affairs and to the study of nature, there lay before humanity the prospect of unlimited improvement in the human condition, materially, morally, and intellectually.

Economics emerged as a distinct social science parallel to the emergence of political science and with it a growing acceptance of "the economy" as something real in its own right, not simply a descriptive abstraction. In retrospect, John Law was clearly an economist, someone who formulated economic policies, though they were not policies deduced from a formal economic theory. David Hume wrote a number of essays on economic matters, especially monetary issues, arguing that economic freedom and political freedom were mutually entailing: no political freedom without economic freedom and no economic freedom without political freedom. Turgot proposed what he claimed were rational policies for the French economy, though these were opposed by Ferdinando Galiani, another proto-economist. Galiani's policies too were rational, based on a theory of value he developed that anticipated the mathematical utility theory of late nineteenth-century economic theory. Credit for the first formal economic theory, however, probably should go to Francois Quesnay.

In his *Tableau Economique* (English title, *The Economical Table*), Quesnay formulated a comprehensive economic theory that was named Physiocracy, and it inspired a school of Physiocrats that included Turgot, another influential aristocratic rational revolutionary the Comte de Mirabeau, and Pierre and Eleuthere du Pont. Settling in America after fleeing the French Revolution, Pierre du Pont negotiated the Louisiana Purchase with Napoleon on behalf of Jefferson's administration, and Eleuthere founded the Dupont Chemical Company, so these people were not obscure "ivory tower" intellectuals. For the Physiocrats, wealth was created only by those engaged in extractive activities, overwhelmingly agriculture for Quesnay but mining, forestry, and fishing would qualify. On the threshold of the Industrial Revolution, which got a later start in France than in England, Quesnay claimed that industry did not create wealth. Industry only transformed wealth created by extractive work into goods and services. This transformation was accomplished by proprietors and by artisans. (Marxists would call these capitalists and workers, respectively.) Ideally, 50% of a society's population should be engaged in agriculture, while the proprietor and artisan classes should be 25% each. Note that this overturns what had been the actual class structure of European societies for over a thousand years, dominated as it was by the aristocracy and the clergy.[6]

Opposing Physiocracy was the economic theory promoted by Adam Smith in *The Wealth of Nations* (1776), a book influenced by the economic and political ideas of Smith's close friend David Hume that proved to be the analog in economics of Newton's *Principia* in physics. According to Smith, industry, in the form of the factory system of manufacturing, with division of labor and a goal of mass production, promised to become the future basis of national prosperity and improved human well-being. This was also the view proposed for America in Alexander Hamilton's 1792 *Report on Manufactures* to the U.S. Congress. Hamilton urged the adoption of policies that would nurture domestic manufacturing enterprises and protect them from foreign competition, at least until they were fully mature, against Jefferson's Physiocracy-like vision of an America dominated by small farmers.

In *The Wealth of Nations*, Smith, who like the French *philosophes* who were his contemporaries favored laissez faire economic policies, defended free trade against the earlier mercantilist policy of maximizing exports and minimizing imports. Strikingly, Smith claimed that an economic climate in which each individual was encouraged to pursue their own economic self-interest would lead to the maximum well-being of society. Entrepreneurial selfishness, morally a vice at the individual level, would somehow, as if guided by an "invisible hand," become virtuous when extrapolated to society as a whole. What was emerging, even at this point in the history of economics, was a conception of "the economy" as a reality, an intellectual abstraction that was at the same time real, independent of the mind (analogous to gravity, photosynthesis, and chemical elements).

For over two thousand years, study of the mind was the province of philosophy. That changed decisively in the eighteenth century as psychology emerged as one of the new social sciences. There is a great deal of speculative and folk psychology in authors from Aristotle to Hobbes and Locke, but in Hume it is clear that the direction of study of the mind has changed. First of all, Hume denied that there was such a "thing" as the mind at all. For Hume, "mind" was a name for the flow of our conscious thoughts and feelings, similar to what William James 125 years later would call the "stream of consciousness." This was quite radical. It dismissed completely the Cartesian "I," which was the essential personal mind/soul that was exposed after all particular thoughts and feelings were stripped away. This "I," which dominated Western philosophy from antiquity to the twentieth century, is precisely analogous to matter in physics and chemistry. That is, both are substances; they "stand under" a stable, coherent cluster of properties. Matter is the bare generic "stuff" that is supposed to be left when you strip away all of the specific properties of some particular instance of matter: an apple, a rose, a person, etc. In and of itself, matter generically has only the properties of occupying space, of inertia, of mass, and of motion.

It was this notion of matter that Berkeley attacked as metaphysical and unacceptable in a science that claimed to be empirical. It follows that only the properties of matter are empirical. Similarly, for Hume, we have no experience of a self stripped of all particular thoughts and feelings. Strip away all thoughts and feelings, and there is nothing to experience, so as a committed empiricist, Hume denied that a Cartesian self existed. The mind simply *was* the flow of thoughts and feelings distinctive of each person. Furthermore, Hume argued, this flow was not determined by the will of an underlying "I" or self. The flow had a *logos*, as Heraclitus would have put it, an order of its own, reflected in Hume's three rules of the association of ideas: resemblance, contiguity in space and time, and causation.

Hume stood at the threshold of the scientific study of mind and its relation to the human nervous system. This required turning away from the historical study of mind as a branch of philosophy to the creation of a specialist science within the new empirical-experimental natural philosophy. Etienne Bonnot Abbe de Condillac and three of the thinkers he inspired—Pierre Cabanis, Destutt de Tracy, and Maine de Biran in the late eighteenth and early nineteenth centuries—illustrate this turn, though all three deviated from the particulars of Condillac's theory.[6]

Beginning with Locke and Hume, Condillac founded what de Tracy called Ideology, the science of the study of ideas. The word "ideology" subsequently acquired a very different meaning, but from the mid-eighteenth to the mid-nineteenth centuries Condillac's meaning prevailed. In his *Essay on the Origins of Human Knowledge*, Condillac argued that the mind was created by the sensations. He proposed a thought experiment. Imagine a human being without functioning sensory organs. Such a being, he claimed, would not have a mind, and would possess no consciousness. Then turn on the sense of touch. This would not only generate tactile sensations but also ideas relevant to tactile sensations. Then turn on, one at a time, the other senses. Again, each one generates its distinctive sensation and ideas relevant to that sensation, but also ideas that link sensations. With all of the senses activated, the full mind is now operative, owing nothing to reason, which is itself an integrative, cross-sensation idea.

Condillac claimed that language and thought were connected and had parallel structures, language being the externalized expression of the experience of sensations. Cabanis, de Tracy, and de Biran rejected the details of Condillac's Ideology but embraced the concept of a science of mind. Cabanis's goal for this science was to link thinking and feeling to the physiology of the brain, to understand how the brain "secretes" thought analogous to the secretion of hormones by the glands. De Tracy and de Biran analyzed consciousness, identifying such structural properties as the selectivity of consciousness, reflecting the influence of remembered past experiences, and the selective activity of sensation, against Condillac's claim that sensation was passive.[7]

These few examples show the intrinsic realism of the new social sciences, in spite of explicit discussion of the problem of claiming knowledge in the strong sense. As we have seen, recognition of the logically problematic character of claiming to produce knowledge of reality by an empirical-experimental method was pervasive in modern natural science from the beginning. It continued in the eighteenth century, as we saw with d'Alembert's *Preliminary Discourse*, but support for the realism of scientific knowledge accumulated, overwhelming this logical "detail."

In 1729, James Bradley, later the Astronomer Royal of Great Britain, announced telescopic proof of the Earth's annual motion around the sun, based on his analysis of what is called stellar aberration: the change in the position of a star produced by observations of it from a moving body. This was the first empirical evidence that the Earth really did move. Bradley also claimed to have observed evidence of the Earth's axial rotation, but it was only Foucault's pendulum experiments one hundred years later that would be taken as definitive evidence of this. Bradley's stellar aberration observations also allowed him to improve on Huygens' estimate of 212,000 kilometers per second for the speed of light to 310,000 kilometers per second, very close to the modern value of 299,000+ kilometers per second. Concurrently, internationally coordinated observations of the transits of Venus in 1761 and 1769 provided much more accurate measurement of the Earth–Sun distance (approximately ninety-three million miles), and with this, and the Kepler–Newton equations of planetary motion, it was possible to calculate the "real" dimensions of the solar system.

In 1783, William Herschel, in collaboration with his sister Caroline, announced the discovery of a new planet, an announcement almost as shocking as Galileo's 1610 announcement of the moons of Jupiter. Herschel's telescope had disclosed a new reality, a planet that had always been there but was invisible to the unaided human eye. This reinforced the claims to the discovery of new realities using instruments like the telescope and microscope, or the new Leyden jar, and calorimeter. The Herschels went on to announce telescopic "proof" of the existence of the asteroids, double stars, and many nebulae, and to identify and track many comets as they passed through the solar system. One comet in particular, though not one of those first observed by the Herschels, played a central role in "proving" that Newton's theory of gravity was "really" true because it corresponded to reality.

The truth of Newton's theory was subjected to three stringent tests in the course of the eighteenth century. One of these was the calculation of the orbit of Halley's comet, the second was the shape of the Earth, and the third was the orbit of the Earth's moon.

In 1705, Edmund Halley, who had underwritten the cost of the initial publication of Newton's *Principia*, published a book on the orbits of comets based on Newton's theory of gravity. While Newton himself was cautious about the very bright comet that appeared in 1686, limiting himself to accepting the probable identity of the comets that appeared, then disappeared "behind" the sun, and then reappeared, Halley argued that this was the same comet that had been observed in 1610 and 1535. Using Newton's theory of gravity, and estimating the influences of Saturn and Jupiter on the comet's motion as well as of the sun, Halley calculated that the comet moved in an elongated elliptical orbit with a period of 75 or 76 years. He thus predicted its reappearance in 1758 and called on astronomers to look for it. In advance of its reappearance, two men and a woman, the mathematician Alexis Clairaut, the astronomer Joseph Lalande, and the astronomer-mathematician Nicole Lepaute, using more accurate data than Halley had and new mathematical tools for analyzing the perturbations produced by Jupiter and Saturn, predicted that the comet would reappear on April 13, 1758, plus or minus thirty days. It did, providing a major public vindication of the truth of Newton's theory (and by association, of Newton's force-based physics generally) and the falsity of Descartes' theory, which had a major problem accounting for closed cometary orbits based on Descartes' postulated cosmic vortices as the cause of gravity.

Twenty years earlier, the truth of Newton's theory of gravity had passed another test that Descartes' theory failed. In 1679, the French astronomer Jean Richer published data he had collected near the equator in French Guyana. These data included observations that an accurate pendulum clock calibrated in Paris and brought with him to French Guyana kept slower time in Guyana than in Paris. Almost immediately, Newton announced that, based on his theory of gravity, Richer's data revealed that the force of gravity was slightly weaker at the equator because the surface of the Earth was further from its center there than it was in Paris! This, Newton claimed, proved that the Earth was not a perfect sphere as had been assumed since antiquity, but bulged slightly at the equator and was slightly flattened at the poles. Meanwhile, Jean Richer's superior, the astronomer G. D. Cassini, who supported Descartes' vortical theory of gravity, announced that Descartes' theory predicted that the Earth was slightly squashed at the equator and slightly tapered at the poles.

These rival claims were soon shown to have a simple empirical test. If Newton's theory were correct, then a degree of longitude measured near the North Pole would be shorter than a degree of longitude measured at Paris while a degree of longitude measured at the equator would be longer than one measured at Paris. If Descartes' theory were true, a degree of longitude would be shorter at the equator than at Paris, and longer at the Pole. In 1755 expeditions to Ecuador and to the Arctic Circle were approved by King Louis XIV to measure a degree of longitude at each of those locations. The northern expedition returned within two years with the result that Newton was right: the degree was shorter than at Paris. The equatorial expedition returned eleven years later, after overcoming extraordinary obstacles, also proving Newton right and Descartes wrong: the equatorial degree was longer. We now "knew," thanks to Newtonian science, that the Earth *really* was an oblate spheroid.

There was a third test in the eighteenth century of the truth of Newton's theory of gravity as corresponding to reality. This had to do with the orbit of the moon. As astronomers from Ptolemy to Newton knew, the orbit of the moon is very complicated to model. The underlying problem is that the moon's orbit is the product of three *mutually* interacting objects: the Earth, the moon, and the sun (ignoring Venus, Jupiter, and Saturn). Newton's gravitational force equation is only for two bodies and is a function of their masses, taken to be concentrated at their centers, and of the inverse square of the distance between those centers. Newton was well aware that this was an idealization and did not correspond to the real orbits of either the planets or the moon. In the case of three mutually interacting bodies—the sun attracting the Earth and also the moon, the Earth attracting the sun and also the moon, the moon attracting the Earth and also the sun—Newton knew that his theory was only an approximation. In order to calculate the actual rather than the idealized motion of the moon as it orbits the Earth, Newton knew that mathematical approximation methods needed to be used to solve his equation.

In 1745, Alexis Clairaut announced to the Academie Royale des Sciences that Newton's theory of gravity did not correctly describe one observed motion of the Earth's moon, the change in its apogee. Either Newton's theory was wrong or his gravitational equation needed to be modified in order to match existing data. To test Clairaut's claim, approximation methods were needed to solve Newton's gravitational equation for the actual orbit of the moon. Fortunately, more sophisticated approximation tools were available to mathematicians in the late eighteenth century than were available to Newton. Clairaut tackled this problem himself, but it was also taken up by d'Alembert and Leonhard Euler, with Clairaut three of the leading mathematicians of the eighteenth century (Euler was, arguably, the greatest mathematician of all time). Each of these mathematicians used different approximation methods, but all agreed that Clairaut was right: Newton's theory failed to account for the moon's orbit! Clairaut was disturbed by this outcome, in spite of having anticipated it. Carefully reviewing all of the calculations, he discovered that he, d'Alembert, and Euler working independently had made the same mistake. In 1749, he informed the Academie that the problem was with the approximation methods he, d'Alenbert, and Euler used, not with Newton's theory. When this was corrected, Newton's theory did indeed produce the observed lunar motions.

The Fallacy of Affirming the Consequent notwithstanding, Newton's theory of gravity was taken as having been proven true of reality, corresponding to mind-independent laws of nature. So were his mechanics and his theories of light—that light really was corpuscular, made up of minute light particles with well-defined properties, particles that were real, not hypothetical—and of color—that sunlight really was a composite of multiple colors. Furthermore, the dispute with Leibniz over the physical reality of mv^2 was also settled. In 1726, the Dutch physicist Willem 's Gravesande published a textbook, *Mathematical Elements of Physics in Accordance with the Principles of Isaac Newton*, that disseminated Newton's method of experiment-based mathematical physics in Europe, where Cartesian physics previously had prevailed. 's Gravesande described a series of experiments to test the reality of mv^2. He dropped an object of known mass from various heights onto a thick slab of clay and measured the depth of the depression for each height, with its attendant varying velocity. He found that the force exerted by the falling body was a function of mv^2, thus proving its physical reality and vindicating Leibniz.

That modern science revealed truths about reality, including unexperienced and unexperienceable realities, was repeatedly reinforced throughout the eighteenth century. We experience light, but the questions of what light really *is* and what colors really *are* were answered by Newton's weightless light corpuscles possessing different degrees of "refrangibility." (Euler continued to defend Huygens' wave theory as what light really was.) Experiments in the course of the eighteenth century using static electricity created new experiences of something that was called electricity but was not directly experienced in the way that light and sounds were. Nevertheless, that this electricity could be generated at will, stored, transmitted, and discharged on demand in replicable experiments implied that electricity was a real "thing" of some sort. It legitimated the question of what electricity *is*, over and above the question of a description of how it behaved.

Experiments revealed that the common air that we experience was not elementary as it had been thought to be since antiquity. It was made up of multiple different "gases," or "airs," with distinctive properties. These could not be
experienced directly; only experiments revealed their existence, but carbon
dioxide, for example, isolated by Joseph Black, was no less real for that nor was
the "air" that was responsible for combustion. What this latter "air" was became a major scientific controversy in the eighteenth century, one not fully resolved for decades. It was either a weightless substance called phlogiston or a
gas newly isolated from ordinary air and named oxygen. One or the other, and
only one, was real, and in the end it was oxygen. It was joined by other "airs"
including hydrogen, nitrous oxide (laughing gas), and chlorine.

Photosynthesis, identified in the late eighteenth century, is an example of
something unexperienced that science tells us is really the cause of our experience of green leaves on plants, shrubs, and trees. The cause of sounds as experienced is revealed by science to be pressure waves that we do not feel or see,
nor do we experience their transformation into sounds by organs inside our
ear sending electrical signals to the brain. We experience heat, and the question of what heat *is* provoked a centuries-long controversy. Is heat the result of
motion, therefore not a thing at all but the byproduct of moving things, or is
heat a thing in its own right, a weightless fluid that Lavoisier named "caloric"?
In spite of a famous experiment by Benjamin Thompson that was claimed to
prove decisively that heat was motion, many of the most eminent scientists of
the early nineteenth century continued to support the caloric theory of heat,
and continued to treat caloric as real, until the 1840s and experiments that led
to the thermodynamic theory of heat.

Another centuries-long controversy over the reality of the products of
modern science involved classifications of plants and animals. Especially
from the seventeenth century on, as thousands of specimens of new plants,
birds, insects, and animals were brought back to Europe from all over the
world, the need to classify these became a central concern of botany and zoology. The issue was the reality of the various systems of classification that
were proposed, and it came to a head in the eighteenth century in the work
of Carl Linnaeus. Is there an ordering of plants and animals that corresponds
to the way nature is ordered? Recall d'Alembert's tree of knowledge and, once
again, Plato's injunction that the (natural) philosopher must "carve nature at
its joints," not chop it up any which way like a clumsy (Sophist) butcher. Is
there a single, uniquely correct way to classify living things that is correct because it corresponds to the way nature is divided up, or are all classification
schemes conventional?

Linnaeus was strongly committed to the Platonic view, and that's what he aimed to accomplish with his taxonomy of classes, orders, genera, and species based on the sexual organs of plants. On this view, species are real and fixed. They correspond to one level of the organization of nature as it really is. Hybridization experiments in the eighteenth century made it clear that individuals and varieties are not fixed; they can vary and routinely are made to vary by plant and animal breeders. Species, however, do not vary; each is perpetual, as are genera, orders, and classes. On the realist view of classification, "mammal" is an example of a real universal, corresponding to a feature of reality that the mind grasps (as in Aristotle) even though we only encounter individuals in our empirical experience.

The counterpoint is that all classification schemes are conventional in spite of employing objective features of the items being classified. It may be easier to see what is at stake here using books to illustrate the point. Given thousands of books to classify, is there a single, uniquely correct way to do this, a way that corresponds to the way books are as real things? It would not be wrong to classify them alphabetically by title or by author, nor would it be wrong to classify them by date of publication, binding, color, height, or market value, but for most people that would make for an inconvenient library to use, even though all of these criteria are objective features of books. The books could be organized by subject matter as in the Dewey decimal system or as in the Library of Congress system, but it is trivially easy to imagine books whose subject matter crosses Deweyan categories and even Library of Congress categories. All book classification schemes are thus conventional and reflect what a given classification scheme is supposed to achieve. Does this hold for taxonomies of individuals in nature as well?[8]

Linnaeus ultimately had to concede that his classification scheme was not natural after all, and the issue of the fixity of species reached a climax in the evolutionary theory of Darwin and Wallace. What is of interest here is that at the turn of the nineteenth century, modern science was deeply realist. Theories were said to be true because they corresponded to realities that, typically, were unexperienced but were revealed by a combination of observation, reasoning, and experiment, increasingly using newly invented specialist instruments characteristic of modern science. The social sciences, too, claimed to be revealing the real causes of social, political, economic, and psychological behavior.

Perhaps the climactic expression of the fusion of epistemology and ontology in modern science was formulated by Pierre-Simon Laplace, one of the most eminent mathematical physicists and mathematicians at the turn of the nineteenth century. In *A Philosophical Essay on Probabilities* (1807), Laplace wrote:

> We may regard the present state of the universe as the effect of its past and the cause of its future. An intellect which at a certain moment would know all forces that set nature in motion, and all positions of all items of which nature is composed, if this intellect were also vast enough to submit these data to analysis, it would embrace in a single formula the movements of the greatest bodies of the universe and those of the tiniest atom; for such an intellect nothing would be uncertain and the future just like the past would be present before its eyes.[9]

Although a committed Newtonian in his experimental and mathematical physics, Laplace adds to his metaphysical Newtonianism Cartesian necessary truth and Leibniz' Principle of Sufficient Reason. Reality is a rigorously deterministic whole, and the goal of science, in principle, is a uniquely correct account of it. This was a proud, if not arrogant, proclamation of the power of science, but the correlation of reasoning in the natural sciences with reality, indeed of deductive logical reasoning itself with reality, would be seriously challenged in the nineteenth century.

7
What Is Science About?

What is a scientific theory about? This doesn't seem a difficult question to answer. On the face of it, a scientific theory is about nature, about some aspect of the world that it describes independently of the way human beings experience the world. The gulf between our experience of the world and scientific descriptions of it, however, is typically underappreciated.

For over four hundred years science has taught us that the Earth moves, and we acquiesce in this truth though it contradicts our direct experience that the Earth does not move. Furthermore, the Earth's motion is complex. There is the Earth's axial rotation of about 1,000 miles/hour at the equator and its orbital rotation about the sun of about 18 miles/second. On top of these two motions (and ignoring wobbling of the direction of the axis as the Earth rotates about it), there is the motion of our entire solar system around the center of the Milky Way, the motion of the Milky Way around the center of mass of the local cluster of galaxies to which it belongs, the motion of the entire local cluster in the direction of the constellation Virgo, and who knows what else? These extremely rapid motions, much faster than the Earth's orbital velocity, occur simultaneously in multiple directions. They are, fortunately, completely unexperienced by us, and yet science tells us that they are real and that our experience is an illusion.

According to the scientists themselves, scientific theories are abstract, typically idealized, depictions of reality. The history of science reveals that it is often the case that two proposed depictions conflict with one another. Even more often, theories change over time, and thus so does the reality they claim to depict. One instance of both of these situations was the seventeenth-century conflict between Newton's particle theory of light and Huygens' spherical wave theory of light. Only one of these theories could be true, in the sense of being a correct depiction of what light was, really, and the scientific community in England and much of Europe initially picked Newton's. In the nineteenth century, the scientific community overall canceled the reality of Newton's corpuscles, deciding that what light really was was a transverse wave. In the course of the century this evolved into Maxwell's theory that what light really was was crossed electric and magnetic transverse waves. This reality in turn became unreal with acceptance of the quantum theory of light in which what light really was was somehow a "corpuscle" and a wave at the same time.

Another example of conflicting depictions of reality involves late eighteenth-century theories of combustion. Joseph Priestley defended the phlogiston theory, in which burning occurs when phlogiston, a weightless fluid conceived to be in all flammable matter, leaves a body. Antoine Lavoisier's theory was that combustion involved combining with, not losing, something. Lavoisier identified this "something" as a gas he called "oxygen" and cited the increase in weight of the mercury on which he and Priestley conducted their experiments as evidence for combination. Priestley retorted that phlogiston was so light that losing it led to an increase in weight. The scientific community slowly shifted from Priestley's theory to Lavoisier's. In the process, phlogiston became unreal, and oxygen became real, though not as Lavoisier initially defined it.

The atomic theory of matter illustrates how reality changes as the *same* theory evolves. In 1808, John Dalton published *A New System of Chemical Philosophy*, a modern version of an ancient Greek theory of matter as being composed of discrete, invisibly small, indivisible, and indestructible solid particles, called atoms. In the modern version, all atoms of a given element were identical, and the atoms of the different elements differed only in mass and size. Dalton's depiction of what matter really was conflicted with the prevailing Cartesian view that matter really was continuous and indefinitely divisible. In the course of the nineteenth and early twentieth centuries, the reality of atoms was slowly accepted by the scientific community, first by chemists and then by physicists. But by the 1890s, Dalton's solid atoms had lost their reality, replaced by atoms that had some sort of an internal structure involving the distribution of positively and negatively electrically charged forms of matter, the former particulate and the latter continuous. But then these atoms were rendered unreal by a theory that depicted matter as made up of atoms that were overwhelmingly empty space, in which a fixed number of tiny electrons orbited a tiny nucleus composed of electrons and positively charged matter, later identified as particles called protons. Furthermore, these atoms were not at all indestructible. In subsequent updatings of atomic reality, the nucleus became home to protons and neutrons losing its electrons; electrons lost their particulateness and became wavelike; then nuclei became home to protons, neutrons and mesons; and then protons and neutrons lost their elementarity, becoming combinations of quarks and gluons, which is what atoms *really* are, today.

These illustrations are of what is the norm in science, but on what grounds do we choose between conflicting depictions of reality? What criteria are there for pronouncing one theory, and one theory only, a true depiction of some phenomenon? Francis Bacon adapted a biblical criterion: 'By their fruits ye shall know them.' If a theory makes predictions that are confirmed, if it matches experimental results and explains them in a logically consistent way, if it has "fruits," leading to new technologies, then it is a true depiction of reality. This is manifestly false. A theory can explain, predict, and give us control over nature and still be replaced as a picture of reality by a newer theory that does all that and more. The history of science shows us that *every* theory that was once embraced as true because it corresponded with reality was embraced because it explained and predicted and gave some control over nature, but nevertheless it was replaced by a later theory judged to do those three things "better."

It follows from the history of science that judging a theory to be true epistemologically is separate from the ontological judgment that a theory corresponds to reality. The ontological judgment cannot be justified logically, and this was well known from the beginning of modern science. Reason cannot with certainty connect our ideas with a reality external to the mind. Nevertheless, by something like what d'Alembert called an "irresistible impulse" to believe that a world external to the mind exists, scientists, like everyone else, are realists. We all are certain that there is such a world—if we didn't, we could not function existentially—and it would seem to follow that at least in principle it can only have one correct depiction. Without a way of stepping outside our minds to compare a theory with reality, however, can we know that a particular theory is the one correct description? The answer to this question must be "no," but scientists typically do not settle for claiming that a theory just works better on epistemological grounds than any other theory available at a given time. They add the ontological claim that the theory is a correct depiction of reality and that the objects associated with the theory are real (electrons, neutrinos, quarks, black holes, dark matter and dark energy, prions, etc.). A seminal nineteenth-century development in modern science, a mathematical theory of heat, illustrates this point and what is at stake.

What is heat? Is it no *thing* at all but only the result of the motion of the small parts (not necessarily atoms) of the matter of which some object is composed, or is heat a thing in its own right, separate from matter? If the latter, then it must be a weightless fluid that is in all matter, and when it is driven out of a given material object, that piece becomes hot. This question vexed natural philosophers for centuries, but especially from Francis Bacon on because he made it an example of his experimental method. It was the only natural phenomenon to which he applied his experimental method. Bacon concluded that empirical evidence revealed that heat was motion, but this did not resolve the controversy. Many eminent natural philosophers in the seventeenth and eighteenth centuries held that heat was a weightless fluid, given the name "caloric" by Lavoisier, who defended that theory.

At the end of the eighteenth century, Benjamin Thompson—an American living in Europe who was married, briefly, to Lavoisier's widow and later was ennobled as Count Rumford—conducted a Baconian "crucial experiment," as he thought it, to decide the ontological question of what heat really was once and for all. In his *New Organon*, Bacon had claimed that when the available evidence could be explained by two or more different hypotheses, it was always possible to do an experiment that would decide which was true and which was false. Remember: the nature of reality is at stake here.

Thomson arranged the following experiment and claimed that it was "crucial." (Many physics textbooks, and some histories of science, have treated it as such.) A solid cast iron cylinder was immersed in a tub of water together with a boring bar of the sort used to hollow out cast iron cylinders to make cannons. This whole apparatus was in a closed container with a coupling such that a draft animal could turn the boring bar. As the boring bar turned, the hole in the cylinder grew in diameter, and the cylinder became hotter and hotter, eventually causing the water to boil. It stayed boiling for as long as the boring bar kept turning. According to Thompson, this proved once and for all that heat was friction caused by motion; after all, there could not be an infinite supply of caloric in the cylinder. It followed that caloric was not real; only the motion of the small parts of matter was real.

The defenders of caloric, however—and the great Laplace was numbered among them as well as Sadi Carnot, whose studies of heat led to the formulation of the new science of thermodynamics—interpreted the outcome of the boring bar experiment in a way consistent with the reality of caloric. Lavoisier is honored in the history of science as the discoverer of oxygen, but in the case of heat, Lavoisier defended an erroneous theory that explained heat as loss of caloric. In one case he was on the "winning" side in the history of science, in the other on the "losing" side. As we have already seen in the case of Galileo—right about some things, wrong about others—and will see again, being on the winning side is not the result of reasoning in the right way, and being on the losing side is not the result of reasoning the wrong way. The reasoning by Galileo and by Lavoisier was the same in both cases.

The persistence of the caloric theory of heat shows that Thompson's experiment was not "crucial" at all. It failed to decide between the motion and caloric theories of heat as true of reality, because, as we have seen with Newton's theory of colors, the data generated by an experiment cannot speak for themselves, Newton's claim to the contrary notwithstanding. The data and the experimental context need to be interpreted, and interpretation is intrinsically pluralistic, which entails that there cannot be such a thing as a crucial experiment. John Losee's *Theories on the Scrap Heap* offers a large number of mini case studies drawn from the history of science in support of this conclusion.

In 1807, in the wake of the cannon-boring experiment and the continuing controversy over what heat really was, the French mathematical physicist Joseph Fourier, who had been a student both of Laplace and of Joseph-Louis Lagrange, two of the most eminent mathematical physicists and mathematicians at the turn of the nineteenth century, read a paper to the French Academy of the Sciences entitled "On the Propagation of Heat in Solid Bodies." This was a highly innovative mathematical description of the flow of heat. Over the next fifteen years, Fourier's attempts to publish this paper were blocked by Laplace and Lagrange, and by Jean-Baptiste Biot and Simeon-Denis Poisson, also eminent physicists and mathematicians. Ostensibly, the objections were to the mathematics Fourier used (subsequently vindicated and used routinely today in science and engineering). Finally, in 1822, after Fourier had been elected Secretary to the Academy of the Sciences, the Academy published what by then had become a book, *An Analytical Theory of Heat* ("analytical" here meaning algebraic including calculus).

What was striking about this theory was that Fourier explicitly dismissed the ontological question of what heat *was* as irrelevant. A scientific theory need not disclose reality. In a move that is reminiscent of the dispute between Newton and Leibniz about what gravity was, Fourier said that his theory accurately described how heat *behaved* independently of what it *was*. What it was could be set aside as a separate, metaphysical, question. Quite possibly it was this that Laplace most objected to in Fourier's original paper, because Laplace was a committed Newtonian realist. The job of scientific theories for Laplace was to tell us something about reality, ultimately, to tell us everything about reality. Fourier's theory correctly described our empirical experience of heat, and it led to experimentally confirmed predictions of its behavior, but it did not *explain* heat for Laplace, because it did not identify its ontological causes. (Einstein had a similar objection to quantum mechanics.)

Must a scientific theory be ontological in order to be considered explanatory? Before Fourier, the consensus answer was 'yes.' Fourier's move, however, offered an alternative definition of "explain." A scientific theory could be explanatory if it provided an empirically adequate description of a phenomenon, made confirmed predictions, and gave control over the phenomenon. Fourier's theory made the questions of what scientific theories are about and what constitutes a scientific explanation an explicit concern of nineteenth-century science. His move, separating mathematical description from ontological causation, was repeated in the creation of thermodynamics in the mid-century, and then some fifteen years later in James Clerk Maxwell's *Dynamical Theory of the Electro-Magnetic Field.*

Thermodynamics was nominally a theory of heat, but in fact it was a theory of a new reality, energy. The nature of matter and the relationship between matter and energy were issues that the creators of thermodynamics had to deal with. Recall that the dispute over what heat was was centuries old before Fourier. The dispute over what matter was was much older, extending back to ancient Greece and Rome. The earliest (Western) atomists were Anaxagoras, Democritus, Epicurus, and Lucretius. Aristotle, on the other hand, vehemently rejected atomism, primarily on the grounds that it entailed the existence of a vacuum in nature—the spaces between atoms—which he considered impossible. Some medieval Christian philosophers, influenced by Plato's dialogue *Timaeus*, and some Muslim philosophers were atomists, but Descartes dismissed it as totally wrong. For Descartes, matter and space were coincident, and as space was continuous and infinitely divisible, so must matter be. Dalton reintroduced the atomic theory of matter as chemistry was becoming a theory-based science instead of a body of collected know-how. Chemists increasingly adopted it, but physicists did not.

Faced with an unresolved ontological dispute over the nature of matter, the creators of thermodynamics made the same move as Fourier. Thermodynamics, in its classical form, sets the question of what matter *is* aside. Classical thermodynamics is independent of what matter is: it works whether matter is continuous or discontinuous. At the same time, thermodynamics posited a new reality, energy, whose properties and behavior are described by the theory's equations. But this new reality was not a *thing* at all. Energy was an immaterial reality that took many specific forms, and according to thermodynamics these forms are interchangeable in accordance with determinate laws, but there is no such *thing* as energy per se, energy not in a specific form. Energy, like matter, was conserved in nineteenth-century physics, but the special theory of relativity made energy and matter interconvertible in accordance with the equation $E = mc^2$, and then quantum field theory made energy *alone* the ultimate reality.

Thermodynamics illustrates the self-consciousness that surfaced in the nineteenth century about the problem of knowledge that lies at the intersection of epistemology and ontology. While classical thermodynamics is explicitly independent of what matter really is, statistical thermodynamics, developed by Ludwig Boltzmann and others, explicitly adopted the atomic theory of matter. When, in the 1890s, Max Planck set out to solve an important problem in the physics of the day—the black body radiation problem—he tried to avoid using statistical thermodynamics because if he did, then the truth of his solution would be dependent on the uncertain reality of atoms. In the end, as will be described at length in a later chapter, he was forced to use it, and the ontological implications of his solution became central to quantum theory.

Classical thermodynamics reinforced Fourier's move. Further reinforcement came from Maxwell's theory of the electromagnetic field. The experimental work of Michael Faraday, and the growth of theories of current electricity enabled by the invention of the electrical "pile" or battery by Alessandro Volta, led a growing number of physicists to accept the reality of electric and magnetic "fields" through which the forces associated with electric and magnetic charges were propagated. The material constitution of these fields became a critical problem for nineteenth-century physics. The fields had to be space filling, and they had to have very specific properties to enable magnets and electrically charged particles to act at a distance on magnetic and electrically charged objects. Further, it was well established experimentally that electricity and magnetism were interconnected. As Michael Faraday demonstrated, an electrical current creates a magnetic field around it, and moving magnets generate electrical currents in conductors. This is the basis of the dynamo from which almost all of our electricity is produced. Finally, to avoid mystical action at a distance (another echo of the Newton-Leibniz dispute over what gravity was), it was assumed that the proposed fields must be made of some form of matter, given the name "aether," a term borrowed from Aristotle, who had argued back in the fourth century BCE that in addition to earth, air, fire, and water there was a fifth form of matter, eternally unchangeable, of which the heavens were composed.

For decades, in fact well into the twentieth century, the greatest physicists of the day, including Maxwell, exerted themselves to explain electromagnetic action via a physically real aether with mechanical properties and failed. The properties that the aether had to have to transmit electric, magnetic, and electromagnetic waves and also allow the planets to move freely seemed impossible to reconcile. In 1865, Maxwell repeated Fourier's move. He published *A Dynamical Theory of the Electro-Magnetic Field*. This was a purely mathematical theory incorporating physical forces that omitted the question of what the electromagnetic field *was* in favor of equations that described how it acted. His theory explained how electrical and magnetic forces propagated through the aether, how they acted on objects, and how they interacted with one another to produce new effects, but without identifying either the physical nature of the aether or the causal mechanism by which it produced its effects. Maxwell believed that there *was* a physically real aether, but he decided that there were too many ways that God could have made it for us to identify the one that He chose.

It was not necessary to know what the aether was in order to have an explanatory theory of its action, and Maxwell's theory was extraordinarily fertile. It united electricity, magnetism, and optics into one set of equations; explained light as *really* an electromagnetic wave phenomenon; and "predicted" (that is, had as a logical consequence) the existence of freely propagating electromagnetic waves. The "reality" of these waves was confirmed independently by Heinrich Hertz and Oliver Lodge in the 1880s and became the basis for radio technologies in the 1890s at the hands of Guglielmo Marconi and others.

Fourier's mathematical theory of heat, classical thermodynamics, and Maxwell's mathematical theory of the electromagnetic field all described the Heraclitean *logoi* of real phenomena—heat, energy, and electromagnetism, respectively. They identified the patterns underlying the behaviors of these phenomena while eschewing identification of the material causes of those patterns. Each shifted the focus of the questions 'What are scientific theories about?" and "How does a scientific theory explain?' from reality to empirical-experimental experience. This challenged the prevailing and historical relationship of epistemology and ontology in modern science, which claimed knowledge of reality, and the reality of the objects associated with scientific theories. Less than twenty years after Fourier's move, the connection between deductive reasoning and reality was severed decisively.

Recall that for Plato, mathematics, employing solely deductive reasoning, showed that we can have knowledge in the strong sense of reality, albeit Plato's ideal reality, not material nature. Because of the reliance of its proofs on geometric figures, mathematics was for Plato a not quite perfect form of knowledge, but it was very close, and it validated deductive reasoning as a model for acquiring knowledge in the strong sense in all other areas of human inquiry. A generation after Plato, Euclid organized two hundred years of Greek geometry into a single deductive system. Given the putatively self-evidently true definitions, axioms, and postulates that Euclid laid down, the necessary truth of hundreds of theorems followed because these theorems were the conclusions of deductive logical arguments. We knew the truth of these theorems with certainty and at the same time that Euclidean geometry was ontological—the geometry of space "out there"—because it was the product of deductive reasoning.

A problem with Euclid's version of Greek geometry had been recognized early on: the status of one of his postulates. This was the postulate about parallel lines, and the issue was whether it should be a postulate because it was self-evidently true, or a theorem, thus needing to be proven true by deducing it from some combination of Euclid's definitions, axioms, other postulates, and theorems already proven deductively to be necessarily true. Envision a perfectly flat surface of infinite extent and a straight line extending to infinity in both directions drawn on that surface. Now envision a point not on the line anywhere on that surface. According to Euclid, it is self-evidently true that one and only one line can be drawn through that point that never intersects the given line. Taken as true, this postulate plays a crucial role in Euclidean geometry—for example, without it, you could not prove that the sum of the interior angles of every triangle is 180 degrees—but is it *self-evidently* true? Many critics thought not, but no one could find a proof that it was true, though many tried, so it lived on as a postulate for some 2,200 years.

Finally, in the nineteenth century, at least three mathematicians—the Hungarian Janos Bolyai, the German Carl Friedrich Gauss (with Leonhard Euler a candidate for the greatest mathematician of all time), and the Russian Nikolai Ivanovitch Lobachevski—independently thought of the same idea. Bolyai was the son of a mathematician who, against his father's advice, wrestled with proving the parallel line postulate for almost ten years. At that point, he decided to assume that the reason the postulate resisted proof was because it was independent of the other axioms and postulates. In that case, it was not a self-evidently true postulate but a stipulation and thus could be replaced by some other stipulation about the behavior of lines. What would happen to Euclidean geometry if he did that?

Bolyai chose to adopt as a stipulation that through a point on a two-dimensional surface outside a given line, no lines could be drawn that would never intersect the given line. That would be the case if the surface were positively curved, for example, an idealized version of the surface of the Earth. What Bolyai discovered was that his stipulation generated a geometry that was just as deductive as Euclid's. It generated theorems that were necessarily true because they were valid deductions from the assumed definitions, axioms, and postulates, including his new postulate. But these theorems conflicted with the equally deductively proven theorems of Euclidean geometry. For example, in Bolyai's geometry, the sum of the interior angles of every triangle (whose sides are now curved line segments) is proven to be *more* than 180 degrees. This is proven by deductive logical reasoning to be a universal, necessary, and certain truth within Bolyai's version of geometry.

Meanwhile, three years earlier than Bolyai (though Bolyai only learned of it much later), Lobachevski came to the same conclusion that Bolyai had. He adopted in place of Euclid's postulate the stipulation that through a point on a surface, an infinite number of lines could be drawn that would never intersect a given line. This implies a negatively curved, or hyperbolic, surface.

Lobachevski then showed that his stipulation led to a deductively valid geometry with theorems that conflicted with Euclid's theorems (as well as with Bolyai's when the two were compared). In Lobachevski's geometry, for example, it was proven to be universally, necessarily, and certainly true that the sum of the interior angles of every triangle was always *less* than 180 degrees.

Bolyai's father sent his son's work, published as an appendix to a textbook the father had written, to Gauss, who responded with a dismissive note stating that he had already worked all this out decades earlier but did not bother to publish it. The note was crushing to the younger Bolyai, who accepted Gauss' word for his priority, and abandoned mathematics. Now there were three geometries, each of which was deductively valid but whose logically necessarily true theorems, and their depictions of space "out there," conflicted with one another. Euclidean space was flat, the only space with (intuitively) straight lines; Bolyai's space was positively curved (the geometry of this space was developed systematically by Riemann two decades after Bolyai); and Lobachevski's space was negatively curved. There are no Euclidean straight lines in either Bolyai's or Lobachevski's geometries. Instead, there are curved lines that meet a new generalized definition of "straight line": the shortest distance between any two points on a surface, whether it is flat or curved.

Which of these three geometries is true of the space in the physical world? This question cannot be answered logically because all three geometries produce theorems that are equally true *logically*. The question can only be answered empirically, by doing experiments, for example, to discover how many degrees there are in physical triangles. But that means that geometry, and by extension all of mathematics, is subject to the limitations of induction when mathematics is applied to the world. The connection of deductive reasoning, which alone can give us knowledge in the strong sense, with the physical world is broken. In a 1921 lecture "Geometry and Experience," Einstein put the issue this way:

> At this point an enigma presents itself which in all ages has agitated inquiring minds. How can it be that mathematics, being after all a product of human thought which is independent of experience, is so admirably appropriate to the objects of reality? Is human reason, then, without experience, merely by taking thought, able to fathom the properties of real things. In my opinion the answer to this question is, briefly, this: as far as the laws of mathematics refer to reality, they are not certain; and as far as they are certain, they do not refer to reality.[1]

And yet physics, couched in the language of mathematics, not only *refers* to reality, for Einstein himself its goal was to *reveal* reality. The "enigma" that

Einstein addresses here echoes Simplicio's question to Salviati in Galileo's *Two New Sciences*: 'Your mathematical poofs are logically valid, but how do you know that they apply to the real world?'

It took several decades before the consequences of severing the connection between deductive reasoning and what was "out there" were fully appreciated, but when they were, it was an intellectual shock. Reasoning alone can never give us certain knowledge of the world external to the mind. Given the Fallacy of Affirming the Consequent, experiments and the data they produce cannot bridge the logical gulf between deduction and induction. Science is inescapably inductive, inescapably conjectural—requiring nonlogical assumptions—corrigible, and tied to experience. It is a product of the mind that claims to be about the world. Kant had rescued the ability of reasoning to produce universal, necessary, and certain knowledge by making the mind's experience of the world the object of knowledge rather than the world itself. It followed that for Kant we could have no knowledge whatsoever of the world external to mind, not even probable knowledge. This was too much to give up for the overwhelming majority of nineteenth-century scientists. Logically, science might have to be about experience, but in practice, scientists claimed that the knowledge they produced had as its goal a correct description of the world.

Though seemingly a technical development within mathematics, the existence of non-Euclidean geometries had major consequences for any reason-based knowledge claim. These consequences affected the application of the conclusions of reasoning to the external world—the existence of which was itself a logically unjustifiable idea, as Hume and d'Alembert among many others acknowledged—and it exposed the necessity of assumptions in all reasoning: logical, mathematical, and scientific. Classical (Greek) logic itself rested on the Laws of Contradiction and of the Excluded Middle. But these "laws" are not logically necessary. They are stipulations, analogous to Euclid's axioms, definitions, and postulates. Twentieth-century logicians showed that dropping or modifying these "laws" led to the creation of many different nonclassical, but equally deductive, logics, analogous to the invention of non-Euclidean geometries.[2]

The existence of multiple geometries also called into question the traditional conception of mathematical truth as validated by deduction, leading to a fifty-year-long search for the "real" foundations of mathematical truth, which had rarely been questioned in the long history of Western mathematics. (Kant, for one, had formulated a theory of mathematics consistent with his philosophical system and the role played in it by synthetic a priori truths.)

In the closing decades of the nineteenth century, Gottlob Frege transformed Western logic, extending its scope far beyond Aristotelian logic. Based on his innovations in logic, Frege in 1884 published *Foundations of Arithmetic* with the goal of reducing arithmetic to logic. The goal was to show that the foundations of mathematical truth were purely logical and logical truth was not problematic (at the time!). Frege completed this reduction in the second volume of *The Basic Laws of Arithmetic* (1903), but his entire project was vitiated on the threshold of publication by Bertrand Russell, who informed Frege that it contained a logical inconsistency, a self-referential paradox that proved irreparable.

In the same year of 1903, Russell himself took up Frege's extended project in *The Principles of Mathematics*, taking on the challenge of reducing all of mathematics to logic. Again, mathematical truths would in that case become truths of logic, and thus secure. Russell believed that he and his coauthor Alfred North Whitehead had completed this reduction in their *Principia Mathematica* (1910–1913), but it, too, turned out to be flawed. Concurrently, others argued that the foundations of mathematics were empirical, following John Stuart Mill; that mathematics was a closed formal system, that is, a kind of logic-validated game invented by the mind, following David Hilbert; or dependent on a kind of intuition that escaped reduction to logic, following Luitzen Brouwer. Hilbert's approach seemed to have the most promise, but in the 1930s, first Kurt Godel (1931) and then Alan Turing (1937) showed that two of the conditions for the success of Hilbert's project could not be met: the completeness of mathematics as an axiomatic system and the computability of all numbers.[3] With that, the quest for identifying the foundations of mathematical truth was set to the side, while the foundations of logical truth itself became an issue. Note, however, that the practice of mathematics and the production of new mathematical knowledge were completely unaffected by the inconclusiveness of the search for its foundations. As with physical science, ontology and epistemology in mathematics seem logically independent, however entangled they are in representing scientific and mathematical knowledge claims.

I started this chapter by asking what scientific theories were about. Hilbert's project suggests that one can ask the same question of mathematics. What is mathematics about? As with science, the answer to this question lies at the intersection of epistemology and ontology. Do mathematical objects exist independently of the mind, and if so, is discovering these objects and their properties what mathematicians do (Plato's view)? Or are mathematical objects invented by the mind, and what mathematicians do is explicate the properties implicit in the invented definitions of those objects, as every game of chess is an explication of the rules of that game, or as the theorems of Euclidean geometry are explications of what is implicit in its definitions, axioms, and postulates?

Everyone agrees that mathematical objects (circles, squares, etc.) do not exist in empirical experience. Circles and squares and triangles are ideas that have no material realization. If we take the Platonic view of mathematics—that mathematics and its objects are real in their own right, independent of the human mind—how the mind can know objects that are independent both of the mind and its empirical experience is a deeply troubling question. We would seem to have to assume, with Descartes, that the mind has innate knowledge of certain realities or has a special power to intuit mathematical objects independently of experience. On the other hand, if we adopt Hilbert's view that mathematics is invented by the mind, then the fact that mathematics, especially extremely abstract mathematics, works so well in describing the physical world is an equally troubling question. There do not seem to be any alternatives to these two views of mathematics and no way to decide between them.[4]

That assumptions play a pivotal role in scientific reasoning had been implicit in modern science from the beginning; in the nineteenth century it became explicit. It became necessary to address the question of how assumptions are chosen and on what grounds. Self-evident truth has been revealed to be an unreliable criterion even in mathematics, let alone in physical science. On what grounds does the scientific community choose between conflicting assumptions? As we have seen, Newton believed that we could deduce from experimental data axioms that are universal, transcending experience, and not merely Cartesian "feigned" hypotheses. After Bolyai inter alia, Newton's definitions of space, time, mass, and motion were revealed to be a priori epistemological stipulations disguised as ontological truths. D'Alembert, explicitly aware of the epistemological problem posed by claiming knowledge of reality, nevertheless claimed that a universal principle of nature could be derived even from a single judicious observation or experiment![5]

The conservation laws in physics, for example, the conservation of linear and angular momentum and the conservation of matter and energy, are just such assumptions of universal principles on which reasoning in physics is dependent. While the conservation of momentum remains in place today, the conservation of matter and the conservation of energy, in the nineteenth century taken to be foundational truths of nature, were overturned in the twentieth century. It turns out that matter *can* be created and destroyed, created out of energy as in the Breit–Wheeler process in which electron–positron pairs are created out of colliding photons, and matter can be destroyed by converting it into energy, as in fission and fusion. Similarly, energy can be created and destroyed. The "real" conservation law today is the conservation of mass–energy as expressed in the equation $E = mc^2$, which was a logical consequence of Einstein's special theory of relativity.

To understand any scientific knowledge claim, then, we need to understand the work that assumptions do in a theory and that without assumptions scientists cannot explain or make predictions. Without assumptions, they cannot make either deductively certain or inductively probable inferences from a theory. It follows that in order to employ scientific knowledge in formulating a public policy in response to a real-world problem, we need to know what the assumptions are that the relevant knowledge claims are based on and the grounds on which they taken to be true.

8
The Knowledge Problem
in Mature Science

Early in the nineteenth century, how science worked and the nature of scientific knowledge became open questions as scientists experienced the same "narcissistic" moment that mathematics would later in the century: they became self-conscious about the foundations of scientific knowledge. One expression of this self-consciousness was deliberately bracketing ontological claims, as in Fourier's theory of heat, classical thermodynamics, and electromagnetic field theory. A second consequence was the emergence of the philosophy of science as a subject in its own right, a subject just as much for scientists themselves as for philosophers.

Even as the scope and power of scientific theories grew dramatically, along with public awareness of and respect for science, a need was felt in the scientific community to explain how science worked to produce true theories of nature. In 1830, John Herschel, son of the astronomer William Herschel, published *A Preliminary Discourse on the Study of Natural Philosophy*, echoing the title of d'Alembert's *Preliminary Discourse* but with a very different philosophy of science. Herschel was a scientist of the first order, making important contributions to physics, astronomy, chemistry, and photography. In his *Discourse*, he defended Bacon's inductive-experimental method as the basis for scientists' knowledge claims.

Nature is like a vast, dark room filled with diverse objects and mechanisms. Bacon's inductive-experimental method casts a light that steadily grows over time, exposing more of these objects to our view, allowing us to identify them and to study how the mechanisms work. The method creates a kind of positive feedback loop, feeding new observations and experiments back into existing knowledge such that science advances cumulatively toward comprehensive knowledge of nature as it really is. The cornerstone of this method is control of reasoning to prevent any self-activity of the mind from corrupting the collection and analysis of facts, the formation of hypotheses, and the experimental tests of these hypotheses. Recall Bacon's four "Idols" of the mind, examples of how a mind not controlled by method projects prejudices and preconceptions onto experience. As Bacon himself said, reasoning about nature must be done strictly methodically, as if by machine. Facts are transformed into knowledge, revealing "laws" of nature, by means of a specific stepwise process, by what today we would call executing an algorithm.

Herschel was a close friend of William Whewell, who was the first "scientist" because in the early 1830s he invented the word as a substitute for "natural philosopher" or "savant." Whewell made contributions to mathematics, geology, and crystallography, so he had credentials as a working scientist, but he was most famous as an exponent of modern science to the general public. He served for many years as Master of Trinity College (Newton's college) at the University of Cambridge. He wrote about science for the public and cofounded the British Association for the Advancement of Science, an organization highly respected and supported by the scientific community as well as by the general public. Most important for us is Whewell's response to the knowledge problem in science as contained in two works: *History of the Inductive Sciences from the Ancient Times to the Present* (1837) and *Philosophy of the Inductive Sciences Founded upon Their History* (1840). The order of publication, and the titles, reveal one of the distinctive features of Whewell's theory of science: how the sciences work to produce knowledge is revealed in the history of science.

Close friend or not, Whewell's theory of how science works is diametrically opposed to the account in Herschel's *Preliminary Discourse*. For Whewell, too, scientific reasoning is based on induction, but Whewell redefines what "induction" means. Historically, induction meant generalizations from particulars, as in "All swans are white." Logically, the truth of such generalizations can never be certain, but for Baconians empirically and experimentally well-founded inductive inferences are taken to be true. Generalizations from particulars, however, are not what Whewell meant by "induction." For Whewell, "induction" is a name for a process by which the mind creatively identifies Fundamental Ideas that, if taken as true, allow a body of data they are trying to explain to be organized deductively, as in Euclidean geometry. An example of this is the Fundamental Ideas that Kepler and Newton imposed on Tycho Brahe's planetary motion data, resulting in Newton's universal theory of gravity.

Brahe's data, the most accurate available, did not match the inherited assumptions of circular orbits and uniform speeds (though this did not bother Galileo). Kepler tried various shapes for the orbits that matched the data, until the idea of elliptical orbits occurred to him. Elliptical orbits matched the data very well, but only by adding the idea that the speed of each planet varied as a function of its distance from the sun in such a way that the area swept out by a line from the sun to the planet in a given time was the same regardless of the distance from the sun (Kepler's second law). Kepler added a further idea: that each planet's motion was caused by forces emanating from the sun that varied inversely with distance. This last idea did not fit the data, but Newton had an idea that the planetary motions were caused by a force that varied inversely as the *square* of a planet's distance from the sun. This idea did fit, and it allowed Newton to *deduce* that the planetary orbits *must* be ellipses, that their speeds *must* vary in their orbits as Kepler proposed, and that the mathematical relation that Kepler had proposed between the period of a planet's orbit and its distance from the sun (his third law) *must* hold.

For Whewell, ideas such as these were creative products of the mind. They were not simply found in the data by mechanical, Baconian analysis. Kepler and Newton came up with ideas from which they saw that particular data could be *deduced*. This is a profoundly anti-Baconian conception of scientific reasoning. Recall that for Bacon scientific reasoning requires suppressing any creative activity by the mind in order to escape the Idols of the mind. It requires subordinating personal speculation to a strictly impersonal analysis of relevant data, avoiding deduction, and minimizing mathematics. Scientific reasoning proceeds by collecting data and letting the data themselves "tell" the researcher how they are related to one another, and to their underlying cause, by suggesting hypotheses that can be tested experimentally.

For Whewell, creative contributions by the mind are the foundation of scientific knowledge. Whewellian "induction"—later named "abduction" by Charles Sanders Peirce to allow "induction" to retain its historical meaning—refers to the creation of Fundamental Ideas that provide a bridge, albeit a nonlogical creative bridge, between classical induction and deduction. Scientific knowledge thus comprises a subjective element, the Fundamental Ideas, as well as an objective element, deductive reasoning. Whewell called this the "fundamental antithesis" of scientific knowledge, that not only is scientific knowledge simultaneously subjective and objective, but also that the objective is *dependent* on the subjective. Fundamental Ideas alone enable particular empirical experience to be reasoned about deductively.

Note that organizing experience into a deductive system is itself a profoundly anti-Baconian goal. Fundamental Ideas arrived at through Whewellian induction allow scientists to begin with particular facts of experience yet produce knowledge in the strong sense. For Whewell, the universality, necessity, and certainty of such knowledge are only in the mind, only in the *form* in which empirical data are ordered, contingent on Fundamental Ideas that change over time as experience accumulates.

In Whewell's language, Fundamental Ideas allow the "colligation" of facts, the organization of facts into a deductive theory. Fundamental Ideas are historical; they develop over time and change over time in response to new observations, new experiments, and new data from new instruments. Fundamental Ideas are taken to be true a priori, but not because they are necessary features of the human mind as for Kant. Copernicus used as a Fundamental Idea the axial and orbital motions of the Earth but retained the old Fundamental Ideas of circular orbits and uniform orbital speeds. Brahe's Idea for colligating astronomical data retained a stationary Earth but proposed that Mercury and Venus orbited the sun while that triplet orbited the Earth, as did the outer planets. Working with Brahe's data, Kepler returned to Copernicus' Fundamental Idea of a moving Earth but rejected the legacy Ideas in favor of his new Ideas of elliptical orbits and varying speeds. Newton showed that Keplerian astronomy followed necessarily from his new Fundamental Idea of a universal gravitational force that obeyed a particular, inverse-square law, together with Ideas defining space, time, and motion in particular ways. Some 230 years later, Einstein substituted a different set of Fundamental Ideas for Newton's, colligating more and different data than Newton's Ideas could, resulting in his relativity theories.

Scientific reasoning is thus *both* deductive and inductive, deductive *because* inductive, if we accept Whewell's new definition of "inductive." Whewell argued that through its inductive character, scientific knowledge is historical, that theories change for the better as Fundamental Ideas change over time. This is revealed by the history of science, which is really the history of Fundamental Ideas. This why Whewell published his *History of the Inductive Sciences* before publishing his *Philosophy of the Inductive Sciences*. The history of science contains the data confirming Whewell's theory of how scientists produce knowledge and what kind of knowledge they produce. There remains the further problem of what it is they produce knowledge *of*, which Whewell called the "ultimate problem" in epistemology: linking knowledge in the mind to the world "out there" as its object. How, given the a priori foundation of scientific knowledge in Fundamental Ideas, could science produce knowledge of a world beyond experience?

Whewell rejected Kant's synthetic a priori, which entailed the necessary fixity of the construction of experience by the understanding and also Kant's conclusion that an external world was forever beyond our comprehension. Whewell proposed that the test of the truth of a theory lies first in its success in colligating facts, deducing facts from Fundamental Ideas. But because these Ideas must be expected to change over time, a theory's account of experience while logically certain is not ontologically definitive. Even with Whewell's redefinition of "induction" enabling deduction there remains a gulf between deduction and the world as it is in itself. (This is, once again, the question pressed on Salviati by Simplicio in Galileo's *Two New Sciences*.[1]) The deductive aspect of our reasoning about the world is internal to the mind, keyed to the subjective *assumption* of the a priori truth of the Fundamental Ideas. Whewell thought that this gulf could be narrowed, though not completely bridged, by three criteria: predictive success, "consilience," and coherence.

Alas, predictive success cannot prove that a theory is true of the world because of the Fallacy of Affirming the Consequent. The same history of science that Whewell brought in support of his theory of science is rich with examples of theories that were predictively successful in their time but were subsequently dismissed as false. The same is true of the coherence of a theory with other theories that we hold to be true, or with all the relevant data supporting it. This leaves "consilience": a theory's ability to explain diverse kinds of facts. The nineteenth-century atomic theory of matter, for example, explained facts associated first with chemical reactions, then with chemical properties of molecules, then through the kinetic theory of gases with facts associated with properties and behavior of gases, then with facts associated with Brownian motion, and finally with a vast array of unexplained spectroscopic data. But none of those nineteenth and early twentieth century atoms are considered real today.

Whewell claimed that scientific theories, supported by predictive success, consilience, and coherence converged over time onto the one true account of the world that corresponds to how the world really is. Whewell, a religious man, believed that theories would converge on ultimate truths about nature because our minds are so constituted by God that we will eventually think of the very same Fundamental Ideas that God used in creating the world (a distinctly Augustinian belief, echoed by Descartes). But Whewell had no criterion independent of the mind's reasoning for *confirming* that the conclusions of our reasoning corresponded to the way the world is. Scientists may say so, but they can never be certain that it is true.

Not surprisingly, Herschel was not converted to his friend Whewell's theory of science. He rejected Whewell's redefinition of induction as well as Whewell's self-identification as the *real* Baconian. Three years after Whewell's *Philosophy of the Inductive Sciences* was published, John Stuart Mill, a philosopher rather than a scientist, published his *System of Logic* (1843). In it, he defended traditional induction and the "true" Baconian method against Whewell. Deductive reasoning about the world is vacuous; all our information about the world comes from experience, and all our knowledge of the world comes only from (classical) "enumerative" induction.

But what is the basis of our confidence that inductive reasoning leads us to knowledge of the world, given Hume's powerful argument that induction is only an idea in the mind? Mill's answer is tricky. Our confidence in induction is based on our experience that nature acts uniformly. This is equivalent to saying that our confidence in induction is based on induction, a patent circularity. Mill realizes this of course and adds that it is a fact of experience that some phenomena are so constant that we are compelled, involuntarily, by a "natural inclination" (recall d'Alembert's "irresistible impulse") to recognize that they are always constant. We are effectively compelled to generalize them. Mill rejected Whewell's claim that elliptical orbits was a creative idea that Kepler imposed on data. According to Mill, Kepler *saw* the ellipticity *in* Brahe's data. Whewell responded to this in an article "Of Induction, with Special Reference to Mr. John Stuart Mill's System of Logic" (1849), defending his philosophy of science and his claim that the ellipticity was not at all latent in the data, like a yet-to-be-developed latent image in chemical photography. (The consensus view was that data alone cannot be the source of uniquely correct scientific explanations.)

Like earlier empiricists, Mill held that the goal of modern science was to reveal the true causes of natural phenomena in spite of employing inductive reasoning. And it could do so because, he claimed, universal kinds have a real existence in nature, echoing Aristotle, and can be discovered inductively, without invoking an Aristotelian active intellect. Also like earlier empiricists, Mill needed to provide criteria for determining that our inductive reasoning had correctly identified the operative cause of a particular phenomenon. He proposed four criteria: the methods of agreement, difference, residues, and concomitant variation.

> Agreement: If B always occurs when A precedes it, then A causes B.
> Difference: If B does not occur when A is absent, then A causes B.
> Residues: Given several effects, all but one with known causes and one residual cause, then that cause is the cause of the residual effect.
> Concomitant Variation: If an effect varies, for example, in intensity, as a candidate cause varies, then that candidate is the cause.

Mill clearly thought these methods sufficed to identify causes uniquely, but equally clearly (to us) they do not. In addition to assuming monocausality and lacking a logical basis for identifying candidate causes, his four methods are not enough to identify "true" causes.

Mill's thinking about science was influenced by his reading Auguste Comte's *Cours de philosophie positif* (*Course on Positive Philosophy*), published in French in six volumes from 1830 to 1842. Comte studied for two years at the École Polytechnique (he was expelled for political activism), then the best engineering college in the world with a rigorous science, mathematics, and laboratory-based curriculum. Comte was a social and political philosopher, but his plan for a rational society rested on a scientific foundation. This required clarifying the nature of scientific knowledge in order to justify the role he assigned to science in the governance of society. (His plan for a society organized around "Order and Progress" was detailed in his *System of Positive Polity (1851–1854).)*

Though he is little studied today, Comte was one of the most influential thinkers in mid-nineteenth-century Europe. He invented the term "sociology" as the name for the emerging science of society, replacing the eighteenth-century term "social physics," reinforcing the status of sociology as a science in its own right. He also coined "positivism" as a name for the mature stage of human mental and social development. Though Comte used the term "evolution," his theory of human history is developmental because it unfolds necessarily while in the Darwin/Wallace theory, "evolution" has a contingent character as a result of the role attributed to spontaneous variation. Comte's theory of society rests on two "laws": the law of three stages and the encyclopedic law.

The law of three stages holds that the human mind and human societies necessarily pass through three stages. The first stage is theological, the second metaphysical, and the third "positive." In the theological stage, our thinking and the organization of society are dominated by the attribution of causation to anthropomorphic beings. In the metaphysical stage, we search for abstract, experience-transcending causes of experience. The positive stage is the maturity of humanity. Metaphysical thinking is replaced by thinking in which knowledge is based strictly on facts and necessary relationships among facts, abandoning metaphysical ontologies. That scientific concepts should have a correspondence with a reality behind facts was, according to Comte, a metaphysical myth that humanity needed to outgrow.

Comte's second law, the encyclopedic law, is reminiscent of d'Alembert's "tree" of knowledge in the introduction to the *Encyclopédie*. Like the mind and society, science, too, displays a developmental history, moving necessarily from mathematics to astronomy to physics to chemistry to biology to sociology. This movement reflects an inherent relationship, not coincidence or historical accident. Mathematics is the first science and is the most general and least complex. Each of the sciences that follow is dependent on the preceding science, though not reducible to it; is less general than it; and is more complex. The emergence of social science in the eighteenth century was a sign that humanity was moving into the last phase of its positive maturity and was ready for the science-based reorganization of society that he had worked out. More than a decade before Whewell's *Philosophy of the Inductive Sciences*, Comte announced his intention to formulate a "philosophy" of each of these six sciences, but Comte seems never to have realized that his own recourse to necessity—as in the law of the three stages and relationships among facts—was itself metaphysical. Hume, for one, had argued convincingly that necessary connection was not a fact of experience, but an idea projected onto experience, hence metaphysical.

Comte and Mill were not scientists, but they were highly influential thinkers about science. Herschel and Whewell were major figures within the science community. Together, the four are symptomatic of the extent to which that community felt the need to clarify the foundations of scientific knowledge even as that knowledge was growing exponentially. Others in the scientific community shared this feeling. Early in the century, the mathematician and logician Bernard Bolzano had developed a probabilist epistemology for science. He claimed that the concepts and laws of logic are independent of experience and are known non-cognitively by intuition, both clearly contingent assumptions (to us). They enable the mind to reason about the world as if the mind were a spectator. Reasoning within the mind can produce universal, necessary, and certain logical and mathematical knowledge, but this knowledge is empirically empty, being independent of experience. The concepts and laws of empirical statements are dependent on our sense-based experience of the world. Our reasoning about the world begins subjectively, with empirical experience, but with the aid of intuition we can reason to objective knowledge of the world. Because this knowledge ultimately rests on empirical experience, however, it can never be certain, only probable.[2]

Hermann Helmholz was one of the greatest scientists of the century, making important contributions to physiology, thermodynamics, acoustics, psychology, and geology. His *On the Conservation of Force* (1847) was the founding document of the new science of thermodynamics. In the introduction, Helmholz gives us an insight into his philosophy of science and his response to the knowledge problem. (His extensive writings on this subject were collected and published in 1921 by the physicist-philosopher Moritz Schlick and the mathematician-logician Paul Hertz, influencing the logical positivist school of philosophy of science.) For Helmholz, science was fundamentally empirical and experimental, but he was also a realist: science produced knowledge of the world "out there." At the same time, his studies on the physiology and psychology of vision and hearing led him to conclude that the brain, noncognitively, shapes external stimuli into what we see and what we hear. He came to believe that analogous unconscious mental processes shaped the way we reason, including the way scientists reason about the world.

On the one hand, Helmholz rejected metaphysics; on the other, he was a committed materialist and assumed a priori knowledge of features of the world that allowed science to achieve knowledge of the world, not just knowledge of experience. We know a priori, he held, that perpetual motion is impossible, that events are causally connected, and that there are Newtonian forces acting at a distance between material particles. Helmholz was a reductionist in holding that all of the natural sciences ultimately reduced to physics. Experimental physics was a necessary condition for theoretical physics, in which phenomena are deduced from experimentally warranted generalizations. Observational facts only became scientific when they were deduced from laws and causes operating in the world.[3] Helmholz' work on the conservation of force/energy was an example of this. It had a deductive character but was also about the behavior of the real world, revealing truths about the unity of action of Newtonian forces in mechanics, heat, electricity, magnetism, and chemistry. This very soon led to proclaiming energy a new feature of reality.

Helmholtz' close friend and fellow physiologist, Emil du Bois Reymond, adopted the mantra "ignorabimus": the ultimate natures of matter and force, motion, sensations, life, and consciousness are unknowable. Helmholz disagreed: for him, the "final aim of theoretical science is . . . to discover the ultimate unchangeable causes of natural phenomena."[4] Unresolved, however, was the problem of moving in a logically consistent way from his subjectivist epistemology to an objective ontology!

Near the end of the century, Heinrich Hertz published *The Principles of Mechanics in a New Form* (1894 in German, English translation 1899, both editions with an introduction by Helmholtz). In the introduction Hertz explained that in this book he presented a version of Newtonian mechanics in which the motions of material particles are derived from forces exerted by invisible masses. The resulting mechanics was logically consistent, consistent with all available experimental data, and reproduced all the results of traditional Newtonian mechanics, yet it rested on an assumption in whose reality Hertz did not believe for a moment. The lesson was that the object of scientific knowledge claims cannot be a reality independent of the mind. The object of science is experience in the form of experimentally acquired data, and successful prediction is the sole criterion of truth. As quoted earlier, he wrote that "As a matter of fact, we do not know, nor have we any means of knowing, whether our conceptions of things are in conformity with them [correspond to the way they really are] in any other than this one fundamental respect, anticipation of future events."[5]

Pierre Duhem and Ernst Mach were both prolific physicists, between them publishing hundreds of papers and dozens of books, including works on the history and the philosophy of science, among them Duhem's *The Aim and Structure of Physical Theory* and Mach's *The Science of Mechanics* and *The Analysis of Sensations*. Duhem refused an appointment to a professorship in the history and philosophy of science, insisting that he was a theoretical physicist and his work in the history and philosophy of science was integral to his physics, not an avocation. Mach, on the other hand, accepted an appointment as Professor of the History and Philosophy of the Inductive Sciences at the University of Vienna.

Duhem, a religious Catholic, was an insightful critic of scientific realism and of the claim that scientists produced knowledge of nature in the strong sense of "knowledge." According to Duhem, we cannot know that a theory is true because theories are underdetermined by data and experimental testing can neither confirm nor falsify a theory, or any hypothesis, decisively. It follows that we can never know that a theory corresponds to reality, and it is a mistake to claim that the object of science is reality. The proper objective of science is a "natural" classification of phenomena. (Once again, recall Plato on cutting nature "at its joints.") To this end, scientific theories employ a symbolic language to systematically summarize and classify empirical facts and relationships. Facts and theories are not only correlated; they also cannot be separated because to a greater or lesser degree, what is a fact *for science* already incorporates elements of a theory (at the least, assumptions about relevance), and all theories implicitly or explicitly incorporate definitions of facts.

Theories are underdetermined by data because there are always multiple theories from which any finite data set can be deduced, as Hertz showed with his invisible masses. Theories cannot be deduced from data alone and different assumptions will produce different theories: "If we restrict ourselves to invoking considerations of pure logic [as Hertz did, for example], we cannot prevent a physicist from representing different sets of laws, or even a single group of laws, by several irreconcilable theories." Duhem wrote this in 1893, a year before Hertz did precisely this.[6]

Bacon's claim of crucial experiments to decide between two rival theories is untenable because we cannot know all possible candidate theories. Experiment cannot confirm the truth of a theory because of the Fallacy of Affirming the Consequent. Nor can experiment falsify a theory because a theory is always complex, with many conceptual elements, and one or more of these can always be adjusted to be consistent with any experimental outcome. Duhem argued that a negative experimental outcome, for example. a failed prediction, only threw a theory into doubt. At most, it tells us that *something* in the theory—or in the experiment—is wrong or not quite right, not that the theory as a whole is false. (This claim reappeared in the 1950s in the critique of empiricism by the American logician and philosopher of science W. V. O. Quine and now is known as the Duhem–Quine thesis.)

Examples of such adjustments abound in the history of science. One occurred at the end of the 1920s when experiments on beta decay—the emission of electrons from radioactive nuclei—suggested that the energies of the electrons expelled were not quantized. In order to protect the new and rapidly developing theory of quantum mechanics, Niels Bohr proposed a radical reinterpretation of the experimental outcome: allowing limited violation of the principle of the conservation of energy in nuclear physics. Many physicists supported this move, but Wolfgang Pauli proposed an alternative interpretation. He proposed that a hitherto unknown particle existed—later called the neutrino—that was expelled from the nucleus during beta decay along with an electron and with exactly the energy necessary to conform to the quantum mechanical prediction. The physics community chose this interpretation, and the neutrino became real, built into the further development of quantum mechanics even though it was not detected experimentally until the 1950s. (The expelled particle in beta decay was later identified as an anti-neutrino.)

Examples also abound of defending a theory seemingly disconfirmed experimentally by challenging the validity of the experiment. A recent instance was the announcement in September 2011 that neutrinos traveling faster than the speed of light had been detected by a device named OPERA at a particle physics research facility in Italy. If true, this would call into question the truth of the special theory of relativity. Instead, OPERA was called into question. A separate detector in the same facility, ICARUS, operated by a different team of scientists, measured the speed of neutrinos from the same source as OPERA (CERN's Large Hadron Collider) and found it just *under* the speed of light, consistent with relativity theory. Under intense pressure, the OPERA team painstakingly re-examined their equipment and announced that a loose fiber optic cable had affected their timer.

Duhem's physics was consistent with his philosophy of science. He belonged to a group of physicists and chemists called the Energeticists who believed that all physical and chemical change could be deduced from classical thermodynamics without any assumptions about what matter really was. Duhem went further. He vehemently opposed *all* explanation in science, arguing that explanations invariably involved models that smuggled metaphysics into science. He rejected Newton's and Descartes' methods alike as implicitly explanatory and criticized even Maxwell's electromagnetic wave theory, in spite of Maxwell's description of it as a mathematical theory only. Duhem claimed that it was implicitly metaphysical, requiring the acceptance of the reality of new entities, for example, what Maxwell called "displacement currents."

Instead of explanation, Duhem argued that the goal of science was *representation* of phenomena. A theory was a set of rules for correlating the terms in its equations with physical magnitudes. A theory uses hypotheses and logical arguments from which its equations can be deduced. (Werner Heisenberg's creation of matrix mechanics in 1925 is an example of just such a theory.) In physics, Duhem wrote, "an equation detached from the theory that leads to it, has no meaning."[7] The ultimate goal of any physical theory for Duhem is the "ideal" theory, in which phenomena are represented as they "really" are "out there." As in Whewell, actual theories converge on the ideal over time, and the history of science for Duhem reveals their "trajectory" toward that ideal.[8] The existence of the ideal theory can be inferred from the trajectory, and we can assess new theories by whether they conform to it. The history of a science therefore is essential to forming and evaluating new theories as well as for intuiting first principles and axioms (Whewell's Fundamental Ideas).

It should be obvious that the idea of as-yet-unknown ideal theories that at a certain point of time in the history of a science can serve as a test of the truth of actual theories is completely inconsistent with Duhem's aggressively anti-metaphysical philosophy of science. Subtly, it smuggles ontology, hence metaphysics, into epistemology. Duhemian science seems purely empirical and relational, Giants-like, a symbolic representation of human experience, but in the end Duhem could not give up Gods-like reality as the arbiter of truth and the object of scientific knowledge. Duhem's use of the history of science to link ontology and epistemology explains the time he devoted to historical studies, but more importantly to the use he made of history in his philosophy of science. Unlike Whewell, for whom the history of science served an epistemological function as a source of data, Duhem took continuity with the past ontologically, as a criterion of truth in science. It was historical continuity that revealed the trajectory of a theory to the final ideal theory.

Ernst Mach, like Duhem, integrated the history and philosophy of science into his physics. He, too, was an anti-realist and thus an anti-materialist, holding that there was no place in science for metaphysics. A reality independent of the mind and of human experience was not and could not be the object of science. Scientists, for Mach, pursued what he called "economical" descriptions of nature, which for him was the world as it appears to the human mind via the senses: "It is the object of science to replace, or save, experiences by the reproduction and anticipation of facts. Memory is handier than experience, and often answers the same purpose." The "aim of all scientific research," he wrote, is the "adaptation of thoughts to facts . . . In this, science only deliberately and consciously pursues what in daily life goes on unnoticed and of its own accord."[9] But facts for Mach are not at all objective. By "facts" about the world, Mach means how the world appears in the mind after being processed by our senses and then by the brain in ways that are distinctive of human beings. As it was for Helmholz, such facts are not representations of the stimuli the brain receives.

Mach was perhaps the first exponent of what is today called evolutionary philosophy of science, the view that our reasoning about the world reflects the evolution of our nervous system in that world. Knowledge is a product of human biological evolution, but also, anticipating by a century Richard Dawkins' "memes," cultural evolution. Knowledge is not the product of a transcendent mind neutrally exploring its experience of an external world. Alongside his *Science of Mechanics* (1883), a history of mechanics supporting his philosophy of science, Mach published *The Analysis of Sensations* (1897), which made important contributions to psychology at an early stage in the creation of experimental psychology.

Much more than Whewell, Mach attributed originary activity to the mind, harking back to the ideas of Pierre Cabanis and Destutt de Tracy on the selectivity of perception and consciousness and to Helmholz's studies of perception. Based on his own extensive experimental studies, Mach argued that what we call the external world is actively constructed by the senses and the mind, unconsciously. In vision, hearing, smelling, and touching, the brain and the senses working together construct a world in accordance with response patterns built into our nervous systems. Helmholz' pioneering research in the physiology and psychology of vision and hearing fits in here, as do the studies on the selectivity of sensation by Wilhelm Wundt—who in 1879 created the first dedicated experimental psychology laboratory (at the University of Leipzig)—and by William James on the selectivity of consciousness described in his *The Principles of Psychology* (1890). The object of our conscious reasoning about the world is not the world as it is "out there." We reason about the world "out there" only as that world is refracted through our evolutionarily determined nervous system.

Mach's ideas about the selectivity of sensation and consciousness, and the constructed character of what we experience as "the world," have a Kantian aspect to them. They also imply that something like Bacon's Idols of the Mind is inescapable. These ideas influenced twentieth-century psychology, especially the school of gestalt psychology early in the century and cognitive psychology today, as well as neurophysiology. The orthodox account today of the human visual field, for example, is that the eyes are selectively responsive to the photons they absorb and relay already selectively processed information to the brain, where different groups of neurons respond to shape, to color, to motion and, influenced by memory associations, construct what we see as the world. As the philosopher of science Norwood Russell Hanson put it in the 1960s, 'people not their eyes see.'[10]

Mach's evolutionary conception of knowledge extended to method as well. He thought that something like selection operating on spontaneous variation applied to scientific reasoning as it did in biological evolution. New ideas unconsciously pop up in the minds of some people, analogous to biological mutations. Many of these are dead ends or bad ideas, but some are good, like the ideas underlying Newtonian mechanics, thermodynamics, and electromagnetic wave theory. New ideas drive theory change, which is progressive when new theories are more comprehensive "summaries" of empirical facts than older ones, but not because they converge on the one, true, account of reality.

Mach was the most consistent of all of the scientists and philosophers discussed here in keeping his epistemology free of ontological claims, and perhaps the only one. Mach alone does not conflate the positions of the Gods and the Giants. The object of scientific theories is experience, not reality, and that's the end of the matter. Theories are pragmatically justified "economical" accounts of experience and as experience changes, so must theories. Not surprisingly, Mach vehemently opposed the growing trend by his colleagues to attribute reality to atoms. He was not converted to the reality of atoms as many were by Einstein's 1905 theory of Brownian motion. In 1910, near the end of his life, he engaged in a correspondence with Max Planck over atoms. Planck had written to him that now we knew many truths about the way nature really is, including that color is a function of the frequency of electromagnetic waves and that matter is atomic. Mach replied that what scientists mean by "frequency" in relation to light and color has changed several times since Newton and that the "findings of atomic theory . . . can undergo a variety of convenient reinterpretations, even if we are in no great hurry to take them for realities."[11] He was quite right about this. Between 1910 and his death in 1913 the atom was completely "reinterpreted" first by Ernest Rutherford with his solar system model of the atom, and then by Niels Bohr with his quantized orbital electron model.

One final illustration of the extent to which the foundations of scientific knowledge had become an open issue inside and outside of science: J. B. Stallo was a politically active German American intellectual, a lawyer, then a judge, and ultimately American Ambassador to Italy. In 1883 he published an extraordinary book, *The Concepts and Theories of Modern Physics*. In it, he argued that so-called empirical scientific theories in fact rested on metaphysical assumptions, for example, about space, time, matter, and motion, assumptions that could not be *known* to be true. Mach read and was very impressed with Stallo's book, and so was Bertrand Russell, who cited it several times in his own theory of science. In 1900, the nature of scientific knowledge, how it was acquired, and what it was knowledge of, remained unanswered questions, but not for lack of trying.

9
Scientific Realism and the Romantic Reaction Against Reason

Romanticism was a reaction against the claim that reason and reason alone was the means to human progress. The claim that man is, in Aristotle's words, a "rational animal," has deep roots in the Western philosophical tradition. Reason is what distinguishes humans from all other living things, from which it followed, for Aristotle at least, that happiness was a function of the exercise of reason. The purest form of reasoning, and the ultimate form of happiness, was reasoning about reasoning itself, as in logic.[1]

This view, or something close to it, has dominated the twenty-five-hundred-year history of Western philosophy. In the modern era, Descartes proclaimed the necessary truth of "I think therefore I am, therefore I am a thinking being."[2] Spinoza deduced from his definitions and axioms that the intellectual love of God (Spinoza's name for the cosmic reality) was the highest form of human happiness.[3] For Kant, humans were rational beings, and it was on rationality that he based his moral philosophy.[4] For Hegel, reality *was* reason, Absolute Reason coming to self-consciousness in history.[5] In the midst of this welter of philosophical speculation, modern science claimed to be producing knowledge of reality by means of methodical reasoning testable against experience. And as the practice of science proceeded, the case for this claim became stronger, even as the ambiguity of what "knowledge" meant and the problematic character of claiming reality as its object were widely acknowledged.

We have already discussed the de facto correlation of reason with reality in the eighteenth century, reflected in the continuing claims of revelations of new realities by the natural sciences, but also in the creation of the reason-based sciences of human affairs. In the nineteenth century, Fourier's theory of heat, the creation of thermodynamics, and Maxwell's electromagnetic wave theory reflected an awareness among scientists that claiming knowledge of the constitution of reality was a problem. So, too, did the philosophies of science of Herschel, Whewell, Bolzano, Helmholz, Duhem, and Hertz, among others. All recognized that, logically, reality could not be the object of scientific reasoning, yet each one found some ground for claiming that that was nevertheless the case. For each, scientific reasoning plus some philosophical or psychological loophole produced knowledge of the real. The one exception was Mach, for whom the object of scientific reasoning was empirical experience, with no knowledge possible of what existed independently of that experience.

If the case for the real as the object of scientists' reasoning strengthened in the eighteenth century, in the nineteenth it strengthened exponentially. In the course of the century, John Dalton's proposed chemical atoms became real before the eyes of the scientific community as it was discovered that molecules composed of the same atoms had different properties. The atomic theory explained this by attributing these properties to the spatial arrangement of the atoms within a molecule, and structural chemistry was born, reinforcing the reality of atoms. In 1848 Pasteur showed that chemically identical tartaric acid crystals rotated the plane of polarization of incident light either to the right or to the left, apparently depending on how the atoms were positioned in the crystals. In 1860, Stanislao Cannizaro confirmed an earlier claim by Amadeo Avogadro that the same volume of different gases at the same temperature and pressure contained the same number of atoms, allowing the calculation of relative molecular and atomic weights. In1865, August Kekule showed that closed rings of six carbon atoms formed a family of molecules with different properties depending on the point on the ring where other atoms were attached to one or more of the carbon atoms. The kinetic theory of gases, reinforced by statistical thermodynamics, brought atoms into physics. In 1905, Einstein's explanation of the Brownian motion suggested how to calculate the sizes of atoms and molecules.

In 1838, the astronomer-mathematician Friedrich Bessel finally measured the stellar parallax, proving the Earth's orbital motion, as well as measuring the absolute distance to a star, 61 Cygni, giving the first idea of the size of the universe. In 1843, he applied Newton's theory of gravity to an observed minute periodic motion of Sirius and predicted the existence of a "dark" companion star that was observed telescopically in 1862. (Recall William Herschel's application of Newton's theory to tiny periodic motions of binary stars.) In 1846, Johann Galle observed the new planet Neptune at almost precisely the location in the sky predicted by Urbain Le Verrier, again based on assuming that Newton's theory of gravity was true. The cell theory of life was clearly to be understood realistically, as more powerful microscopes revealed detailed structures not only within cells, but within their nuclei. The germ theory of disease had to be understood as implying that microbes were real disease-causing agents, especially as effective vaccines were developed for rabies, cholera, and typhus. At the turn of the century, genes were accepted as real, the carriers of heredity, especially after 1910 and Thomas Hunt Morgan's experiments correlating specific locations on chromosomes with specific body features of fruit flies.

Meanwhile, theories in physics and chemistry became central to technologies that transformed Western societies: the electric telegraph; transoceanic telegraph cables; long-distance telephony; wireless telegraphy/radio; the generation and distribution of electricity; electric motors; the synthetic dye, perfume, fertilizer, and fabrics industries; plastics; synthetic pharmaceuticals and antibiotics; new, more powerful explosives and propellants; internal combustion engines; and the petroleum industry. How could science, the systematic application of reason to experience, not be producing knowledge of what was real? Our personal and social lives were increasingly oriented around the new realities created by science-based engineering, and these products of reason were improving our lives.

In the face of all this, it was the shared conviction of eighteenth- and nineteenth-century Romantics that the apotheosis of reason by philosophers, social reformers, and scientists was mistaken. For the Romantics, as for the Greek sophists, reason could be useful in human affairs, but it was of limited value for understanding the human condition, which at ground was based on feeling, intuition, will, and history, thus on time. Reason-based philosophy dismissed all of these, particularly time. Time was patently an elemental feature of human experience, yet deterministic modern science—based on the Principle of Sufficient Reason, Laplace's manifesto, and the time-reversible equations of Newtonian mechanics—dismissed the reality of time and thereby the significance of history in epistemology and in human affairs generally.

I have already referred to Jean-Jacques Rousseau as an early figure in what came to be called the Romantic movement, and his initial fame arguing that science and technology had worsened the human condition, not improved it. For Rousseau, will, in individual and general forms, was the central fact about human beings in the world. Earlier than Rousseau, Giambattista Vico's *The New Science* was a more ambitious and scholarly response to modern science. Though *The New Science* was first published in 1725, it became well known and influential only in the last third of the eighteenth century, largely through the writings of the German philosophers J. G. Hamann and Gotffried Herder. Vico objected above all to Cartesianism and its claim to produce knowledge of reality. To understand something, Vico argued, one must know its history, one must put it into the historical context in which it is located, and that made it what it was. He described the historical character of language, customs, and thought from earliest antiquity to the eighteenth century. Over that time, languages changed, customs changed, and the ways that people thought about the world and about the human condition changed. These changes over time, these histories, needed to be appreciated in order to understand how we speak, behave, and reason today, because these, too, are located in an ongoing history: they're not final.

Vico contrasted philosophical and philological approaches to knowledge. The former seeks to identify universal principles of reality while the latter seeks to acquire knowledge (really know-how) beginning with particular human experience. Modern science is philosophical, seeking somehow to snatch deductive certainty out of inductive contingency. According to Vico, this is a complete misunderstanding of what goes on in the thinking mind. Reasoning is not an experience-neutral, timeless faculty within the mind. The way we reason reflects the history of our reasoning, which is specific to Western cultural history (including its interactions, or non-interactions, with other cultures). To "introduce geometrical method into practical life is 'like trying to go mad with the rules of reason', attempting to proceed by a straight line among the tortuosities of life, as though human affairs were not ruled by capriciousness, temerity, opportunity, and chance."[6] True knowledge is the wisdom that follows from the integration of theoretical philosophical reasoning (Greek *episteme*) and practical philological reasoning (Greek *phronesis*). The true, Vico claimed, is and can only be "the made," whether in technology or in epistemology. What our minds conclude is true is the product of how we reason about it and the latter is a function of our location in cultural space-time.

Romanticism's roots are in the eighteenth century, but it became a movement in the nineteenth century, growing in intellectual and cultural stature, strongly influenced by Vico, Hamann, and Herder. Hamann was at the center of German intellectual life in the second half of the eighteenth century. Recall that it was Hamann who in 1761 recommended that Kant reread Hume—rousing Kant "from his dogmatic [epistemological] slumbers"—and introduced Kant to Rousseau, whose writings influenced Kant's moral philosophy. Ironically, Hamann completely rejected the shift in Kant's philosophical direction that resulted from his "awakening." Kant changed from an empirical-skeptical approach to philosophical issues to the ultra-rationalist transcendental idealism for which he is famous. What Hamann rejected in the later Kant was precisely Kant's apotheosis of reason in quest of certainty. The mind, for Hamann, and for his student Herder, does not and cannot take the transcendental position vis-à-vis experience that Kant assigned to it. Our thinking reflects nonlogical influences, such as feelings, instincts, and intuitions, and constraints deriving from the dependence of conscious reasoning on language.

Hamann and Herder extended Vico's ideas about the historical relationship between language and thought with its implications for our understanding of texts, for translation, and for interpretation. Their work laid the foundations for modern linguistic theory and hermeneutics. Herder was particularly influential in arguing not only for historical studies of language and culture, but also for comparative studies, laying the foundation for comparative linguistics and comparative anthropology. Studying non-Western languages gives us insight into non-Western forms of thinking, but it also gives us a deeper insight into the historically and culturally conditioned character of our own thinking and values (recall Montesquieu's *The Persian Letters*). One byproduct of this heightened self-consciousness about our own thinking and its language dependency is to force us to confront the problem of meaning. How do we know, how *can* we know, the meaning of another person's speech/writing? How do we determine the meanings of words, and thus of expressed thoughts, of our own or others? These questions transpose readily to questions about the meanings of the terms used in science.

Early in the nineteenth century, Friedrich Schlegel, influenced by Herder, developed further the implications of the historical and cultural dimensions of the relationship between language and thought. Schlegel rejected the claim that science could give us knowledge of reality. He argued that so-called self-evidently true principles, the assumptions that play an important role in scientific reasoning, can always be doubted and thus the certain truth of deductive arguments always challenged. (He did not live long enough to see this validated by non-Euclidean geometries.) There are no first principles of unassailable truth that can serve philosophers, including natural philosophers, as a foundation for their systems. Philosophy must start in the middle, "like an epic poem," because starting from first principles is a delusion. Inevitably philosophy, including natural philosophy, curves back on itself, the conclusions reached reflecting its contingent starting point.[7] As Nietzsche later put it, philosophical writing is first and foremost about the philosopher doing the writing.[8] (The same can be said of a scientific theory: that it is first and foremost about the scientist and the assumptions he or she makes on which their theory rests.)

Schlegel recognized that describing reality was the objective of modern science, but because this required an unavailable criterion confirming that a theory corresponds with reality, the most that science can hope for is coherence of a theory with data, logical consistency, and acceptance by the scientific community. That's as much universality as scientific truth claims can have. Knowledge in the strong sense is an idea, and may be an ideal, but it cannot be achieved. However inconsistently, Schlegel nevertheless accepted that the object of scientific reasoning was what existed "out there" and not merely patterns in subjective experience. The ideas of Schlegel and of his brother August and Friedrich Schleiermacher, a philosopher and theologian close to the Schlegels, resonate strongly with ideas associated with late twentieth-century postmodernism, which, with its vehement rejection of science's claim to objective knowledge, echoes the response of the Romantics to science.

For the Romantics, nature is a process in Heraclitus' sense of continuous flux, not an unchanging reality already fully there waiting to be uncovered by successive generations of scientists. Nature is an organic, holistic, and creative process that modern science violates and distorts. (Recall Vico on going "mad with reason.") In his poem "The Table Turned," Wordsworth expressed it this way: ". . . Our meddling intellect/Mis-shapes the beauteous forms of things/ We murder to dissect." The "meddling intellect" freezes a changing reality *in order to* apply logical reasoning to it. Logical reasoning cannot comprehend a changing reality—though Hegel tried and believed he had succeeded—but such a reality can be apprehended aesthetically, as Protagoras and Gorgias also held. And for the Romantics, as for Protagoras and Gorgias, poetry, the novel, drama, and music are means to such aesthetic apprehension. It is not that science is worthless, but only that its claim to tell the whole truth about nature is an overstatement. There are truths about reality that can only be apprehended allusively via art, literature, dreams, fantasies, and Plato's "likely tales," all of which are untruths that somehow allow us to feel truths that words cannot directly describe. We cannot fully comprehend reality intellectually because of the limitations on conscious reasoning imposed by language and the ambiguity of linguistic meaning. To the complexity of extracting meaning from texts that Hamann and Herder had discussed, the Schlegels, Schleiermacher, and the poet-philosopher known as Novalis (Friedrich von Hardenburg) added two important theses that have implications for what scientific theories are and mean.

First, authors themselves cannot tell us the definitive meaning of their speech or writing because both of these incorporate unconscious meanings into what is said or written, how it is said—word choices, metaphors, allusions—what is not said, and even the structure itself of a text, the rationale for which may not be given to the reader and may be unknown even to the author. Because interpretation is intrinsically pluralistic, no text, not even the "book" of nature, can be fully understood logically. Interpretation requires setting aside the search for the one true reading of a text in favor of a search for a sympathetic "understanding."

A second thesis is what is called semantic holism. Meaning is defined by the relationship of words to one another within a language and within the text as a whole in which they occur. Meaning is not determined by extralinguistic entities to which words are taken to refer but solely by their relationship to other words within a language, conceived as at any given time a closed system. This thesis was developed further by the linguist Ferdinand de Saussure early in the twentieth century and further still as a theory of reading and textuality by Jacques Derrida and others later in the century.

The implication for science is that scientific theories, too, are closed systems in which the meaning of each of the terms used in the theory is determined within the theory and not by nature, not by what is "out there." A scientific theory always refers to something it *claims* is "out there," for example, atoms, but what atoms *are* is determined by the theory, not by what is independent of the theory. In this respect, science is like language, a thesis that became central to the Science Wars of the 1980 and '90s in the form of the claim that scientific knowledge was socially constructed.

Is discursive reason, whose ideal is purely logical reasoning, the only means available to us to obtain knowledge of reality? For Parmenides, Plato, and Aristotle down the millenia to Descartes, Spinoza, Leibniz, Kant, and Hegel—all allies of the Gods in Plato's "perpetual battle" over the nature of knowledge and Being—the answer is 'yes.' As we have seen, justifying that attribution has often required attributing to the human mind some special faculty that allows the reasoning mind to transcend the process of reasoning in order to have as its object a reality external to and independent of its reasoning. That reason does not have such a privileged position was, of course, the view of the sophists and the skeptics in antiquity, Renaissance magical nature philosophers such as John Dee who made a distinction between knowledge and wisdom, and modern skeptics such as Montaigne and Hume. The latter took a secular, clearly Giants-like position on the limitations of reason, giving up the possibility of certain knowledge of anything. Hume, we saw, argued that reason was subordinate to the passions, incapable even of proving that there *is* a world external to the mind, something that every human being "knows" with certainty without reasoning.

There were literary Romantic parallels to this reasoned philosophical critique of reason. Examples include Rousseau's novels, particularly *Emile*, in which Rousseau proposed a system of education that subordinated training in reason to nurturing moral sentiments; Laurence Sterne's plotless novel *Tristram Shandy*, which openly mocks logical reasoning in the person of Tristram's father; Diderot's *Jacques the Fatalist*, explicitly indebted to *Tristram Shandy*; the novels of the Marquis de Sade; and the new genre of so-called Gothic novels, the most famous of which is Mary Shelley's *Frankenstein, or the Modern Prometheus* (keeping in mind that the "modern Prometheus" of the title is the scientist who bungles the scientific project to which he has dedicated his life with tragic consequences).

In the course of the nineteenth century Romanticism increasingly became identified with poetry, music, art, and literature. This was consistent with the role that the Romantic philosophers—and the sophists—assigned to "aesthetic" sensibility, described philosophically by Friedrich Schiller in his *Letters on Aesthetic Education.* The aesthetic was a non-rational capacity of the mind for apprehending reality. It gave us insight into the true nature of human being in the world, an apprehension that was beyond the scope of discursive reasoning. The aesthetic was a philosophical concept, but it was rationally non-rational as opposed to irrational. As the century unfolded, however, the truly irrational intruded into art, literature, poetry, and music in the guise of the supernatural, the fantastic, and the unconscious.

On the one hand, there were extraordinary nineteenth-century advances in science, in the form of theories and their associated realist claims, together with their role in the creation of new technological "realities." This emboldened many to claim that reason, as exemplified by science, was producing knowledge of reality and was the only route to what deserved to be called knowledge. This claim grew stronger in the course of the century in spite of what we have seen was a growing awareness within the scientific community of how logically problematic this claim was.

On the other hand, and parallel to the aesthetic theme in nineteenth-century Romanticism, a second generation of Romantic philosophers, among them Søren Kierkegaard, Arthur Schopenhauer, Friedrich Nietzsche, and Henri Bergson, rejected the hegemony of reason. All used philosophical reasoning to deny the ability of science, and of reason generally, to produce knowledge of the real. They acknowledged only that at best, science produced empirically justifiable opinions about experience, a view with which Ernst Mach would have heartily agreed.

Kierkegaard, as Protagoras and Gorgias before him, held that as the universe ultimately was not rational, reason cannot grasp it. He contrasted two familiar approaches, or "stages," to human existence. One, which he called the aesthetic, is dominated by feeling. The second, which he called the ethical, is dominated by reason. Neither of these alone can do justice to human being in the world, yet each claims to be capable of doing so and dismisses the other. Kierkegaard urged a synthesis of these two stages into a third, religious, stage by way of the imagination, not reason. In the religious stage, humans have a feeling of connectedness to a source of all being that transcends rationality because that source, God, is not rational in any human sense.

The work for which Kierkegaard is most widely known is *Fear and Trembling*. It retells the biblical story of Abraham being commanded by the God he believes in to offer up his son Isaac as a sacrifice to God, and it explores Abraham's state of mind as he prepares to do so. This command can make absolutely no sense at all to Abraham. Morally, it is in conflict with God's earlier commandment that murder is wrong and must be punished by the death of the murderer. In addition, God has repeatedly told Abraham that his son Isaac will not only inherit the blessings that God has bestowed on Abraham but will also be the progenitor of a people who will inherit those blessings. How can God now ask Abraham to kill that child?

For three days, as they journey together to the place God has chosen for the sacrifice, Abraham lives with the overwhelming burden of both believing in God and obeying a command that is without qualification wrong and makes no sense on God's own terms. But Abraham's faith triumphs over reason, parental feeling, and moral sensibility. He prepares to obey God's command. At the last minute, with Isaac bound on the altar Abraham built, God stays Abraham's hand and reveals that this had been a test, an experience that God subjected Abraham to of what it meant to have faith in a transcendent, hence trans-human, Deity, beyond human reason and beyond human moral sentiment.

Quite apart from the different conclusions to which they came, Schopenhauer was a very different type of philosopher from Kierkegaard. Kierkegaard used discursive reason allusively to somehow express a non-discursive sensibility, and the convoluted construction of his books reflects this. Schopenhauer had no problem with words and with reasoned arguments for claiming the subordination of reason to will, identified as the ultimate and non-rational ground of the universe. By contrast, Hegel, at the time the most prominent philosopher in Europe, argued that the history of the universe was the history of Absolute Reason, the incarnation as it were of the Principle of Sufficient Reason, coming to self-consciousness.

Schopenhauer's philosophical career began with a detailed critique of the "root" of causality, of causes as necessary and sufficient conditions for their effects, as in the Principle of Sufficient Reason. Like Protagoras and Gorgias among the sophists, and like Hamann and Kierkegaard, reason's inability to comprehend reality for Schopenhauer is not the result of an inability get outside of itself to "see" reality directly. Reason is unable to comprehend reality because reality is not reasonable. In *The World as Will and Representation*, it is will that rules the physical universe and will is antithetical to reason. The scope of reason is restricted to the representation of the universe in our minds. The reasoning mind constructs a universe in its own image.

For Schopenhauer, will is real, a cosmic reality, a metaphysical truth, the ground of Being, acting through the material universe but obeying its own drives, not laws of nature. Nietzsche, too, based his philosophy on will, but Nietzsche was an anti-realist, holding that our thinking cannot discover what is independent of our thinking. Will, for Nietzsche, is not an external metaphysical reality, but a personal reality internal to each individual human being's experience of the world, an echo of Leibniz' monads. It is will that underlies all valuing, continually making and remaking the "whole eternally growing world of valuation."[9] No thing, and certainly not nature, has value in itself; all values come from us. We give values to nature "as a present,"[10] and we revalue nature as we choose to. It follows that there are no universally correct values and no good as a value "out there" to which we must conform. Instead, the good is that within us that heightens our feeling of power, of the will to power in us, in the end of power itself, or just sheer willfulness.[11] At the same time, Nietzsche often writes as if the values he propounds are the right ones and the values proposed by other thinkers false!

When it comes to truth, Nietzsche argued that modern science generates a "realization of general untruth and mendaciousness" by its claim to reveal the "real" world behind experience, a world that is fundamentally ahuman and ultimately inhuman and even anti-human.[12] Science teaches us that all our experience of the world is an illusion and erroneous. The real world is independent of us and indifferent to us, as in Matthew Arnold's poem "Dover Beach." A rigorously honest person, Nietzsche held, taking scientific knowledge for the truth, would feel "nausea" at its misrepresentation of the real in experience and even find in it grounds for suicide. (Jean-Paul Sartre in his novel *Nausea* and Albert Camus in his essay "The Myth of Sisyphus" explored both of these ideas.)

What saves us from the "utterly unbearable" consequences of taking science as truth is art. Art, like science, is untruth, but contrary to science, it is a benevolent and life-affirming untruth. Scientific truth denies the meaningfulness of human experience, while art is a form of untruth that enriches meaningfulness. The Impressionist painter Pierre Bonnard, for example, said that a painting was a lot of little lies that added up to a big truth, not a truth in the sense of a revelation of a reality external to the mind, but of a truth internal to the mind's experience. The problem with science, for Nietzsche, is that science's knowledge claims are universal, hence aperspectival and thus false: "There is only a perspectival seeing, only a perspectival knowing . . . the more eyes, different eyes, we know how to bring to bear on one and the same matter, that much more complete" our conception of that matter will be.

Perspectivalism implies that philosophy, including natural philosophy, is essentially pluralistic. We can never conceptualize a reality that is independent of our conceptualizing because "our thoughts grow out of us" the way fruit grows out of a tree. The mind and its unique experience-based perspective on the world cannot escape that perspective. Great philosophies, he wrote, and we can add great scientific theories, have always been "the personal confession of its author and a kind of involuntary and unconscious memoir."[13] Science presents its results as if the scientific method allows scientists to escape perspectivalism, to give us, borrowing a phrase from Thomas Nagel, a "view from nowhere" of reality, but this is impossible.

Nietzsche's philosophizing was cut short in 1889 when he suffered a mental breakdown from which he never recovered. That same year, Henri Bergson's first book was published: titled *Time and Free Will* in its English translation. Bergson would go on to become, for the first third of the twentieth century, the most renowned philosopher in Europe and one of only three philosophers, so far, awarded (in 1927) the Nobel Prize in Literature, the other two being Bertrand Russell (1950) and Jean-Paul Sartre (1964), who declined it.

In *Time and Free Will, Matter and Memory* (1896), *Introduction to Metaphysics* (1903), and *Creative Evolution* (1907) Bergson articulated a systematic refutation of the claim that reason generically, and modern science in particular, was capable of giving us an account of reality. Bergson was unambiguously metaphysical, arguing that the mind, through experience, *could* know the real, but not by way of reason. Reality, for Bergson, manifests itself in the mind in our experience of the irreversible flow of time; reality *is* the irreversible flow of time. By contrast, in the mainstream rationalist Western philosophical tradition, but also in modern science, reality is atemporal, identified in philosophy and in modern science alike with timeless, "spatial" representations of the real.

Bergson championed process metaphysics, in which change is the ultimate fact about reality, as Heraclitus said it was, though for Heraclitus there were underlying laws guiding change. Bergson proposed an ontology in which continual change is the essence of reality, not merely an appearance behind which there is a changeless reality. A corollary of this is that time, too, is real and elementary. Change entails a direction to time and that direction is consequential; it has effects unique to time. It is in the directed flow of time that novelty emerges, unpredictably, as in biological evolution. There is no final cause, no goal that biological evolution is working toward. Biological evolution and the "creative" evolution of Bergson's metaphysics are open-ended.

Process metaphysics is the polar opposite of the substance metaphysics that has dominated Western philosophy and modern science. In substance metaphysics, the ultimately real, echoing Parmenides, is changeless. The ontology of substance metaphysics comprises changeless objects—in one version ideal forms, in another material atoms—each possessing unchanging intrinsic properties. In substance metaphysics, reality is already made, complete as it ever was and ever will be, always there for the mind to discover. For process metaphysics, on the other hand, reality is a continual making, never complete, not already "there." It is tempting to see in this contrast between substance metaphysics and process metaphysics the battle between the Gods and the Giants, but this is too simple. Leibniz was a process philosopher but a rationalist, hence on the side of the Gods. Hegel attempted to integrate process and substance metaphysics, but he was clearly on the side of the Gods, arguing that the history of the world was a process that moved necessarily to a foreordained goal, the self-knowledge of Absolute Reason.[14]

Bergson cannot be assigned either to the Gods or to the Giants. His metaphysics is realist—the mind can know reality—but he is neither a rationalist nor a skeptic, nor is he an empiricist in the traditional sense of that term. Rationalism, he wrote, is as much the "dupe of the same illusion" as traditional empiricism is: the illusion that reason can disclose reality. The "true empiricism" for Bergson "is that which proposes to get as near to the original self as possible . . . and this true empiricism is the true metaphysics."[15] Traditional empiricism is oriented outward, toward the world, striving to correlate the data of conscious perception and the external world causing that perception. Bergson's empiricism is oriented inward, to discover the "original" self behind conscious perception with its unconscious perception of reality. "Absolute knowledge" of this self is possible, but only through memory and intuition, not reason.

According to Bergson, the immediate data for consciousness are not spatially arrayed atemporal sensations—what philosophers later called sense data—and not spatiotemporal sensations either, as in Kant's philosophy or Einstein's relativity theories. What is immediately given to consciousness by the world is temporality, the experience of sensations embedded in directed time. There is nothing more fundamental that is given to consciousness than the directed flow of time, which Bergson calls "duration," and the content of our experience of duration. This experience is immediate in the sense of being pre-reflective and non-discursive. It is at once the most elementary feature of human consciousness and the point of contact between the mind and reality. Reason plays no role in the exploration of this opening to the real in the mind enabled by its experience of duration. As a response to conscious perception, which is necessarily utilitarian, reason is practical and action oriented: "We seek knowledge to satisfy an interest."[16] (The American pragmatist philosophers would say the same, but without the metaphysics Bergson attached to duration.) Reason "freezes" flowing time in order to deal with the world pragmatically, because accepting continual change would be overwhelming. To that end, reason populates the world with more or less stable, spatially juxtaposed objects possessing more or less fixed properties.

It followed that modern science was only what Nietzsche called a perspective on reality, one partial perspective that was erroneously judged universal because of its usefulness. Bergson acknowledged that reason was successful as a response to the physical environment in which we find ourselves. If that were not the case, we would not have survived as a life form! But reason fundamentally misrepresents reality because it suppresses the temporal dimension of our experience of the world in favor of simpler, atemporal, spatial representations. The practical success of this strategy has "duped" both rationalist and empiricist philosophers into concluding that these representations are at least partial representations of reality.

Where static spatial experience finds expression in discursive-logical reasoning, dynamic temporal experience, the experience of duration, is given to us in what Bergson calls a non-discursive "intuition." Because duration mirrors the true, temporal, nature of Being, it follows that intuition is a "truer" means of connecting the mind to reality than reason is with its timeless Parmenidean-Platonic reality. Note that Bergson is not calling for space and time to be taken as complementary and equally real. That's what Kant and Einstein did in their respective theories. Bergson's philosophy is based on the thesis that the experience of duration is the *only* basis for the mind's apprehension of reality. Reasoning reaches its heights in mathematics, philosophical systems, and modern scientific theories with their deductive logical structures. In logical reasoning and in science, reason hypostatizes simultaneity-spatiality. Reason is the "frozen memory" of duration from which it abstracts what it then calls reality. For Bergson, the "really" real is recognized by its continual variability, while reason, with its analytic methodology, is recognized by the invariability of its explanatory elements.[17]

Reason is analytic; intuition is holistic. Analytical reasoning dismembers experience, treating it as an assembly of separable objects (hence the "spatiality" of reasoning). In analysis, reason seeks homogeneity, a common denominator to these separate objects, and when it succeeds in synthesizing experience (stripped of duration) out of reason-generated unexperienced objects, it proclaims its understanding of that experience. Reason reduces reality to a series of atemporal snapshots, but no number of snapshots of a city can substitute for the experience of living in a city.[18] Echoing Bergson, William James would make a similar point. The perception of succession, for example, the perception of a melody, is fundamentally different from a succession of perceptions.[19] Indeed, the melody has a reality of its own, as the same melody can be played using different notes (in another key).

Intuition, unlike reason, invokes memory and the unconscious because the past for a living being is never gone. All our previous temporal experiences are *in* each present experience of duration without displacing earlier experiences. What we as temporal beings call our past is, in fact, consciously and unconsciously actually present. It affects how we experience the present, and it influences how we move into the future, but not deterministically. Marcel Proust's novel *Remembrance of Things Past* (more accurately, *In Search of Lost Time)* is based on Bergson's theory of memory. (Bergson was married to a cousin of Proust's, Louise Neuburg.) At a certain point in his life, Proust's narrator finds that, quite suddenly, his mind is flooded with vivid memories of his childhood at his grandmother's house, memories that he was unaware still existed within him. These memories were not evoked deliberately and by reason, but spontaneously and unconsciously by the experience of dipping a particular kind of cookie into tea, as he used to do long ago at his grandmother's house. He realizes that his entire past, in exquisite detail, is still there in his mind and accessible in the present, but only if he can learn how to unlock it deliberately.[20]

Bergson's philosophy is Romantic. He shared with such figures as Vico, Rousseau, Hamann, Herder, the Schlegels, Schleiermacher, Kierkegaard, Schopenhauer, and Nietzsche a rejection of the Enlightenment claim that reason is the sole means available to us to know reality and to increase human well-being. His philosophy, in which an "inner absolute knowledge of the self by the self is possible"[21] but only by transcending reason, overlaps and develops in a systematic way the roles the Romantics assigned to history, instinct, intuition, will, and the dependence of discursive reasoning on language. Bergson extended the role that history plays for the Romantics into an ontology of temporality that was beyond reason to explore, an echo of those Renaissance philosophers who contrasted reason-based knowledge and wisdom. And he extended the dependence of reason on language to a broader dependence on concepts, inevitably derived from a utilitarian response to sensory experience. Bergson wrote that it is a mistake when philosophizing to believe "that all knowledge must necessarily start from concepts with fixed outlines in order to grasp with them the reality which flows." Instead, it is necessary to enter *into* that reality by means of an "intellectual sympathy" enabled by intuition, a sympathy that draws support from memory and the unconscious. Intuition "attains the absolute"; analytical science cannot. Science professes to be "a vast mathematic, a single closed-in system of relations imprisoning the whole of reality in a network [of concepts] prepared in advance," but this is precisely why it cannot do what it professes to do.[22]

One final point before we turn in the next chapter to twentieth-century attempts at an explanation of how scientists produce knowledge. In spite of Bergson's realist metaphysics and his claim that the mind can grasp the real, there is a parallel in Bergson to Nietzsche's anti-realist perspectivalism, which denies just that. Bergson claimed that the "true empiricism" aimed to re-cover the "original self" behind conscious perception, the self that through its unconscious perception of duration already apprehended reality. The use of intuition to explore this unconscious perception consciously reveals to the "original self" the role of its own memories in shaping its experience of the world. It follows that philosophizing via intuition is self-centered. In intuiting duration, we experience our connectedness to reality, but this connectedness is inescapably our own, a function of our own unique, personal experience of the world (again reminiscent of Leibnizian monads). We cannot know reality comprehensively. What we can have is a perspective on reality that radiates outward from our personal experience, but this cannot encompass reality as it is apprehended by others. It follows that no one can have a definitive intuition of reality.

10
Early Twentieth-Century Philosophy of Science

Henri Poincaré was a mathematician and mathematical physicist of the first rank, a prolific theoretician who was also actively involved in practical applications of theory. He incorporated lectures on the philosophy of science into his teaching of physics and mathematics at the Sorbonne, publishing them in three slim volumes: *Science and Hypothesis* (1902), *The Value of Science* (1905), and *Science and Method* (1908).

For Poincaré, scientific knowledge is limited to relational structures discovered in our experience of the world: "That which science captures are not the things themselves, but simply relationships between them. Beyond these relations, there is no knowable reality."[1] And again:

> The purpose of mathematical theories [in science] is not to reveal the true natures of things. Such a claim would be unreasonable. Their only goal is to coordinate the physical laws that experiment reveals to us, but that we could not even state without the help of mathematics.[2]

This is more radical than what I earlier called "Fourier's move." Fourier's theory of heat flow, classical thermodynamics, and Maxwell's theory of the electromagnetic field all have as their object a reality independent of our experience. They are presented as uniquely correct descriptions of how that reality *behaves,* only bracketing the underlying physical causes of that behavior.

Poincaré, like Mach and Plato's Giants, denied the possibility of the mind having as its object a reality independent of experience: ". . . beyond a doubt a reality completely independent of the mind which conceives it, sees it, or feels it, is an impossibility. A world as exterior [to the mind] as that, even if it existed, would for us be forever inaccessible."[3] All our theories are inescapably dependent on definitions, assumptions, concepts, and metrics that are not logically necessary, nor are they derived logically from data. They rest on conventions that are freely adopted by the scientific community on the basis of convenience or advantage: "Experience . . . guides us by helping us to discern the most convenient path to follow."[4] The objectivity of scientific knowledge is thus really intersubjectivity, deriving not from reality but from tacit or explicit agreement to share assumptions, concepts and metrics in order to formulate a theory that the community explains to the community's satisfaction. Given such agreement, theories are objective but only relative to adopted conventions. There can be no uniquely correct account of relational structures imposed on our empirical experience.

Émile Meyerson abandoned a career in chemistry for the history and philosophy of science. He was perhaps the only philosopher of science that Einstein openly admired and cultivated as a personal acquaintance. Meyerson's books, primarily *Identity and Reality* (1908), *Explanation in the Sciences* (1921), and *The Relativistic Deduction* (1925), influenced many of the leading thinkers of the day, including Henri Bergson. According to Meyerson, scientific reasoning reflects the operation of two principles that are innate to the human mind: lawfulness and causality. The world as we experience it is predetermined by our imposition upon experience of expectations of orderliness and strict causal connection. All of our actions, he wrote, "are performed in view of an end which we foresee; but this foresight would be entirely impossible if we did not have the absolute conviction that nature is well-ordered, that certain antecedents determine and will always determine certain consequences."[5] Like Kant, what is "out there" is unknowable, but unlike Kant, for Meyerson the *ways* that these two principles are employed by scientists are not fixed. The history of science reveals that scientific explanation has taken different forms in the face of changing experience, but it always incorporates some combination of a generic fixed a priori element—lawfulness and causality— and a specific variable a posteriori element, namely, the content of empirical experience.

In science, the innate need to explain experience in terms of laws and causal connections takes the form of universal laws of nature incorporating strict causality: causes must be necessary and sufficient conditions for effects. This is Leibniz' Principle of Sufficient Reason, and it entails a strictly deterministic ontology and the identity of cause with effect, which, as already noted, eliminates time from reality. Without the flow of time introducing novelties, nothing happens because, as Parmenides had asserted, nothing that is real changes; cosmic preformationism prevails. Furthermore, nothing that we experience is real because human experience is inescapably temporal, thus an illusion. Meyerson argued that, logically, universal laws and strict causality lead to the reduction of matter to space, as Descartes had claimed and as Einstein proposed in the general theory of relativity, a conclusion of Meyerson's with which Einstein agreed. With the reduction of matter to a property of space, there is no *thing* at all "out there," only space! Science escaped this absurd Parmenidean conclusion, according to Meyerson, only with the creation of thermodynamics in which causes *cannot* be recovered from effects, thus reintroducing time into science's account of reality.

Percy Bridgman was a Nobel Prize–winning experimental physicist whose entire academic career was spent at Harvard, from undergraduate student to holder of the Chair in Natural Philosophy. Bridgman presented his theory of science, called "operationalism," in *The Logic of Modern Physics* (1927). "It is becoming increasingly recognized," he wrote, after the introduction of the relativity theories and quantum mechanics, that "the world of experiment is not understandable without some examination of the purpose of physics and of the nature of its fundamental concepts."[6]

As a working physicist, he felt 'compelled' to step back and formulate a critique of the "interpretative fundamentals" of physical science. This

is forced upon us . . . by a rapidly increasing array of cold experimental facts . . . Whatever may be one's opinion as to our permanent acceptance of the analytical details of Einstein's restricted and general theories of relativity, there can be no doubt that through these theories physics is permanently changed. It was a great shock to discover that classical concepts, accepted unquestioningly, were inadequate to meet the actual situation, and the shock of this discovery has resulted in a critical attitude toward our whole conceptual structure which must at least in part be permanent. Reflection on the situation after the event shows that it should not have needed the new experimental facts which led to relativity to convince us of the inadequacy of our previous concepts, but that a sufficiently shrewd analysis should have prepared us for at least the possibility of what Einstein did.[7]

The relativity and quantum theories grew out of a common insight: that the meaning of scientific concepts is a function of how they are to be measured: "In general, we mean by any concept nothing more than a set of operations; the concept is synonymous with the corresponding set of operations."[8] If you want to know what an electron is, what energy is, what space and time are, identify the instrumental operations by which they are measured. What "electron" means is the outcome of specified experimental setups that measure mass, velocity, size, charge, and spin. Each setup must correspond uniquely with its associated concept/object in order for the concept and the operations to be equivalent. What cannot be measured, absolute space, time, and motion for example, is meaningless.[9]

"Pragmatism" was the name given by Charles Sanders Peirce—American philosopher, logician, and physicist—to his theory of knowledge, to distinguish it from rationalism, idealism, and empiricism. Peirce's friend, the philosopher-psychologist William James, adopted pragmatism in 1898 and disseminated it widely in his very popular books, articles, public lectures, and teaching at Harvard University. (Peirce rejected James' interpretation of his ideas and changed the name of his philosophy to pragmaticism to make sure that no one confused his ideas with James'!)

Peirce introduced the core ideas of pragmatism in 1877 and 1878 in two articles that he published in the magazine *Popular Science Monthly*: "The Fixation of Belief" and "How to Make Our Ideas Clear." Reasoning, he explained was an instrumental response to feelings of anxiety and doubt generated by the need to respond to concrete problems posed by experience. This kind of existential doubt had nothing at all to do with the celebrated doubt that Descartes had made the basis of his rationalist theory of knowledge. For Peirce, Cartesian doubting of everything was an intellectual fiction, a ploy in an abstract intellectual game. Real doubt emerges out of confronting real-life problems requiring action in order to solve them. Peirce used "belief" as a name for actions that we anticipate will solve particular problems. Beliefs that repeatedly lead to outcomes we desire no longer need to be thought about before acting on them. Peirce called such routinely effective beliefs "habits."

Scientific reasoning systematizes the formulation of effective action-generating beliefs. It requires that we specify clearly the problem to which we are responding, the outcome we desire, and the means to be employed to achieve that outcome. It privileges those beliefs, and the ideas that enter into them, the consequences of whose actions meet our expectations, while dismissing those beliefs that do not. Scientific reasoning is thus characterized by a deliberate feedback loop between beliefs and ideas on the one hand and the experiential consequences of actions based on those beliefs and ideas on the other. This makes scientific knowledge instrumental, a means for solving practical problems in particular contexts and its conclusions are true just to the extent that those problems are judged, in context, to be satisfactorily solved. Although in practice it is decontextualized by employing abstract universal concepts, scientific knowledge is rooted in contextual, practical know-how and justified in terms of success in experience, not correspondence with a transcendental reality: "Consider what effects which might conceivably have practical bearings that we conceive the object of our conception to have. Then, our conception of those effects is the whole of our conception of the object."[10]

Peirce claimed that scientific reasoning was based on a form of logical reasoning that he called "abduction." This is a process by which the mind somehow invents theory-anchoring hypotheses and assumptions, very much like Whewell's redefinition of "induction" as the creative production of Fundamental Ideas. For Peirce, as for Whewell, these assumptions/Ideas are stimulated by, but not deducible from, empirical data. They serve an epistemological-explanatory function, but they do not entail a coordinated ontology.

Everything so far is incorporated into Peirce–James pragmatism. Peirce's *pragmaticism*, however, evolved into an increasingly metaphysical philosophy with Kantian and Hegelian overtones. It had an ontology in which chance played a fundamental role and in which, over a long period of time, as human experience widened and along with it our reasoning about experience, scientific knowledge converged on *the* truth about the world.

Like Peirce before him and John Dewey after, for William James science alone was knowledge-producing reasoning. In a series of public lectures that he gave at the Lowell Institute in Boston in 1906, James began with his "positively most important point." Modern science had so permeated Western thought that "our children, one may say, are almost born scientific."[11] He was acutely aware of the Gods–Giants battle in Western philosophy and that it carried over into conflicting philosophies of science. It was a "perpetual" battle because it was in principle not winnable by either side. The combatants possessed one or the other of two exclusive temperaments that determined the way that they reasoned: tough-mindedness or tender-mindedness. Tough-minded reasoning was Giant-like: experiential (a posteriori), positivistic, materialistic, pessimistic, irreligious, and fatalistic. Tender-minded reasoning was God-like: a priori, deductive, idealistic, religious, and optimistic. The pursuit of certainty and knowledge of reality was the passion of the tender minded. The denial of the possibility of certain knowledge of the world was the passion of the tough minded, "tough" because such thinkers accepted the unknowability, and perhaps even the irrationality, of the universe.[12]

James' tough–tender dichotomy links reasoning to personal temperament, a factor innate in individuals. This manifests itself in irreconcilable conceptions of what it means to explain and to understand, and of what the words knowledge, truth, reality, and rationality mean. This fits the history of Western philosophy to some degree, but not the history of modern science. The history of modern science reveals the compresence of *both* tough-mindedness and tender-mindedness in the reasoning about nature of individual scientists, as well as in their reasoning about scientific reasoning itself. With the exception of Mach and Poincaré, the philosophies of science we have examined so far all reflect some combination of a priori and a posteriori forms of reasoning, of induction and deduction, of positivism and empiricism, but also of rationalism and idealism, of certainty and probabilism, and of the goal of science as knowledge of reality and recognition that knowledge of experience is what scientists actually produce at a given time.

James also called his theory of knowledge "radical empiricism." In *The Meaning of Truth* he wrote that "the only things that shall be debatable among philosophers shall be things definable in terms drawn from experience," which consists of "disconnected entities" and connections among them, both immediately given to the mind.[13] There is, he argued, no need at all to refer in our reasoning about experience to "trans-empirical" factors behind experience, because experience already "possesses in its own right a concatenated or continuous structure." In his *Essays on Radical Empiricism* he wrote,

... I maintain [that] a given undivided portion of experience, taken in one context of [associations], play[s] the part of a knower, of a state of mind ... while in a different context the same undivided bit of experience plays the part of a thing known. At one and the same time a portion of experience is a thought and an object, and "we have the right to speak of it as subjective and objective, both at once.[14]

James distinguished pragmatism, for which reality is a continual process of making, from rationalism, in which reality is "ready-made and complete from all eternity."[15] What makes James' empiricism "radical" is his claim that our experience of the world not only already includes relations among sensory facts but also includes values and meanings. These are just as objective as sensory facts themselves, not subjective interpretations projected onto sensory experience. Subjective and objective are not opposed to one another, either epistemologically or ontologically. Humans *make* truth, carving "out everything, just as we carve out constellations [in the night sky], to serve our human purposes."[16] (Recall that Vico had said that we can understand only what we have ourselves made.)

This would seem to entail a radically relativistic, ontology-free epistemology, but there are "resisting factors in every experience of truth making" that constrain our "carving up" experience. This is the objective facet of subjective experience, two sides of the same coin. There are facts given to all of us in our experience of the world, as well as the mutual relations of those facts including values and meanings, together with the constraint imposed on our experience-based reasoning by the need for confirmation by others. James' denial that he was an epistemological relativist reflects his (pragmatic?) belief in some kind of trans-empirical ontology, "out there," that is the ultimate reference of knowledge and truth, again slipping reality into an ostensibly experience-based theory of knowledge.

It was John Dewey, acknowledging his indebtedness to Peirce and James, who developed pragmatism into a comprehensive theory of science-based knowledge that he called "prospective empiricism." Dewey was a process philosopher, one for whom the real comprises processes of endless, open-ended interaction and change. He argued that "mind" and "world" must be understood as verbs, thus processes, not nouns, names for self-contained things. As things, independent of one another, knowledge of the world by the mind creates an unsolvable problem: how to connect experience of the world inside the mind with the world as it is outside the mind. For Dewey, experience is not inside the mind because there is no *thing* as mind for experience to be inside of. Instead, "experience" is a name for a continuously evolving transactional process that produces mind and world as distinguishable, but not distinct, facets of experience.

Mind and world are *of* experience in the same way that mountains are *of* the Earth, not things plopped down on the Earth's surface. Mountains are so visible and stable that we give them names *as if* they were things in their own right, instead of being expressions of primal tectonic processes shaping the Earth's surface. Similarly, trees are highly visible expressions of transactional processes involving root systems, bacteria, soil, water, atmosphere, and sunlight. Conscious experience, too, is a transactional process in which two features—mind and world—stand out so sharply that we are led to give them names as if they were entities in their own right, separable from one another and from experience itself, though this is profoundly misleading.

History suggests that humans have a need for snatching stability from constantly changing experience. We prefer nouns to verbs, things to processes, equilibrium to change, substance metaphysics to process metaphysics. In novels no less than in fairy tales some form of ". . . and they lived happily ever after" is the ideal ending to a conflict-filled story. Analogously, in nineteenth-century science the equilibrium state is the assumed normal state of a material system. Equations describe how a system behaves when it is disturbed from its equilibrium state by an external force and how it returns to that state. (Nonequilibrium thermodynamics only came into its own in the last third of the twentieth century.) For process philosophy, on the other hand, equilibrium is a fairy tale because everything is constantly changing. Some things, such as flowing water and clouds, change faster; others, among them mountains and trees, change more slowly, but everything is constantly changing.

For Parmenides and most Western philosophers, as well as for almost all modern scientists, the objective of reasoning is the changeless behind change. Dewey urges us to suppress this inclination to reify, to substitute things for processes, to substitute an ideal of equilibrium for real change. For Peirce,

James, and Dewey alike, the Darwin/Wallace theory of evolution revealed that a transactional process was the ultimate reality underlying life. All of the "forms most beautiful" that for Darwin comprise the vast diverse life-world were expressions of a single process characterized by constant transformational change. This process had no goal, no final cause, no end. It was driven by one factor internal to each life form, spontaneous variation, and one factor external to it, a constantly changing environment to which survival demanded adaptation. For Dewey, conscious experience was just such a transactional process.

We experience an existential need to adapt to features of our experience that are pleasing, challenging, or threatening, or offer perceived opportunities. The temptation to reify features of changing experience into a mind-thing "in here" and a world-thing "out there" is functionally justified. Most of the time, we adapt to our environment better when we do this, but reason can become abstracted from experience. It may then produce conceptual and logical fictions—philosophical systems and scientific theories—that impede effective adaptation to complex experiences because these fictions claim to be more real than experience. Reasoning, for Dewey, as for Peirce and James, is first and foremost a problem-solving response to experience, but it need not only be a *reactive* response to experienced problems; it can also be *proactive*. Dewey called his version of pragmatism *prospective* empiricism because thinking can anticipate new kinds of experiences, imagine new goals for future experiences, and then invent means for achieving those goals. For Dewey, traditional philosophy had hijacked reason, dismissing its existential, practical function and replacing it with an abstract intellectualized function: to "understand" experience in terms of an experience-transcending reality that was knowable only to the self-consciously reasoning mind.

The goal of pragmatism was to return reason to experience and return philosophy to its "proper" function: the pursuit of what Aristotle called practical as opposed to theoretical wisdom, which for Aristotle was superior to practical wisdom. Philosophy should be what it once was, an "education for living," a reflection on experience that aims at making life better. Isocrates, a contemporary of Socrates and Plato, had railed in his essay "Antidosis" against Plato's intellectualization of philosophy, making philosophy irrelevant to improving one's mortal life. This was, as well, the view of the sophists. For Epicurus, for the skeptics, and for the Stoics, among others, the whole point of philosophy was to teach us how to live better lives, to be better people, not to reach an idealized understanding of an abstract reality.

"Knowledge," Dewey wrote, "as an abstract term is a name for the product of competent inquiries. Apart from this relation, its meaning is so empty that any content or filling may be arbitrarily poured in."[17] The "natural," pre-intellectualized, role of reason is making "inquiries" into experience. Logic is the study of inquiries. Its aim is to identify the forms of inquiry that are effective in responding reactively to current problems in our experience and proactively to opportunities (imagined problems). One "pattern" of successful problem-solving inquiry jumps out for Dewey as particularly successful: the method of modern science, in which experience and our reflection on it are "cooperative instrumentalities for economical dealing with the problem of the maintenance of the integrity of experience." This, for Dewey, is the pattern of reasoning that defines the kind of "competent inquiry" that reliably produces knowledge, not just in science, but in all areas of life, including philosophy. In "scientific inquiry," he wrote, "the criterion of what is taken to be settled, or to be knowledge, is being so settled that it is available as a resource in further inquiry; not settled in such a way as not to be subject to revision in further inquiry."[18]

This definition of "knowledge" returns us to Peirce's definition of "habit": a belief that has been so reinforced by acting on it that we no longer have to think about it before acting on it. In both cases, nothing is guaranteed in the face of future experience; knowledge and habits are corrigible, however well they may have worked in the past. Where does truth fit into this picture? Dewey joins Peirce and James, and Poincaré on truth:

> Truth, in the final analysis, is the statement of things 'as they are,' not as they are in the inane and desolate void of isolation from human concern, but as they are in a shared and progressive [future oriented] experience . . . Truth is having things in common.[19]

Every account of experience, of nature as it enters into experience and of the sociocultural world as it enters into experience, must be empirical, hypothetical, perspectival, and evolutionary.

The most influential philosophy of science from the 1920s into the 1960s was not pragmatism, but logical positivism and the related logical empiricism. Both of the latter were products of "circles" of philosophers, scientists, and mathematicians at the Universities of Vienna and Berlin, respectively, who met regularly for the specific purpose of formulating, finally, after one hundred years of inconclusive efforts, a correct theory of how science produces knowledge.

The founder of logical positivism was Moritz Schlick, a philosopher with a deep understanding of contemporary physics, having studied under Max Planck and having written well-received articles and books on logic and on the special and general theories of relativity. In 1922 Schlick was appointed to the Chair in Natural Philosophy at the University of Vienna that had previously been held by Ludwig Boltzmann and before Boltzmann by Ernst Mach. In 1925, Schlick organized a weekly discussion group to address philosophical issues raised by the relativity and quantum theories and by the ongoing search by mathematicians for the foundations of mathematical truth. Kurt Godel—whose extraordinary "incompleteness" theorems published in 1931 and 1932 effectively terminated this search—was an early member of the "Vienna Circle," as Schlick's group came to be called; so was Ludwig Wittgenstein. Besides Schlick, members of the Circle who made a lasting impact on the philosophy of science included Otto Neurath and Rudolf Carnap.

Three years later, Hans Reichenbach, a member of the physics faculty at the University of Berlin, who, like Schlick, had advanced training in both physics and philosophy, organized a Society for Empirical Philosophy. This soon morphed into a Berlin Circle analog of the Vienna Circle. It included among its members, in addition to physicists and philosophers, leading mathematicians such as David Hilbert and Richard von Mises. Reichenbach named the Berlin Circle's theory of science "logical empiricism" to differentiate it from Schlick's logical positivism, though the two are closely related. Among Reichenbach's many books on physics and philosophy, three give a sense of the agenda of both Circles: *Experience and Prediction: An Analysis of the Foundations and the Structure of Knowledge* (1938), *The Philosophical Foundations of Quantum Mechanics* (1944), and *The Rise of Scientific Philosophy* (1951).

Schlick's aim in creating logical positivism was to defend the view that scientific knowledge was universal, necessary, and certain and was about the world "out there." Logical positivism set aside the process of *discovery*, the question of how scientists came to formulate theories, as a subject for psychology, focusing instead on the process of *justification*: what made a scientific theory true. The logical positivists were influenced by Helmholtz's epistemological writings, by Mach's phenomenalism, by Bertrand Russell's attempts at reducing mathematical truth to logical truth, and by the logical atomism of Ludwig Wittgenstein. Of these, the most important influences were Russell and Wittgenstein.

As discussed earlier, Russell had attempted to prove that mathematics could be reduced to logic, more specifically, that truth in mathematics, which was being questioned at the time, could be reduced to logical truth, which was considered unquestionable. This attempt failed, but reading Russell's *The Principles of Mathematics* led Wittgenstein to become fascinated by the question of the foundations of mathematical and logical truth. He first went to see Frege, who recommended that he study with Russell, and in 1911 Wittgenstein, unheralded, went to meet Russell at Cambridge University. In a very short time, Russell recognized, and even deferred to, Wittgenstein's genius.

Russell's 1918 *Philosophy of Logical Atomism* blends Wittgenstein's ideas on reasoning about the world and Russell's commitment, shared by Wittgenstein, to logical truth as the only unquestionable truth. While openly acknowledging his indebtedness to Russell for shifting his focus from engineering to logic, Wittgenstein was also open that Russell misunderstood his (Wittgenstein's) theory of logical atomism. In 1921 Wittgenstein published his own version of his ideas, the extremely influential *Tractatus Logico-Philosophicus*. This became, along with works by Frege and Russell, a central text for the logical positivists. The Vienna Circle studied the *Tractatus* very carefully, initially together with Wittgenstein, who dropped out, upset at having to defend his ideas against challenges by members of the Circle. The members persevered, however, taking logical atomism very seriously even after Wittgenstein repudiated it in the late 1920s.

In *The Philosophy of Logical Atomism* Russell argued that the goal of philosophers must be the creation of a logically ideal language capable of mirroring the nature of the world. Ordinary language is incapable of doing this because it is irremediably afflicted with misleading connotations. Ordinary language is "totally unsuited for expressing what physics really asserts [about the world]"; only mathematics and mathematical logic can do that. The uncritical use of ordinary language by philosophers is why "from the earliest times [philosophers] have made greater claims and achieved fewer results than any other branch of learning." This "unsatisfactory state of affairs can be brought to an end" only "if the logical analytical method" is used by philosophers.[20]

In his introduction to Wittgenstein's *Tractatus*, Russell wrote that starting with "the relations which are necessary between words and things in any language" Wittgenstein shows in this book that the failure of traditional philosophy to solve any of its problems "arises out of ignorance of the principles of [logical] Symbolism and misuse of language."[21] For his part, Wittgenstein wrote that his "book deals with the problems of philosophy and shows, as I believe, that the [traditional] method of formulating these problems rests on the misunderstanding of the logic of our language."[22] There is a limit to which any language can express our thoughts, and "what lies on the other side of the limit" is nonsense. These ideas about language were seminal for what was called the "linguistic turn" in twentieth-century philosophy, a focus on logic, language, and the form of linguistic expressions as the central concern of philosophy, as defining the conditions for formulating and solving philosophical problems.

The world, according to both Russell and the early Wittgenstein, is capable of being mirrored by a properly constructed logical language because the world is the sum of a vast number of "atomic" facts, individual things and their mutual relations, all of which can be expressed using logical connectives. In the *Tractatus*, Wittgenstein wrote that the world "is the totality of [atomic] facts not [the totality of] things."[23] An atomic fact is a combination of objects (entities, things). It is essential to be a thing [of any kind] that it can be a "constituent part of an atomic fact." Furthermore, if "things can occur in atomic facts, this possibility must already lie in them." This last is necessary if the "picture" of the world we construct using a logical language is to mirror the real world: "things" must form atomic facts in just one way, the "right" way.

Atomic facts combine to form more complex "molecular" facts, and these combine to form the objects familiar to us in sense experience. The result is that objects in empirical experience are, ultimately, logical constructs, so we can reverse the process. We can analyze empirical experience and using only "sensibilia," immediate sense data, identify first the molecular and then the atomic facts of which perceptual objects are composed, along with their relations, for which we can substitute logical connectives. We can thus construct, using a sense-based logical language, a "picture" of the world that correctly "mirrors" the world. Statements about the world so "pictured" are logically true, hence universal, necessary, and certain. This reverse engineering of experience to formulate true scientific theories became the "holy grail" of logical positivism. (In the late 1920s, Wittgenstein reversed his position, undercutting logical atomism by denying the possibility of a sense-based, unambiguous logical language. Only a public language, he now argued, that is, only ordinary language with all its flaws, can be used to describe experience!)

For the logical positivists, theories in physics needed to be shown to fit into this Russell-Wittgenstein conception of mirroring the world. This required that scientific knowledge begin with immediately given, uninterpreted, and non-inferential observed facts about the world. For Schlick, there must be "an unshakable point of contact between knowledge and reality" because without "the good old expression 'agreement with reality,'" scientific theories are reduced to myths or fairy tales.[24] In order for observations to provide such an "unshakable point of contact between knowledge and reality," they must be completely theory free, unambiguous, and indisputable, hence self-evidently true. Scientific theories are then a kind of unambiguous logical language from which observational, "atomic" facts, including their relations, can be deduced. If a theory can do this using agreed upon rules, "correspondence rules," that link theory statements to observation statements logically, not empirically, then the theory is logically true. And the theory is about the world because it is logically tethered to facts about the world.

The theory neutrality of observations together with its corollary, the absolute distinction of empirical, so-called synthetic statements and analytical (logically true) theory statements, is one of the two pillars of logical positivism. The other is the verifiability criterion of meaning: theory statements are meaningful only if they can be cashed out, to use a Jamesian expression, in terms of observation statements. The meaningfulness of any term in a non-observation statement is a function of its verifiability in observation statement terms. This was intended to rule metaphysics and values out of science and to render traditional philosophy meaningless. (Nevertheless, Schlick, for one, wrote extensively on ethical and religious questions.)

The relativity and quantum theories seemed perfect examples of the logical positivist theory of scientific knowledge. The special and general relativity theories were both strictly deductive, and measurement operations played a key role in defining the meaning of such central terms as space, time, matter, and motion. Quantum mechanics, too, had a deductive logical structure, even in Schrodinger's version according to the interpretation of the terms in his wave mechanics as probability distributions, because these distributions developed deterministically in time. But quantum mechanics referred to the world in terms of objects and properties that were in principle *not* directly observable. It seemed inescapable that a looser definition of "observable" was necessary, as adopted by the Berlin Circle logical empiricists, for including observations of the outcomes of typically complex experiments.

So, on the one hand we have theories with a deductive logical structure, all of whose terms purportedly meet the verifiability criterion of meaning. On the other hand, we have a body of purportedly theory-neutral, indisputable facts about the world, if we accept the output of instruments as facts about the world. Knowledge of the world, it is claimed, follows from deducing the observation facts from the theory. But to do that, to deduce the relevant facts from a theory, we need to have some logical rules—called correspondence rules or protocol statements—that link observational data to terms in the theory. The rules cannot have empirical content because that would undermine the deduction, making it dependent on particular facts. The rules cannot be defined solely in terms of the theory, however, or the deduction would be circular. The logical positivists wrestled with this problem but never solved it.

Meanwhile, it became clear even *within* the Vienna and Berlin Circles, that the theory neutrality of observation statements was false. Observation statements, especially those based on indirect observation, always incorporate elements of theory, implicitly or explicitly. But if observation statements are theory infected, the verification criterion fails because verification becomes circular. It follows that as scientific theories cannot meet the verifiability criterion, deductions from a theory cannot be taken to be about the world "out there." Schlick himself came to this realization as early as 1930, and in time so did just about everyone else in both Circles. Reichenbach's logical empiricist version of this, in his *Axiomatization of the Theory of Relativity* (1924), was that "Unfortunately . . . every factual statement, even the simplest one, contains more than an immediate perceptual experience . . . it is already an interpretation and therefore itself a theory . . . the most elementary factual statements therefore contain some measure of theory"[25]. This is yet another confirmation of the claim that data cannot speak for themselves.

Reichenbach nevertheless continued to believe that scientific knowledge was objective and about the world, but he argued that it was only probable, not certain, though *objectively* probable. The justification of scientific theories required a probability function reflecting empirically observed frequencies of events. This function itself, unfortunately, was not empirically verifiable. Hume had argued that there was no logical justification for the principle of induction and Reichenbach agreed. He made accepting induction a transcendental feature of reasoning about the world: it's the way we humans think, inescapably.

Rudolf Carnap, who came to America in 1935, had been a member of both the Vienna and the Berlin Circles. While responding to the evolving criticisms of the logical positivist agenda, Carnap attempted to remain true to its core tenet: the logical justification of scientific knowledge as knowledge of the world. Accordingly, he attempted to create a formal (logical) theory of induction that would allow for the kind of correspondence rules linking theory and observation that logical positivism required. Although the effort produced important insights into logic and language, he failed. Induction resisted formalization and still does. Carnap collaborated with several members of the Berlin Circle, most notably with Carl Hempel, who was a graduate student in the early 1930s and an active member of the Berlin Circle. Carnap arranged for Hempel to be his assistant at the University of Chicago when Hempel emigrated to America in 1937. Hempel went on to teach at City College of New York, then at Yale, and finally at Princeton, where Thomas Kuhn was a colleague before Kuhn's move to MIT.

Logical positivism was committed to the elimination of metaphysics from science and philosophy because metaphysics was meaningless, failing the verification test. Nevertheless, Hempel (correctly) saw implicit metaphysics in the Comtean origins of the term "positivism," and so he identified himself as a logical empiricist rather than a positivist. Hempel is best known for his "covering law," or "hypothetico-deductive," theory of scientific explanation. A theory explains when its premises, a combination of general laws and specific conditions, allow the deduction of what is to be explained. These premises must have empirical content that is testable and true, either deductively, inductively, or statistically. Testability was Hempel's substitute for the verification principle. By 1950 at the latest he had concluded that while universal statements were necessary for a scientific theory, they were not verifiable. At the same time, he recognized that testability could not be reduced to logic. Logically, theory statements cannot be confirmed or disconfirmed. Hempel proposed a suite of reasonable but not strictly logical criteria for determining the truth of a theory, among them, precision of terms, explanatory power, experimental confirmation, and predictive success. These may be necessary, but they are clearly not sufficient for determining the unique truth of a given theory.

In 1952, the American logician-philosopher Willard van Ormond Quine, in "The Two Dogmas of Empiricism," argued that the distinction of synthetic and analytic statements required by the logical empiricists did not exist. Analytic statements, even when they are apparently simply definitional, always have some synthetic content. A common example given of an analytic truth was, and is, "A bachelor is an unmarried man," said to be necessarily true because "unmarried man" is defined to be what the word "bachelor" means. But Quine argued that the terms "bachelor," "unmarried," and "man" all contain empirical content reflecting prevailing word usages. Quine also argued that while observations could not confirm a theory, because of the Fallacy of Affirming the Consequent, neither could they disconfirm a theory. Quine revived Pierre Duhem's argument that as a theory is always a combination of theory statements and assumptions, any apparently disconfirming observation could be negated by modifying one or more terms in the theory. On strictly logical grounds, then, a theory cannot be confirmed or disconfirmed by observations. Quine also revived Duhem's claim that theories are always underdetermined by data. It can never be the case that there can be just one "mapping" of a data set onto a theory.

The logical positivists initially attempted to formulate a theory of scientific knowledge in the God-like sense of "knowledge" (universal, necessary, and certain) based on a correspondence criterion of truth—a statement is true if it corresponds to the way the world is. What they produced, in fact, was a Giant-like theory in the weak sense of "knowledge" (contextual, contingent, and probable) resting on a coherence criterion of truth: logical and conceptual coherence internally and with other related theories, and coherence with a relevant body of data. Theories have a deductive logical structure such that they possess universality, necessity, and certainty but only *within* the theory. The logical form of a theory may be deductive, but it's connection to the world is only inductive, so what it says about the world is only probable, not certain.

Note an important semantic point about scientific theories. It is a commonplace to say that theories make predictions. If these predictions are confirmed empirically, the truth of the theory is thereby confirmed or at least made more probable. But theories do not make predictions; theories have logical consequences. The distinction is important. That electromagnetic waves travel freely through space is a logical consequence of Maxwell's theory of the electromagnetic field, not a prediction. The word "prediction" carries ontological, and psychological, connotations that implicitly link the theory to reality. The logical consequences of a theory are necessary and certain, but the *existence* of those consequences "out there," however strongly confirmed observationally or experimentally, is not necessary and certain. It follows that the form of a theory may be deductive, but the connection of that theory to world is only inductive. Furthermore, the premises that give a theory its deductive form change over time, so that the connection to the world is again contingent. (This is, once again, Simplicio's question to Salviati: How do you know that your deductively arrived at conclusions correspond to the world?)

This is another reminder that the assertive realism of scientists is not a consequence of logical reasoning. On logical grounds alone it is not possible to reason from data to just one theory from which those data can be deduced, it is not possible to confirm a theory based on predictive success or to disconfirm it based on predictive failure, and it is not possible to justify logically a claim that a theory corresponds to reality. The fact that scientists routinely make these kinds of claims reveals the conscious and unconscious incorporation of nonlogical elements into scientific reasoning. This is the conclusion reached, ultimately, on one ground or another, by every philosopher of science we have discussed so far, including Mach, Russell, Wittgenstein, and the logical positivists, who tried hard to avoid it.

11
Einstein Versus Bohr on Reality

The inconclusiveness of a century-long effort to clarify how science works once again brings to mind the "mysterious" intersection of epistemology and ontology. Knowledge claims and existence claims clearly intersected at the heart of nineteenth- and twentieth-century science, as they had for nineteenth- and twentieth-century mathematics. The mystery lies in the failure of scientists, logicians, and philosophers to find a logical justification for doing what almost all scientists do: link epistemological claims to ontological claims. It is interesting that this widely appreciated failure had little if any impact on the continued practice of science, just as the parallel failure to find the foundations of mathematical truth had little impact on the practice of mathematics.

Ontology was a central issue for the relativity and quantum theories from their beginnings. Recall that Planck had introduced the idea of quantizing electromagnetic energy in order to solve the "black body" radiation problem. Physicists had found one equation that described the distribution of electromagnetic energy within a so-called black body—one that absorbed all incident radiation—for high frequencies (Wien's law) but not low, and another that worked for low frequencies (Rayleigh–Jeans law), but not for high. Planck came up with a single equation that worked for all frequencies, but at a price.

In Maxwell's theory, electromagnetic energy takes the form of *continuous* waves with a continuous distribution of energies that propagate through space at the speed of light. The reality of these waves was considered established based on independent experiments in the 1880s by Oliver Lodge and Hertz, followed by their application in the form of radio wave technologies. Planck's solution to the black body radiation problem, however, required that these continuous waves could only be absorbed or emitted *discontinuously*, in discrete packets of energy he called quanta. Epistemologically, this is fine, but it is a radical proposal if taken ontologically. Is this quantization of electromagnetic energy real? If so, what are the implications for the reality of electromagnetic waves? Is quantization real but limited to the processes of absorption and emission? Perhaps it isn't real, but only an artifact of the mathematics in Plank's equation with no physical correlate. (Recall the

Newton–Leibniz controversy over the reality of mv and mv².) Planck himself took the latter view and worked for years to derive his equation without quantization. Meanwhile Einstein, in a 1905 paper "On a Heuristic Point of View Concerning the Production and Transmission of Light," claimed that electromagnetic waves themselves were quantified, really.

Certain metals had been observed to emit electrons when exposed to ultraviolet light and this was called the photoelectric effect. The energy of the emitted electrons, however, did not match the energy associated with incident light waves. The energy of a continuous wave is measured by its amplitude (the "height" of the wave), and it was observed that the stronger the light shone on a metal, the greater the *number* of electrons emitted from the metal, but the *energies* of the emitted electrons did not increase. Experiments in 1902 by Philip Lenard showed that the energy of the emitted electrons was a function of the *frequency* of the incident light waves, *not* their amplitude.

Einstein used Planck's black body radiation equation in an explanation of the photoelectric effect. He concluded that light behaved as if it were a very dilute gas, hence composed of *particles*, that knocked electrons out of a metal's surface:

> According to the assumption considered here, when a light ray starting from a point
> is propagated, the energy is not continuously distributed over an ever-increasing
> volume, but it consists of a finite number of energy quanta, localized in space,
> which move without being divided and which can be absorbed or emitted only as
> a whole.[1]

The energy of these quanta was a function of their frequency as Lenard's experiments showed, and it was correctly described by Planck's equation. The *number* of emitted electrons, however, was correlated with the amplitude of the incident light as if it were a continuous Maxwellian wave. Light behaved as if in some sense it really was both continuous *and* quantized, both a wave *and* a particle.

Ontology is thus central to the birth of quantum theory. If, following Mach or even Duhem or Poincaré, mathematical physics is about experience only and essentially conventional, there is no problem at all here. The equations involved match the available data, make correct predictions about data from new experiments, and even suggest practical applications. There is only a problem if you claim to be revealing a reality that is behind and the cause of experience. For years Planck denied the reality of Einstein's light quanta. So did Robert Millikan, whose Nobel Prize was for establishing the reality of the unit charge on the electron. So did Niels Bohr, right up to his 1922 Nobel Prize award address to the Swedish Academy, the same year that Einstein received the (delayed) 1921 award for his theory of the photoelectric effect. Bohr was opposed to the reality of light quanta because of the ontological as well as conceptual contradictions entailed by quanta being both waves and particles.

Right up to the experimental confirmation in 1923 of the existence of what were by then called "photons," Bohr argued that the phenomena Einstein's theory explained could be explained without attributing reality to quanta. (This echoed Planck's resistance to an ontological interpretation of his initial quantization of electromagnetic energy.) Instead of real quanta, Bohr, in collaboration with Hendrik Kramers and John Slater, proposed the existence of a real "virtual radiation field" that acted at a distance and controlled the emission and absorption of radiation by orbital electrons. Even though accepting the reality of this field meant giving up strict causality and strict conservation of energy and momentum, a number of physicists, including Max Born and Erwin Schrodinger, preferred this to accepting the reality of light quanta and its corollary and giving up the reality of the continuous waves in Maxwell's electromagnetic wave theory.

Clearly, the nature of reality is what is at stake here, and the opening paragraph of Einstein's photoelectric effect paper is revealing. "There is a profound formal difference between the theoretical conceptions physicists have formed about gases and other ponderable bodies [matter], and Maxwell's theories of electromagnetic processes in so-called empty space."[2] The "theoretical conceptions" about matter that Einstein refers to include the atomic theory of matter, as employed in the kinetic theory of gases and Einstein's theory of Brownian motion (also 1905). The "theoretical conceptions" employed in Maxwell's theory include the continuous nature of electromagnetic energy. Einstein continues: the state of a material body "is completely determined by the positions and velocities of a very large but nevertheless finite number of atoms and electrons." By contrast, "we use continuous spatial functions to determine the electromagnetic state of a space, so that a finite number of quantities cannot be considered as sufficient for the complete description of the electromagnetic state of a space."[3] Furthermore, the energy of a body composed of atoms cannot be "broken up" into parts smaller than the energy of a single atom, while there is no lower limit to the energy of a wave radiating out from a point source.

Einstein is proposing a theory of what light really *is*, and he writes that he cannot base such a claim on "conceptions" selectively drawn from logically and ontologically exclusive *theories* of the real. This was precisely what Planck had done in arriving at his 1900 equation. Although Planck was not then convinced of the reality of atoms, he found, to his regret, that in order to arrive at his black body radiation equation he needed to use concepts drawn from classical thermodynamics, which is ontologically neutral, from Maxwell's electromagnetic theory with its real continuous waves, and from the Maxwell–Boltzmann kinetic theory of gases, which assumes that matter really is atomic. For Einstein, a theory that rests on such an eclectic conflation of contradictory concepts cannot give us a description of reality, only a theory that works empirically, and that was not enough for Einstein. At the same time, he knew that he needed Planck's equation in his photoelectric effect theory. Einstein thus devotes the first part of his paper to re-deriving Planck's equation but without reference to the atomic theory of matter (though Einstein then believed the atoms were real), using only physically consistent "conceptions."

Einstein proceeded to an extended argument that not only did light behave *as if* it were a highly dilute gas composed of particles, but that it really *was* composed of particles. He claimed that not only the photoelectric effect, but also black body radiation, photoluminescence, the production of cathode rays [electrons], "and other groups of phenomena associated with the production or conversion of light can be understood better if one assumes that the energy of light is discontinuously distributed in space."[4] Epistemologically, his reasoning leads to a conclusion that is ontologically paradoxical, a paradox that was for a long time referred to as the paradox of wave–particle "duality," but he arrived at this conclusion using physically consistent reasoning.

In yet a third 1905 paper, "On the Electrodynamics of Moving Bodies," Einstein presented his special theory of relativity. In addition to the universally accepted principle that the laws of physics are the same for all observers in uniform (non-accelerated) motion, Einstein stipulated that the speed of light in a vacuum was the same for all observers regardless of their motion with respect to the light. There was suggestive experimental evidence in the late nineteenth century that this was the case, but Einstein made it into a principle of nature as an axiom of his theory. With this principle in place, Newtonian mechanics morphs into relativistic mechanics. In the process, Newton's absolute space and absolute time lose their reality in favor of a new reality: unified space-time.

All measurements of space, time, motion, and mass now become relative to a freely chosen frame of reference: there is no longer a universal frame of reference for all such measurements. Mass, defined by Newton as indifferent to motion, becomes a function of the velocity of a moving body. These changes define a new ontology, not merely new conventions that are explanatorily more powerful than the Newtonian definitions. The equation $E = mc^2$, for example, is a logical consequence of the special theory of relativity, and it required acknowledging a new truth about the physical world, that the separate "laws" of the conservation of matter and of energy were wrong.

In November of 1915, Einstein published his general theory of relativity, in which the laws of physics are the same for all observers even if they are in accelerated motion with respect to one another. The general theory, even more than the special theory, was explicitly ontological. A cornerstone was another ontological move by Einstein, making into a principle of nature the equivalence of inertial and gravitational mass (weight). There was suggestive experimental evidence for this, especially extremely precise measurements made by the Hungarian physicist Lorand Eötvös, but Einstein took the equivalence axiomatically, as a universal fact about the world. His theory's field equations integrated space, time, matter, energy, and motion in an entirely new way, a way that profoundly altered our "picture" of reality. Space and time, already linked in the special theory, now had properties that varied as a function of the local concentration of matter and energy. The curved path of light rays passing by stars and galaxies (creating the galactic lensing effect), the existence of black holes and gravitational waves, Mercury's anomalous orbit, and the time dilation effect are all logical consequences ("predictions") of the general theory of relativity. A realist understanding of the general theory is inescapable.

Space, time, matter, and energy really are entangled for Einstein. Relativistic space-time is local, not universal. Space really is curved in the presence of matter and so is the rate at which time "flows." The existence of black holes was seen as a logical consequence of the general theory of relativity early, and today black hole studies is an active branch of both theoretical and experimental physics, linked to thermodynamics, astrophysics, astronomy, and cosmology. Gravity waves, also a logical consequence of the general theory (according to most physicists), are considered to have been observed and are now real. With the invention of atomic clocks, the measurement of time dilation became routine, and today it is built into the satellite GPS systems of position location. Contrary to classical (Newtonian) physics, all motion, uniform and accelerated, is relative, and matter is not indifferent to motion. Logic notwithstanding, these changes are treated within physics as revealing new realities.

The general theory of relativity is deductive, hence deterministic. Understood ontologically, it follows that the universe, too, is deterministic, and that is what Einstein believed. From the late 1920s until his death in 1953 he had a running dispute with Bohr, a dispute conducted via papers and correspondence, not face to face, over the completeness of quantum mechanics. From his 1905 photoelectric effect paper until 1925, Einstein had made important contributions to what is now called the "old quantum theory," pre-quantum mechanics. When Bohr first launched the old theory in 1912 with his proposed quantization of the energy of orbital electrons in atoms, he credited Einstein's ontological application of Planck's hypothesis to light as having inspired his theory of the atom, a theory even more radical vis-à-vis classical physics than Einstein's 1905 theory of light was. Because of the paradox of wave–particle duality, Bohr rejected the reality of Einstein's light quanta right up to their experimental confirmation in 1923, but his own quantum theory of the atom was unequivocally ontological. Atoms were real, nuclei were real, orbital electrons were real, as were the laws governing their orbits, and the presence of electrons inside nuclei was considered real.

The rise of quantum mechanics in 1925 proved to be a turning point. Over the next five years, this theory led Max Born, Bohr, Werner Heisenberg, and many others to adopt a probabilistic interpretation of the behavior of reality at the quantum level, the so-called Copenhagen interpretation of quantum mechanics. This was incompatible with the claim that reality was deterministic "all the way down." With Born and Pascual Jordan, Heisenberg had created a theory, called matrix mechanics, that Mach would have applauded. It correlated observable electron energies with spectroscopic observations without reference to unobservable causal mechanisms and without ontological claims. Independently, Erwin Schrodinger created a theory of quantized electromagnetic waves modeled on classical nineteenth-century physics. Schrodinger's wave mechanics solved all of the problems that Heisenberg's matrix mechanics solved, plus it was ontological and deterministic. It was a surprise to both Heisenberg and Schrodinger to learn that their theories were mathematically equivalent, but Born, who coined the name "quantum mechanics," showed that this was the case if the terms in the wave equation were taken as probability distributions. Schrodinger was deeply upset by this equivalence. In 1926, in a letter to Born, Einstein wrote:

Quantum mechanics is certainly imposing. But an inner voice tells me that it is not yet the real thing. The theory says a lot, but does not really bring us any closer to the secret of the "Old One." I, at any rate, am convinced that He is not playing at dice.[5]

To the end of his life Einstein remained a critic of quantum mechanics under the Copenhagen interpretation, that ultimately, natural phenomena *were* probabilistic, really.

Quantum mechanics worked empirically and Einstein acknowledged that. Because it was not deterministic, however, because even in principle it excluded an underlying deterministic causal mechanism responsible for the probabilistic quantum level, he could not accept it as an account of reality. For some twenty years, Einstein and Bohr argued about this, neither succeeding in convincing the other that he was right. The reason for this failure did not lie in physics, but in metaphysics. For Einstein, the intelligibility of experience required causal-deterministic explanations. When Einstein said that the most wonderful thing to him was that the universe should be intelligible to the human mind, he was expressing an ontological commitment. What was wonderful was that the mind "in here" could understand the way the world "out there" really was. A theory that only described the world "out there" probabilistically failed to make experience intelligible. Bohr, on the other hand, had no problem at all with a probabilistic theory of the world as providing intelligible explanations of experience.

Is the world *really* deterministic "all the way down," or is it *really*, deep down, random and statistical? Quantum mechanics made this into a pressing ontological question for physicists. Probability and randomness as features of nature had entered more and more deeply into science in the course of the nineteenth century. The Belgian astronomer-mathematician-sociologist Adolphe Quetelet had pioneered the use of statistics to describe human behaviors in ways that would allow public policies to be based on a rational "social physics." From the late 1820s to 1870, Quetelet published statistics he collected about behavior such as crime and suicide and their correlation with such factors as age, poverty, alcohol consumption, and gender. He showed that one could make useful predictions of group behavior without knowing its underlying causes.

Quetelet did not doubt that the world *was* deterministic, but he showed that useful information about the world could be had without knowing determining causes. Maxwell took Quetelet's work seriously and saw in it a solution to the moral-philosophical problem of the compatibility of human free will and ontological determinism. Individuals really did have free will, but the free choices of large numbers of people were predictable, as if they were determined. In the 1860s Maxwell, along with Rudolf Clausius and Boltzmann, formulated the kinetic theory of gases. It was based on a statistical account of the behavior of gases assuming that such phenomena as temperature, pressure, and diffusion were the aggregate result of the motions of vast numbers of atoms/molecules. These motions are not at all random, but we cannot know the positions and momenta of each of these particles, and so we must use statistical averages to describe their behavior. As it did for Quetelet, in the kinetic theory of gases probability becomes a functional surrogate for an underlying determinism. This was the case as well for the subsequent creation late in the century of statistical mechanics and statistical thermodynamics, in which Maxwell, Boltzmann, and J. Willard Gibbs played prominent roles.

The Darwin/Wallace theory of evolution was more radical than this, how-ever. It was the first modern scientific theory to attribute randomness to na-ture itself, in the form of intrinsically unpredictable "spontaneous variations" (later called random mutations) between parents and offspring of all living things. In 1896, radioactivity was discovered, and within a very few years it was explained to be a random process, random at the level of individual atoms, but highly predictable in the aggregate as each radioactive element has a unique rate at which it decays (half-life). Randomness was now established as a feature of physical reality, and it became more deeply entrenched with the growing acceptance of Bohr's quantum theory of Rutherford's "solar system" model of the atom.

Bohr's theory aimed at explaining a vast body of puzzling spectroscopic data that had accumulated in the course of the nineteenth and early twen-tieth centuries. It was discovered that each chemical element radiated light at a distinctive set of frequencies, constituting a "lightprint" as it were, for each element. By the end of the century, it was widely speculated that the source of each element's lightprint was a process internal to the atoms of which each element was composed. (We see in this the convergence of an ontological un-derstanding of the electromagnetic wave theory of light and of the atomic theory of matter.) The discovery of the electron in 1897 and then Rutherford's proposed "solar system" model of the atom in 1910 suggested to Bohr, who had spent most of 1911 in Rutherford's lab, an explanation of spectroscopic data: that light was absorbed or emitted by electrons orbiting each element's distinctive atomic nucleus when those electrons moved between orbits.

First, however, Bohr had to resolve a serious problem with Rutherford's model. If, as Rutherford proposed, atoms were composed of very light nega-tively charged particles (electrons) circling a much heavier positively charged nucleus, then according to Maxwell's electromagnetic wave theory, those electrons would spiral into the nucleus almost instantaneously. To avoid this, Bohr proposed making the following assumptions: Maxwell's theory does not apply inside atoms; the mechanical energy of orbital electrons is quan-tized consistent with Einstein's application of Planck's quantization to light; electrons only emit or absorb light when they *change* orbits; and only cer-tain orbits are permissible for the electrons in a given atom, the simplest case being hydrogen with one electron and a simple positively charged nucleus. With these assumptions, Bohr was able to deduce hydrogen's lightprint and predicted the same would be possible, in principle, for all elements.

Note the ontological character of these assumptions, and note, too, that they are completely ad hoc, creative inventions of Bohr's as in Peirce's abductive rea-soning or Whewell's induction of Fundamental Ideas. In effect, Bohr worked backward from a problem as he defined it to what would allow a solution (the procedure recommended by Galileo). The assumptions must have seemed "off the wall" to Bohr's contemporaries, certainly to the older ones with careers anchored in classical physics. (An exception was Arnold Sommerfeld, who embraced them.) The payoff for accepting these assumptions, Bohr claimed, was worth it: explaining all spectroscopic data. Major conceptual and empirical challenges arose and were overcome between 1913 and 1925 as Bohr's initial theory was extended and refined (primarily by young physicists) to match existing and new spectroscopic data. Bohr's quantization assumptions became embedded in physics and morphed into quantum mechanics with its radically new "picture" of reality compared with what was real in classical physics. However much the relativity and quantum theories owed to the concepts and mathematics of classical physics; between them these two theories generated new, not altogether consistent, ontologies, new physical realities.

The "old" quantum theory took the reality not only of atoms but of their internal structure and constitution as established: electrons as independent particles, discrete electron orbits with permitted and forbidden transitions, and nuclei. Between 1920 and 1932, an internal structure was attributed to the nucleus: first a mixture of electrons and protons, then no electrons but a mixture of protons and neutrons, then protons, neutrons and neutrinos, all as independent particles with neutrons described as a proton plus an electron plus a neutrino (later identified as an anti-neutrino). In one form of radioactive decay, a neutron inside the nucleus randomly "decays" into a proton and emits from the atom an electron and an (anti-)neutrino. In another form of decay, one or more neutrons are emitted from the atom when the nucleus randomly splits into two smaller nuclei, with a very large release of energy, as described by Lisa Meitner and Otto Hahn in 1938–1939.

It is virtually impossible to talk about these phenomena without attributing reality to intrinsically unobservable atoms, nuclei, electrons, protons, neutrons, neutrinos, orbits, and orbital changes—later also mesons and a host of other subatomic particles and later still to quarks and antiquarks—along with their properties and behaviors: all this, plus electric, magnetic, electromagnetic, gravitational, and quantum energy fields with families of force carriers. All of these particles and fields are described as real, possessing properties and engaging in processes that are indirectly observable, though requiring very complex instruments to do so.

Quantum mechanics revealed another ontological dimension of scientific knowledge: a direct linkage between mathematics and reality. The ancient Pythagoreans taught that mathematical forms determined physical reality, and the influence of this teaching can be traced from the Greek and Roman period through medieval natural philosophy to Renaissance natural philosophy to early modern science (Galileo, for one) to the twentieth century. In a 1960 essay titled "The Unreasonable Effectiveness of Mathematics in the Natural Sciences," the Nobel Physics Laureate Eugene Wigner wrote, "The enormous usefulness of mathematics in the natural sciences is something bordering on the mysterious and . . . there is no rational explanation for it."[6] In 1915, the mathematician Emmy Noether, for example, proved a theorem that for every differentiable mathematical symmetry possessed by a mathematical structure called a group that describes the action of a physical system—the action also being a mathematical expression—there will be found a conservation law or an invariance principle in that physical system. If this suggests that one can discover truths about nature solely by studying the properties of mathematical structures, that is indeed the case.

Schrodinger's 1926 wave equation, whose probabilistic interpretation was and remains at the heart of quantum mechanics, did not incorporate relativistic effects associated with the motions of the orbital electrons. In 1928 Paul Dirac remedied this, producing a version of the equation that incorporated the special theory of relativity. Dirac's theory was immediately recognized as important and impressive, but its correctness was challenged because his equation had negative solutions for electron energies. This seemed absurd and if taken as true would make atoms unstable, but Dirac stood by his theory. He proposed that the negative solutions implied the existence of as yet unobserved particles exactly like electrons but with a positive charge. Because of the opposite charge, the energy of this particle—later named "positron"—appears in Dirac's equation as having a negative value. Nature imitates mathematics!

It is an understatement to say that this was a bold claim by Dirac, made at virtually the same time as Pauli's rescue of quantum mechanics by proposing the existence of the neutrino, but in 1932 Carl Anderson observed signs of positively charged electrons in his cosmic ray experiments. The positron became real and with it, antimatter. There are other examples of the linkage of quantum theory to ontology that are worth a close look. In 1927, Heisenberg published a paper announcing an inescapable uncertainty in the simultaneous measurement of certain pairs of physical properties. One pair was the position and the momentum of a particle; another, the energy of an event and the time of its occurrence. It was a consequence of quantum mechanics that to know the position of a particle exactly is to lose all knowledge of its momentum and to know the momentum precisely is to lose all knowledge of position; the same holds for energy and time. The implications of Heisenberg's principle for classical physics' deterministic picture of reality were profound. Recall Laplace's formulation of this picture: from knowing the precise position and momentum of each particle of matter in the universe at a single instant of time—in principle possible in Newtonian mechanics—all future states of the universe could be predicted and all past states retrodicted. But for quantum mechanics, position and momentum *cannot* both be known precisely at a single instant of time. The uncertainty principle thus pulls the rug out from under determinism, making it a metaphysical assumption.

Heisenberg sent a draft of his paper to Bohr, who read the paper and wrote back to Heisenberg that he (Bohr) had just come up with what he called his principle of complementarity. Bohr was seeking an explanation of the ontologically puzzling wave–particle duality: photons behave as if they were simultaneously particles and waves, and, as recent experiments revealed, particles also behave as if they were simultaneously waves and particles. Bohr's solution to this problem was conceptual, bracketing the question of what was real "out there." Wave–particle duality is not a paradox about nature, he said; it is an artifact of how we conceptualize nature. Limited as we are to concepts deriving from empirical experience, the only way we can describe the quantum level of nature is in terms of complementary concepts derived from empirical experience, for example, waves and particles.

Complementarity *seems* to bracket ontology. This is supported by books about science that Bohr and Heisenberg wrote claiming that the object of knowledge in physics was our *experience* of reality, not reality as it was in itself.[7] In Bohr's words: "There is no quantum world. There is only an abstract physical description. It is wrong to think that the task of physics is to find out how nature is. Physics concerns what we can say about nature."[8] This Machian phenomenalism is belied by the fact that from 1927 on Bohr and Heisenberg worked as physicists as if quantum mechanics were an ontological theory, not something we had to settle for because of epistemological limitations. Both the Heisenberg uncertainty principle, which began as an artifact of experimental measurement, and Schrodinger's wave equation were increasingly interpreted ontologically, and this continues to be the case today.

Instead of the uncertainty relations reflecting operational limitations, or in Bohr's formulation, epistemological limitations, the uncertainty relations are said to imply that there cannot *be* such a thing as a true vacuum, for example. A vacuum requires a region of space in which there *is* precisely nothing. The value of the quantum field energy in that region would then be precisely zero, and this is interpreted as a violation of the uncertainty relations. It must be the case, therefore, that the quantum field energy continually fluctuates around a zero value. Thus, what seems to be a field-free vacuum is actually only a statistical average of a fluctuating energy field. These fluctuations take the form of particles popping into and out of existence, mutually annihilating one another so quickly that some region of space *seems* empty but at the quantum level is not. We know this must be the case because quantum field theory and the uncertainty principle tell us something about reality: a precise field energy value is not permitted. And this bizarre-sounding prediction has been confirmed experimentally, beginning with the work of the Dutch physicists Hendrik Casimir and Dirk Polder in the late 1940s.

In the seventeenth century, Aristotle was routinely mocked by the founders of modern science for maintaining that a vacuum was impossible because 'nature abhors a vacuum.' The reality of a vacuum became a hallmark of modern science from the seventeenth century to the twentieth. Today, the vacuum has returned to the ranks of the unreal because nature "abhors" a zero value for the energy of quantum fields. Perhaps even more bizarre is John Wheeler's 1955 "quantum foam" hypothesis. Just as the uncertainty relations do not allow a quantum energy field-free vacuum, they do not allow space-time to *be* continuous. At the quantum level, space-time also must fluctuate, it must "really" display minute discontinuities, and there is some experimental evidence that this may indeed be the case.

Multiple universe theories are yet another example of the ontological interpretation of the mathematics of quantum mechanics. Schrodinger's wave equation is, as described earlier, at the heart of quantum mechanics and is experimentally confirmed to a very high degree. Schrodinger considered his equation to be both deterministic logically because it is deductive, and also physically real. As in classical physics, the parameters of the equation corresponded, he believed, to physical realities. His wave mechanics had as its object the world, while Heisenberg's matrix mechanics only had as its object correlations among a body of experimental data.

Unfortunately, the probabilistic Copenhagen interpretation of the wave equation undermined a correspondence with a deterministic world. The ostensibly physical parameters of the equation, for example, the position of a particle, were now probability functions. These functions developed in time deterministically, but they remained probabilities and their physical interpretation was problematic. One solution was to claim that the probability waves were physically real, with the wave function for an as yet unmeasured parameter having a nonzero value everywhere in the universe. If you want to calculate the position of a particle after an earlier measurement, the wave equation gives a distribution of probabilities over the entire universe. Once you measure the position, the wave "collapses" to zero everywhere except for where the particle is discovered to be when the measurement is made.

The instantaneous cosmic "collapse" of a physically real probabilistic wave function was deeply problematic, violating the special theory of relativity. The waves remained central to the Copenhagen interpretation of quantum mechanics, but their reality was ambiguous. (David Bohm's deterministic subquantum theory of matter and energy, and subsequent Bohm-like theories, don't have this problem, though they have others.) In the 1950s, Hugh Everett III, a graduate student of John Wheeler's at Princeton, made a proposal far more radical than Dirac's positron. For his doctoral dissertation, Everett proposed an ontological interpretation of the probabilities in Schrodinger's wave equation–based version of quantum mechanics. At every measurement, he argued, equally real, alternate noncommunicating universes come into existence, each one a realization of one of the possible outcomes of a measurement.

The wave function does not collapse, and therefore there is no problem of how a measurement causes its instantaneous cosmic collapse. What are abstract probabilities in the Copenhagen interpretation of quantum mechanics become physical realities in Everett's interpretation. The famous 'Schrodinger's cat' problem is easily solved by Everett. When the box is opened, the universe bifurcates. In one, the observer finds the cat alive and events unfold accordingly. In the other, the same observer finds the cat dead, and events unfold accordingly. The observer does not *feel* his/her bifurcation and does not feel that they are living in two different universes, having different experiences, but that is not evidence against its happening. After all, the fact that we don't feel the Earth moving is not evidence against its motion!

Everett's interpretation of quantum mechanics was almost universally mocked when he proposed it, not least by Bohr and his circle, who were wholly committed to the Copenhagen interpretation. Wheeler was supportive of Everett (though not of multiple universes) and worked with him to modify the dissertation so that it was approved and Everett received his doctorate. Unfortunately, the revisions Wheeler urged made Everett's ideas more vulnerable to criticism than the original draft. The ridicule and rejection were too much for Everett, who abandoned theoretical physics for a career in applied mathematical physics, achieving some notable success. However mad Everett's theory may seem at first hearing, note that logically his "move" in interpreting the mathematics of wave mechanics ontologically is no different from Dirac's ontological interpretation of the mathematics of his relativistic theory of the electron. Dirac correctly predicted the existence of antimatter on no other grounds than his faith in the physical reality of the negative solutions to his equation. Today, Everett's ontological interpretation of the wave equation is no longer mocked.

Everett's thesis was revived in its original, pre-Wheeler-edited form and promoted by Bryce DeWitt, who had become a friend of Everett's during a year DeWitt spent in Princeton at the Institute for Advanced Study. Later, at the University of North Carolina, DeWitt developed a solid reputation as a theoretical physicist and in 1973 co-edited a volume of essays that took Everett's theory very seriously, *The Many-Worlds Interpretation of Quantum Mechanics* (MWI). Subsequently, the theory was adopted or defended by a number of highly respected physicists, among them David Deutsch and Max Tegmark. Deutsch, a pioneer of quantum computation, incorporated MWI into his explicitly realist theory of multiple coexisting universes, described in *The Fabric of Reality* (1997). Tegmark defended MWI as consistent with his belief that all mathematical possibilities are physically real in *Our Mathematical Universe: My Quest for the Ultimate Nature of Physical Reality*. Today, MWI is a respectable, mainstream alternative to the single-universe Big Bang as a cosmological theory and to the Copenhagen interpretation of quantum mechanics. Brian Greene's *The Hidden Reality* surveys a range of multiple-universe theories currently being pursued by physicists.

Wheeler remained indifferent to MWI: one universe was enough for him, he said. But Wheeler came to champion an ontology at least as radical as Everett's: that "really" the universe was a binary information structure whose existence, and behavior, incorporates human thinking: "The universe does not exist 'out there,' independent of us . . . We are inescapably involved in bringing about that which appears to be happening. We are not only observers [of the universe]. We are participators. In some strange sense, this is a participatory universe."[9] Richard Feynman, also a Wheeler Ph.D. student, dismissed MWI out of hand. He saw that it entailed an infinite number of branching universes, which he felt was a physical impossibility. At the same time, Feynman was an uncritical realist about science. Science described the way the world "out there" really behaved, and that's that. He had no patience for what to him was the philosophical question of why it behaved that way: "All I'm interested in is trying to find a set of rules which would agree with the behavior of nature [not experience] and not try to go very far beyond that."[10] (In an eerie anticipation of MWI, in 1941 the Argentine writer Jorge Luis Borges published a short story, "The Garden of the Forking Paths," in which at every instant the universe branches into multiple universes such that all possibilities are realized.)

Philosophically and culturally, the most dramatic re-ontologization of nature has been wrought by quantum field theory. In the late 1920s, Dirac created a theory he called quantum electrodynamics (QED) that united Maxwell's electromagnetic theory, quantum mechanics, and the special theory of relativity (but not the general theory). In QED, now incorporated into the so-called Standard Model of quantum theory, the ultimate reality is no longer matter, but immaterial energy fields. From a cultural perspective, this is far more shocking than the dethroning of the unique truth of Euclidean geometry and therefore of the connection between deductive reasoning and reality. That nature is ultimately material, that natural phenomena are all the result of matter in motion, was an almost biblical dogma of the secular Western conceptualization of the real. A crack developed when in the nineteenth century immaterial electric, magnetic, and electromagnetic fields were pronounced real *alongside* matter. But in quantum field theory, energy fields *replace* matter as the ultimately real. What we call matter is only "excited states" of quantized immaterial energy fields, a byproduct of the varying outcomes of the interaction of the Higgs field, by way of the Higgs boson, with other energy fields.

Physics, and science generally, seems to have an ineluctable ontological dimension and not just with respect to mathematics. By the early 1960s, the number of elementary subatomic particles had exploded from three in the early 1930s, electrons, protons, and neutrons, to scores, and according to some counts, hundreds. In 1962, Murray Gell-Mann and Yuval Ne'eman proposed a taxonomy they called the Eightfold Way, organizing the meson and baryon families of particles into octuplets. (Mendeleev's periodic table of the elements was also based on patterns of eight.) There was a gap in one family, and Gell-Mann predicted the existence and properties of a particle he called Omega-minus to fill that gap. Subsequently, this was experimentally confirmed. (Mendeleev also made existence and property predictions based on gaps in his periodic table.) This suggested that the Eightfold Way was more than a conventional taxonomy, that it was "natural." Nevertheless, in 1964 Gell-Mann (and independently, George Zweig) replaced it with a new theory, quantum chromodynamics (QCD), in which all particles except electrons and the family to which they belong (leptons) *really* are combinations of quarks and/or antiquarks held together by gluons.

Science tells us what is real and what is not real, but science routinely redefines what is real and redefines the objects that comprise reality. Science may intend to *reveal* the real definitively—Einstein liked to say that a true theory 'lifted a corner of the veil'—but it clearly does not do that. Whatever the real *really* is, in and of itself, we should not accept as definitive what science says it is at any given time. When scientific theories change, for example, when the relativity theories redefine space, time, matter, energy, and motion, *reality as defined by science* changes. Reality as it is in itself—if indeed there is such a thing, and Poincaré among many others denied that there was—independent of our reasoning about it, does not change, of course.

But that there is anything independent of us is somewhere between an intuitive or instinctive belief and a useful metaphysical fiction (like matter?). The only reality we can speak of *knowledgeably*, based on reasoning, is the one that science tells us about based on reasoning about experience. And, however counterintuitive it may sound, in that sense reality does change when scientific theories change. What may make it sound counterintuitive is that the word "reality" is used equivocally. In can mean what exists independently of the human mind, or it can mean what people have *concluded* exists independently of the human mind. We cannot *know* what exists independently of our reasoning about it. Typically, we *believe* that science refers to that, so that when science changes its description, it changes what we are expected, on rational grounds, to believe to be real.

The American philosopher of science Nelson Goodman argued that scientific knowledge claims were true 'under a description.' In *Ways of Worldmaking*, he described theories as languages for describing experience subject to specified definitions, assumptions, and rules of reasoning. This is reminiscent of Nietzsche's perspectivalism and Poincaré's version of conventionalism. Scientific descriptions of reality are *necessarily* contingent, which sounds paradoxical, so perhaps "inescapably contingent" is better. The ontological dimension of science manifests itself in the announcement of the existence of unexperienced objects and their orderly behaviors. Did these objects and behaviors exist before scientists identified them? Were there electrons before J. J. Thomson, quarks before Murray Gell-Mann, local spaces and times before Einstein, and moving continents before plate tectonics theory?

These questions surely sound silly, but think about them for a moment. Dinosaurs were real before their fossilized bones were discovered only in the metaphysical sense of "real," and of course after the fact. Before their discovery, dinosaur talk was meaningless; it became meaningful only when dinosaurs became scientific objects, and as such, what dinosaurs *really* were continues to be redefined as new fossils are discovered. Similarly, what Neanderthals really were—what they looked like, how they lived, what their relations to Homo sapiens were—has changed dramatically since the discovery of their bones in 1854 and continues to change. So, too, have our accounts of the real lineage of Homo sapiens, as well as of genes, DNA, mountains, and continents.

As a scientific object, the universe has been repeatedly made, unmade, and remade in the course of the twentieth century. As recently as 1920 many astronomers, apparently most, thought the Milky Way *was* the universe. A few years later, Edwin Hubble, using the then-new 100″ telescope on Mt. Wilson, was able to see stars in the Andromeda Nebula, revealing that it was a galaxy, not a nebula. Soon a vast number of galaxies were revealed, changing what the "real" universe was, and this changed again when Hubble in 1929 announced the expanding universe. This universe was redefined beginning in 1935 when Karl Jansky initiated radio-telescopy and was redefined again with acceptance of the Big Bang cosmogony in the early 1960s and then again with inflation theory. With dark matter and dark energy, what we thought was the universe, the universe that had existed before the late twentieth century, ceased to exist and was replaced by a universe in which the old universe is no more than 4% of what is "really" out there.

Nelson Goodman had a point. In *Ways of Worldmaking*, he argued that the real is an artifact of the language you use to describe it. Beyond that lies metaphysics, which seems to be as seductive for scientists as the call of the sirens was to Ulysses.

12

In Quest of the Thinker of Science

Who thinks scientifically?

On the face of it, this is even sillier than asking, "Was DNA a double helix before Watson and Crick?" The obvious response is that the thinking that produces science takes place in the minds/brains of individual scientists because thinking can *only* take place in the mind/brain of an individual. Agreed, but is thinking *scientifically* under the control of the individual thinker? Does each person decide what they are going to think about scientifically and how they are going to think about it? Keeping in mind that the thinking that produces science is a consciously chosen form of discursive-linguistic thinking, the answer to "*Who* thinks science?" is not all that obvious.

Does the linguistic nature of discursive reasoning have implications for what an individual scientist thinks when "doing" science? Is discursive reasoning under the control of the individual thinker? These are important questions. Wittgenstein, after all, abandoned the program of his *Tractatus* because he realized the importance for both epistemology and ontology of how we conceive of language as the medium of thought. The *Tractatus* had been based on a representational conception of language, in which the truth of statements about the world is determined by extralinguistic correspondences between words and things outside the mind. Wittgenstein decided that this view of language was untenable, and he abandoned it in favor of exploring the implications of formulating philosophical questions in ordinary language, in which meaning and truth depend on the usage of words and hence are internal to language. Selections of these explorations were published posthumously as his *Philosophical Investigations*, and it is this work that made Wittgenstein famous. Everything changed philosophically when he changed his conception of language.

There are two responses to the question "Who thinks science?" that are noteworthy with respect to the issue of what scientists know. The dominant response by far in Western intellectual history is that each individual is in control of, and responsible for, their own thinking. An opposing view is that a mind of one's own is a myth. This view has been argued increasingly forcefully since the eighteenth century and underlies the Science Wars of the late twentieth century.

One of the most prominent themes in the history of Western philosophy is the lone thinker struggling to know themselves in order to get beyond opinions and beliefs to truth and wisdom. The tacit assumption is that each individual can think their way to knowledge because they are in control of their thinking. The Socrates of Plato's dialogues (and perhaps even the real Socrates) is one example of this. Descartes, who took it as given that he was in control of his thinking and that he alone was responsible for and in control of the content of his mind, is another. He proclaimed the necessary truth of "*I* think, therefore *I* am" and inferred from it that the "*I*" is quintessentially a thinking thing.[1] It is the "*I*" within me that thinks. It is critical for Descartes' philosophy that he know this "*I*" in order to know anything else because he discovers in his "*I*" the conditions for truth and knowledge: intuition, the criterion of clarity and distinctness, and innate ideas. Spinoza shared this view, and so did Kant, with his transcendental subject able to "look down" on its experience and to discover in it, through logical reasoning, how it itself determines the structure and form of its experience.

It is no less true in modern science that we credit individual thinkers as the source and author of their reasoning. Certain individuals—Galileo, Newton, Einstein—are said to have a "genius" for thinking creatively. This reflects the Renaissance invention of the idea of genius as a property of individuals who are gifted with creativity, as well as a still older value placed on the self in Western culture. It also reflects the growing influence in the modern era of individualism in Western culture. Individualism is as fundamental to democratic politics as it is to economic theory. Both are anchored in social atomism, the view that society is the sum of independent individuals who must be free to decide who they "really" are and should be free to choose how to live accordingly. The highest praise goes to individuals as uniquely responsible for their achievements. They achieve what they do, whether as athletes, entrepreneurs, writers, poets, composers, political leaders, intellectuals, scientists, or mathematicians because of the unique self within them whose potential they have realized through their own efforts. One of the most prominent themes in all of Western literature is the journey of an individual to discover who they really are. One example is Hermann Hesse's *Siddartha*, the journey to enlightenment of the Buddha; another is James Joyce's *A Portrait of the Artist as a Young Man*. Joyce's hero, Stephen Dedalus, declares that he must leave Ireland in order to discover how to live in accordance with who he really is, regardless of how others, for their own reasons, want him to live. "I will not serve that in which I no longer believe," he proclaims, "whether it call itself my home, my fatherland or my church."[2] He acknowledges that he, and others, may have to pay a heavy price for his self-centered choices, but he is prepared to pay it, because living as others want him to live will exact an even greater price.

In science, the products of individual genius are attributed to self-consciously disciplined discursive reasoning. This has not always been the notion of how genius works within the individual. The Renaissance notion of genius, echoing Plato's dialogue *Ion*, was that the products of genius were the result of something non-rational and external that happened to the individual. The ability to know the truth about the real in magical nature philosophy required tapping into a non-discursive ability that the mind possessed to get outside itself and "see" the real, or it required acquiring the knowledge needed to gain the assistance of an external power: God, angel, or demon. (Christopher Marlowe's play *Dr. Faustus* (1590) explores the demonic route to knowledge.) Recall that Bacon, Galileo, and Descartes all said that methodical reasoning should be done 'as if by machine.' Hobbes went further, arguing that all thinking was the product of a machine-like, hence deterministic, process in the brain. While Descartes had exempted the mind from his mechanistic-deterministic account of nature, so that our thinking is under our deliberate control, many of his followers—among them Pierre de la Mettrie and the Baron d'Holbach—included the mind in that account. For them, what we call thinking happens to us with the same necessity by which an object falls under the force of gravity.

Hume denied that there was any such thing as Descartes' self. Strip away all specific content and nothing is left, no *thing* at all. In a sense that Dewey would share, "mind" for Hume is a name for the continually changing flow of thoughts and feelings of which we are aware.[3] Furthermore, we are not in control of this flow. Our ideas and feelings are directed by the "laws" of association that Hume identified: resemblance, contiguity in space and time, and causality. (Edgar Allan Poe gave a brilliant example of this at the beginning of his tale "The Murders in the Rue Morgue.") Hume concluded that reasoning was a sharply limited faculty. Vico, Montesquieu, Hamann, Herder, and Comte all argued that some combination of history, language, and culture played formative roles in determining the thinking of individuals. All foregrounded the fact that language is a social phenomenon so that in acquiring a language an individual internalizes their social situation.

Kierkegaard, Schopenhauer, Nietzsche, and Bergson argued on one ground or another that reasoning was intrinsically incapable of grasping the real. At Columbia University in the early twentieth century, the comparative anthropologist and linguist Franz Boas trained generations of cultural relativist anthropologists, among them A. L. Kroeber, Ruth Benedict, and Margaret Meade. They and their students after them promoted the view that there were no universal cultural forms and values, and no one correct way of doing anything, including reasoning. One of Boas' students was Edward Sapir, who went on to become a professor of comparative linguistics at Yale. Sapir, on his own and then together with Benjamin Lee Whorf, formulated what became known as the Sapir–Whorf hypothesis. Based on their wide-ranging comparative linguistic studies, they claimed that the language a person acquires affects the way that they experience the world and thus the way that they think about it. Many language families have such different noun and verb forms, for example, that their speakers perceive the world differently from speakers of other languages. It follows that there is no one objectively correct way of experiencing the world or of thinking about it. Thinking science reflects, to some degree, the language family in which a person reasons about the world they experience through language and, as language changes over time, when they think it.

The Sapir–Whorf hypothesis, initially very popular, became highly controversial in the 1940s and '50s. Linguists piled up lots of counterexamples and dismissed it as nonsense. It was denounced in the wake of Noam Chomsky's theory of universal linguistic structures built into the human brain, but since the 1980s it has made a comeback. As Protagoras' dictum "Man is the measure(r) of all things" was parodied and dismissed by Plato, Aristotle, and their rationalist philosopher descendants, the Sapir–Whorf hypothesis was often misrepresented by its opponents. Careful study of what they actually wrote shows that they did not claim that world perception was *incommensurable* for speakers of different language families, but only that meanings of world-referring words were different, and thus perception of the world was as well. A speaker of one language needed to take that into account in reflecting on their own reasoning and when communicating with speakers of other languages who are describing their perceptions of (ostensibly) the same world. The truth of this weaker but still very important claim is known to everyone who has attempted to translate a text from one language to another. It is effectively impossible to do so without change of meaning.

At the turn of the nineteenth century, researches by Condillac's students led to recognition of the (unconscious) selectivity of consciousness and to the claim that the brain "secretes" thought as glands secrete hormones, deterministically. This implies that the mind and its thinking are not under the deliberate control of the individual thinker. In the course of the nineteenth century, theories of unconscious mental activity were proposed, with the unconscious affecting and even determining conscious thinking, feeling, and behavior. The most famous of these theories in the twentieth century were those of Sigmund Freud and Carl Jung, but they had significant predecessors.

These included theories of mesmerism-hypnosis and phrenology in the late eighteenth century, and clinical studies in the mid-to-late nineteenth century by Pierre-Martin Charcot and his students, among them Pierre Janet and Sigmund Freud.

Charcot was a highly influential neurologist. One area of his research involved the clinical treatment of hysteria, in men as well as in women. He moved from identifying hysteria as a neurological disorder to a psychological disorder, and he used hypnosis to study and treat it. There was, he concluded, some deep connection between what happened in the mind under hypnosis and the disorder called hysteria. Janet's studies led him to a scientific theory of the mind that included a role for the subconscious in affecting the content of consciousness. Freud's theory of the mind went far beyond Janet in attributing a formative role to the unconscious in thoughts, feelings, behaviors, and physical disorders. The concepts of repression, the id, and the libido are relevant to this, along with the role of (repressed) memories of physical and emotional traumas. Freud emphasized sexual traumas, especially those experienced in childhood.

Freud's contemporary, George Groddeck, largely forgotten today, was in his time a prominent psychiatrist, respected even by Freud, who rarely respected those who disagreed with him: Groddeck had dismissed Freud's focus on sexuality. Groddeck's best-selling *The Book of the It* (1923) is a series of letters to a woman to whom Groddeck explains his theory that our conscious lives are shadowed by an unconscious alter ego that he called the "It." If the desires of the It and the conscious self are discordant, the It causes mental, emotional, or physical disorders, and even apparent accidental injuries, until the discord is resolved. Carl Jung is much better known today than Groddeck. Jung, too, proposed a major role for the unconscious in consciousness, including in addition to a personal unconscious an inherited collective unconscious of archetypal forms that influence our conscious mental, emotional, and intellectual lives. (Wolfgang Pauli was an admirer of Jung, and a patient.)

The cumulative effect of attributing reality to the unconscious seriously undermines the Cartesian self. The mind cannot fully know itself or know why it thinks and feels as it does. An active unconscious puts limits on the value of Socrates' injunction to "know thyself," and on the claim that discursive consciousness is perspicuous, that the mind can "see" clearly what there is to be "seen" in its experience in order to know it fully by deliberate reasoning about it. But there is much more than this to answering the question "Who thinks science?"

In 1879 at the University of Leipzig, Wilhelm Wundt created the first dedicated experimental psychology laboratory, but he was not the first experimental psychologist. The program of extending physics to the study of the mind was begun earlier in the century, notably by Ernst Weber, by Gustav Fechner, and above all by Hermann Helmholz. In 1860 Fechner published a book titled *Elements of Psychophysics*, a term he had invented. It included the first mathematical law linking body and mind: that there is a logarithmic relationship between changes in the strength of a sensory stimulus and its perception, and this applies across all five senses. (Textbooks notwithstanding, Fechner's "law" does not hold either for small or for large stimuli.) Contemporary with Fechner's work were the researches of Helmholz on the physics of the nervous system and of perception, especially of vision and hearing. After his epochal 1847 book *On the Conservation of Force* (later changed to "energy"), Helmholz turned to neurophysiology, a field he effectively created. He measured the speed of transmission of electrical impulses traveling along nerves, and, using instruments of his invention, among them the ophthalmoscope, he studied the neurophysiology of vision, color perception, and hearing. He concluded that the world we perceive and about which we reason is a construct of the nervous system.

Wundt was one of Helmholz's students and like Helmholz had a wide range of interests. He published a staggering number of books and papers on philosophy, logic, ethics, and cultural anthropology in addition to psychology. His researches in experimental psychology centered on quantifying how our senses respond to varying stimuli and on mapping out sensory fields, for example, the "field" of differential tactile responses to stimuli at different locations on the body. Wundt and his students made experimental psychology a prominent and popular field of research. In the United States, his work was widely disseminated by William James, who instituted a Wundt-style laboratory at Harvard. The upshot of all these studies of perception was to establish that the senses delivered to the mind a necessarily selective and limited experience of the world, one that was a function of the human nervous system. This undermined the seventeenth-century distinction of primary and secondary sensations. As Berkeley and Hume had argued, there are no primary sensations in the sense of trustworthy inputs to the mind from the external world. Neuroscience revealed that the mind's empirical experience is a preconscious construct of the brain. In response to the flood of electrical signals arriving from sensory nerves, the brain, in accordance with its own (relatively) fixed architecture, constructs the world that the mind sees, hears, smells, tastes, and feels.

The poet Wallace Stevens caught the implications of this in his poem "The Man with the Blue Guitar":

They said, "You have a blue guitar/you do not play things as they are."/The man replied, "Things as they are, are changed upon the blue guitar."/And they said then, "But play, you must,/A tune beyond us, yet ourselves,/A tune upon the blue guitar/ of things exactly as they are."

The guitarist goes on to explain that this is impossible. There is no instrument that can "play things as they are," only as refracted in ways expressive of the limitations of each instrument. The novelist Howard Jacobson put it this way: "We are unsubtly made. We lack the strings to pluck the tunes we hear."[4]

Early in the twentieth century, Ivan Pavlov, in a further blow to the autonomous mind, published his animal studies on conditioned responses to symbolic stimuli. Shortly thereafter, John Watson initiated behavioral psychology, based on the principle that human behavior was to be explained as reactive responses to external stimuli, not as actions freely chosen by the individual. His more famous successor B. F. Skinner made behaviorism into the dominant mid-century school of psychology, at least in the Anglo-American academic world. For Skinner, the mind is an epiphenomenal byproduct of the total nervous system, of no causal consequence at all. *All* action, including intellectual and artistic "action," is reaction. He decried credit for creativity or genius, claiming that poets "had" poems the way they "had" a cold. Applied to science, Einstein "had" the theory of relativity; he did not create it.

An earlier, rival, theory of the mind seemed to have been overwhelmed by Skinner and his followers: gestalt psychology. Gestalt psychology has been revived with the dethroning of behaviorism by cognitive psychology, for which the mind is real in its own right and a causal agent in behavior. Gestalt psychology is the polar opposite of the experimental psychology of Helmholz, Wundt, and many others. Their research programs sought to identify the smallest units of sensation out of which perceptual experience, and then conscious thinking, is built up, as if out of LEGO blocks. By contrast, for the founders of gestalt psychology—Max Wertheimer, Kurt Koffka, and Wolfgang Kohler (like Schlick a former student of Max Planck's)—consciousness is composed of context-sensitive perceptual wholes that are greater than and different from the sum of their component parts.

As with the perception of melody, what we see when we look at the world are whole visual contexts composed of objects, typically in motion, in relation to one another, with foreground distinguished from background. The same is true for hearing and the other senses. Gestalt wholes are what are now called emergent entities, entities that have properties that are different from the properties of their constituents and cannot be derived from them. Consider table salt. It is composed of molecules of sodium chloride, each of which is composed of one sodium atom ionically bonded to one chlorine atom. The properties of salt are totally different from the properties of sodium, a white metal that reacts explosively with water, and of chlorine, a toxic green gas. Instead of being toxic, salt is necessary for the healthy functioning of the body. Furthermore, dissolved in the blood, the sodium atoms and the chlorine atoms separate (the chemical bond between them is ionic, not covalent), but in the context of the blood they do not revert to their respective individual properties even though separated.

The gestalt psychologists argued not only that the whole is greater than the sum of its parts because it has emergent properties that the parts do not have, but the whole *changes* the behavior of the parts themselves. (Today this idea is a fundamental principle of systems thinking.) The context-dependent sense-based wholes that constitute consciousness can be decomposed by the analytical mind into units out of which we can recompose the original wholes, but those units are intellectual abstractions only; they are not part of our perceptual experience. Experiments showed that the component parts of these wholes are perceived differently in different contexts, and one dimension of these contexts is personal experience itself, which selectively incorporates memories of an individual's past experiences. Truly, as Norwood Russell Hanson put it, "seeing is an experience, people not their eyes see."[5] (Recall Nietzsche and Bergson on perspectival "seeing." For Descartes, "the soul not the eyes see."[6])

Following Hanson, an art historian and a novice viewer looking at a painting do not "see" the same painting. The art historian sees the painting in the context of his/her knowledge of the history of art, of theories of art criticism, and of their own experience seeing, reading, and talking about other works of art. The art historian does not first see a painting as the novice viewer does and only then, consciously, put the painting into the context of their knowledge. Someone with a knowledge of classical music or of jazz hears a piece of music that is different from what a casual listener hears even though the same pattern of sound waves is registered by their auditory nerves. The same pattern of photons may strike the eyes of a physician and a patient looking at a CT scan or an X-ray, but they do not at all see the same thing.

How we conceive the world through our experience is a function of the context of present experiences into which, Bergson-like, memories of past experiences are incorporated. The chemist and philosopher of science Michael Polanyi emphasized the role that *tacit* knowledge plays in thinking science. Unconsciously, experimental and theoretical experiences influence our conceptualization of present experiences. Gestalt psychology brings to mind Francis Bacon's fear of Idols of the Tribe and Idols of the Cave. These were (unconscious) prejudices innate in human nature that influenced our reasoning about the world and misled us, or they were idiosyncratic prejudices, like James' distinction of tough-minded versus tender-minded philosophical temperaments. And then there are the Idols of the Marketplace and of the Schools, which amount to the (unconscious) influence on thinking of the internalization of misinformation. Bacon was confident that the Idols could be overcome by a consciously methodical, subjectivity-suppressing form of reasoning that would allow us to know the world as it really is. The gestalt psychologists disagreed: our conceptualization of experience is inescapably Idol ridden.

One final complication of "*Who* thinks scientifically?": Karl Marx and Friedrich Engels redefined "ideology," the term that had been coined by Condillac for his theory of ideas. Condillac's theory explained that ideas are produced by the activity of our senses, whose inputs are registered, cross-connected, and correlated by the brain preconsciously and determine the content of our consciousness, including our thinking. "Ideology" was a name for the science of ideas as "geology" was for the science of the Earth and "psychology" for the science of the mind. Marx and Engels redefined "ideology," making it their name for the unconsciously values-pervaded thinking of individuals in a class conflict–driven society. Economic relations determine the sociopolitical structure of a society, but they also determine the cultural and intellectual "superstructure" of society, the thinking by which individuals experience the world and judge it. This thinking invariably reflects their position in society. The ideas of the ruling class, for example, are the ideas that prevail in a society; the members of the ruling class "see" the world, unreflectively, consistent with those ideas. "Ideology" becomes a pejorative term for consciously or unconsciously inauthentic, distorted reasoning about social reality.

In the early twentieth century, the German sociologist Max Weber developed a less polemical notion of ideology, arguing, in *The Protestant Ethic and Capitalism*, that there was a close connection between Protestantism and the rise of capitalism. In the 1920s, the sociologist Karl Mannheim, a student of the philosopher and philosophical anthropologist Max Scheler, extended Weber's use of "ideology." In *Ideology and Utopia* (1929) Mannheim founded the field known as sociology of knowledge. He argued that all our ideas, all of our thinking, and all of what is called knowledge at any time are ideological in the sense of being the product of the social reality in which we find ourselves. Contrary to Marx, there is nothing conspiratorial or manipulative about this. Consistent with Nietzsche's perspectivalism, knowledge claims cannot be truly objective, neutral, and universal because no one can think that way, not even in a classless society. The products of our thinking and reasoning are "socially constructed," a term introduced by the American sociologists Peter L. Berger and Thomas Luckman in *The Construction of Social Reality* (1966), in which they argued that social reality *itself* was socially constructed.

Mannheim made one exception to the totalitarian grip of ideology on our thinking: science and mathematics. Scientific and mathematical knowledge were indeed objective because of the universal and impersonal logic of scientific and mathematical reasoning. This exception was incorporated into an influential follow-on work applying the sociology of knowledge to science, Robert K. Merton's *Science, Technology and Society in Seventeenth Century England* (1938). Merton applied Weber's thesis connecting Protestantism to the rise of capitalism to the rise of modern science. His analysis of the relevant sources showed that Protestants made up a far larger percentage of the people who created modern science in England than Catholics, and his book explored the implications of this. Merton went on to become an eminent figure in the sociology of science. He made pioneering studies of the social dimension of the *practice* of science—who becomes a scientist, how and why they choose their research topics, what the values are that motivate and constrain scientists (priority, honesty, trust, sharing data, and ideas)—but, like Mannheim, he bracketed the content of scientific knowledge claims from sociological analysis.

Merton avoided the radical claim that not just scientific practice but scientific knowledge itself was socially constructed. That claim would become fighting talk in the 1970s and '80s, igniting the Science Wars of the time. But the claim was already in the "air" in the 1930s. At an international congress of historians of science held in London in 1931, Boris Hessen presented a paper titled "The Social and Economic Roots of Newton's *Principia*." Hessen was a Soviet physicist and historian of science and, consistent with Marxism, argued that the ideas in Newton's *Principia*, one of the founding texts of modern mathematical physics, were not objective. They reflected the ideas and values of the British ruling class at that time. (Hessen was executed in 1936 in a Stalin-ordered purge. One of his crimes was defending the theory of relativity, which Stalin considered anti-Marxist. He was posthumously exonerated after Stalin's death.) Hessen's 1931 presentation was controversial and his specific analysis largely rejected, but beginning in the 1960s his methodology transformed the history of science and of technology. The Irish chemist J. D. Bernal, for one, took Hessen seriously, publishing a four-volume Marxist history of modern science, *Science in History* (1954).

Prior to Hessen's paper, the small community of historians of science studied science from the inside, so to speak. They focused on detailed descriptions of the content of scientific theories and knowledge, and the chronology of their development over time. What Hessen did was to look at science from the outside, to open the history of science to the study of science in its full social and cultural context. How did contemporary philosophical, religious, and political ideas influence Newton? How did Newton's science reflect such influences? Studying only the ideas in Galileo's *Dialogues Concerning the Two Great World Systems* of 1632, for example, ignores what after Hessen seems the obvious influence on his ideas of Galileo's social and economic situation, his personality, his relationships with individual scientists and nonscientists, and his complex relationships with Church authorities, as well as the fact that the book was published in the middle of the Thirty Years War.

The preceding critiques of the nature of the thinking that goes on in individual minds converged in the 1930s on the ideas about scientific thinking of Ludwig Fleck. Fleck was an internationally respected Polish immunologist, famous particularly for his researches on typhus. In 1935, he published a monograph on the philosophy of science, *The Genesis and Development of a Scientific Fact*. The original German language edition was only translated into English in 1979, after Fleck's death. It was then recognized as seminal, anticipating many of the ideas in Thomas Kuhn's *The Structure of Scientific Revolutions* as well as the subsequent rise of the social construction of scientific knowledge movement (with which Kuhn himself strongly disagreed).

Fleck's answer to the question "Who thinks science?" is that it is not individual scientists who think science, but scientific thought collectives that think *through* individual scientists.[7] Fleck cites the Polish sociologist Ludwig Gumplowicz: "The greatest error of individualistic psychology is the assumption that a person thinks . . . What actually thinks within a person is not the individual himself but his social community." It follows for Fleck that the thinking of individual scientists is framed, and constrained, by having internalized the thought style associated with a particular scientific thought collective. In effect, you cannot be recognized as a scientist, and your thinking will not be accepted as scientific, unless you are recognized by the relevant collective as thinking consistently with that collective's prevailing thought style.

Fleck begins his analysis of scientific thinking by asking, "What is a fact?" He was quite familiar with the logical positivism of the Vienna Circle, having spent 1927 in Vienna, and his own conception of scientific reasoning and knowledge is a decisive rejection of logical positivism. "A fact," he wrote, "is supposed to be distinguished from transient theories as something definite, permanent, and independent of any subjective interpretation by the scientist."[8] His monograph uses the history of the conceptualization and diagnosis of syphilis, a subject on which he was a recognized expert, to show that no such facts exist. All theories of knowledge that base knowledge on reasoning about veridical, self-given facts are mistaken. As the logical positivists and logical empiricists themselves came to admit in the course of the 1930s, there are no theory-neutral observation statements that can serve as the criterion by which theory statements can be confirmed. Fleck went further in his critique of logical positivism. Not only are there no reason-independent scientific facts about which to reason scientifically, but neither can there be protocol statements—what I earlier called correspondence rules—that would allow observation statements to be deduced uniquely from theory statements. Theories of scientific knowledge show too much respect for logic as the essence of scientific reasoning!

Fleck describes the history of the conceptualization and diagnosis of syphilis as it moved from a mystical-ethical stage in the sixteenth century, when it first became a serious medical problem for Europeans, through an empirical-therapeutic stage, to pathogenic and etiological stages in the late nineteenth and early twentieth centuries. In these changes we see the socially constructed character of syphilis as a medical fact changing from a punishment by God to a disease of the blood to identification as a well-defined disease entity with a well-defined cause, *spirochaeta pallida*, and a well-defined diagnostic test, the Wasserman reaction. In the process, Fleck exposes the action on the thinking of individual scientists of changing social values and medical "thought collectives." He defines the latter as "a community of persons mutually exchanging ideas or maintaining intellectual interaction." A thought collective in a field always provides a "special carrier" that Fleck calls a "thought style" that is distinctive of that field, embodying its historical development and its current stock of knowledge.[9]

Fleck defines "thought style" as "the readiness for directed perception and appropriate assimilation of what has been perceived."[10] Thought styles have a specific character reflecting their employment in specific scientific inquiries, but they always also have a generic character. Every thought style inherits "the common mood in which the thought collective of modern natural science lives its life. This mood "is expressed as a common reverence for an ideal—the ideal of objective truth, clarity, and accuracy." It also expresses itself as a belief that what is revered can be achieved, though it may take much time to do so; that dedicating oneself to its pursuit is glorious; in hero worshiping of perceived great pursuers; and in treating the pursuit as a tradition in which the individual scientist locates him/herself.

Fleck's thought styles are similar to what Kuhn later called "paradigms," concepts that frame and constrain the thinking of individual scientists about specific scientific problems. It follows, for Fleck, that cognition is "not an individual process of a particular 'theoretical consciousness.' It is the result of a social activity, since the existing stock of knowledge exceeds the range available to any one individual."[11] For example, "once a statement is published [in a scientific journal] it constitutes part of the social forces which form concepts and create habits of thought. Together with all other statements it determines 'what cannot be thought in any other way.'"[12] The germ theory of disease initiated a thought style as had the earlier Galenic theory of the four "humors." In the germ theory thought style, scientists seek to identify particular invisible entities—germs, microbes, bacteria, viruses, recently prions—as the specific causal agents of particular diseases. According to the "humoral" thought style, by contrast, the cause of the disease was an imbalance among four bodily "humors": blood, phlegm, black bile, and green bile. The task of the physician in that thought style was to identify the imbalance and prescribe a treatment to restore it.

Newtonian physics initiated a thought style, as did the oxygen theory of combustion, the cell theory of life, the atomic theory of matter, the relativity and quantum theories, the chemical bond theory, and plate tectonics. A thought style is the "carrier" of the thinking of a thought collective. It can only determine the thinking of individual scientists after a thought collective has formed that promotes that thought style. It does this through the publication of books and journal articles, the creation of new journals, new scientific societies, and conferences. Journals publish more and more papers in the new style and fewer and fewer in the abandoned style. Before plate tectonics became entrenched in the 1960s, it was effectively impossible to get a paper published in American geology journals arguing that the continents moved. After plate tectonics, it was impossible to get an article published that denied that continents moved. Especially since the nineteenth century, teaching has become a powerful disseminator of the thought style of a collective. In order to become accepted as a member of a scientific community, and thus have your ideas given a hearing, you need to be credentialed, and you need to have a degree. This requires passing through an extended process of education that requires demonstrating that you have mastered the prevailing thought style.

Given the hold of thought styles, how they ever change once entrenched becomes a pressing question, and the history of science shows that they clearly do change. This question is central to Kuhn's theory of science, as we will soon see. Fleck's goal was to use the "genesis and development" over centuries of the scientific "fact" called "syphilis" to expose the nature of scientific reasoning generally. His argument is that at the macro level, so to speak, the history of syphilis reveals changing conceptualizations of what syphilis *is*, of what is factual about syphilis. It follows that scientific facts are not fixed datapoints that transcend theories. They are, rather, the products of the operation of a thought collective and an associated thought style. Scientific facts are constructed in the sense of being defined within the thought style of a thought collective. As we have seen before, and as the history of science amply documents, the outcomes of experiments and the readings of instruments are not theory-neutral facts. They become scientific facts only when they are interpreted within the context of a scientific explanation, and what constitutes an explanation is a function of some thought style, typically the thought style of a thought collective.

The macro level of scientific thinking leads to the micro level of the reasoning of individual scientists. How scientists formulate problems, how they seek solutions to them experimentally and theoretically, and how they distinguish "correct" from "incorrect" solutions all reflect internalization of a collective thought style. They are not original to the individual scientist, and it is in this sense that the collective thinks *through* the individual scientist, who thus does not have a mind of his/her own vis-à-vis thinking scientifically. According to Fleck, reasoning by individual scientists is structured around two poles: "active associations" and "passive associations."

Reasoning scientifically about a phenomenon begins with actively chosen, Protagorean metrics: correlated concepts, assumptions, and conventions. These are contingent, similar to Poincaré's freely chosen "convenient" metrics, and define a thought style. The organization and application of active associations into a theory or explanation of phenomena in empirical experience generates passive associations. These can be understood as the responses of nature to our reasoning insofar as passive associations always incorporate an element of what Fleck called "resistance" to our active association choices: "This is how a fact arises. *At first there is a signal of resistance in the chaotic initial thinking, then a definite thought constraint, and finally a form to be directly perceived.*"[13] Suppose we have adopted the atomic theory of matter and have assumed that there are forms of matter that are elementary. We assume further that each element is composed of atoms unique to that element, that these are unchangeable, always and everywhere the same, retaining their identity regardless of their combination with other elements into molecules. We have isolated what, consistent with our assumptions, are different elements, among them two that we have named hydrogen and oxygen. We decide that atomic weight is a property that is unique to each atom and is conserved unchanged in all interactions. If we adopt the convention that the atomic weight of oxygen is 16 units of weight—everything so far is active association—then the atomic weight of hydrogen *must be* 1.008. This value is a passive association, reflecting nature's "resistance" to its being any other value. Once the choice of 16 for oxygen is made, the atomic weight of hydrogen for everyone, everywhere, at all times, in all situations, must be 1.008, but only if the scientific community's rules are followed for how atomic weights are to be measured, another active association. The crucial point is that the passive elements of knowledge can never be exhibited on their own, only in a context of active elements, and these are to be chosen so as to maximize "thought constraint" with a minimum of "thought caprice."[14]

It follows from Fleck's theory of science, that while theories and explanations in science are not arbitrary, neither can they be uniquely correct accounts of the world "out there." We can no more announce that a theory is true or an explanation is correct because it pleases us than we can define the value of pi to be 3.0 to make computation easier (though the Indiana state legislature once tried to do just that). Fleck insisted that he was not a relativist: "The views outlined here should not be construed as relativism. We are certainly capable of knowing a great deal," but we cannot know everything, because "everything is, ultimately a meaningless term in science. . . . An 'ultimate'. . . set of fundamental first principles from which [a theory of everything] could be constructed is just as nonexistent as this 'everything.'"[15]

Scientific knowledge is objective, and it is about nature, but only relative to a logically consistent set of (contingent) active associations. Most importantly, the armature around which scientific knowledge is woven is made up of nonlogical cognitive choices. "The concept of absolutely emotionless," strictly logical thinking as in the logical positivist theory of science, "is meaningless."[16] Scientific reasoning can be cast into a logical form such that *scientific* facts can be deduced from a theory, but scientific facts are not independent of our reasoning about them. There is within such facts *something* that resists logical inferences that are not consistent with *some* external state of affairs and its relations with other external states of affairs. The upshot is that at the micro level, Fleck's account of scientific knowledge overlaps Poincaré's relationalism and his interpretation of objectivity as convention-dependent intersubjectivity. There is nothing fixed in scientific reasoning, not on the part of scientists and not on the part of nature: "Both thinking and facts are changeable, if only because changes in thinking [changes in our cognitive choices] manifest themselves in changed facts."[17]

A final point before turning to Thomas Kuhn: As already reflected in his title, *The Genesis and Development of a Scientific Fact*, Fleck's argument for his theory of science begins with examining *changes* in thought styles, including the possible coexistence for some time of two different thought styles employed by different scientists in the same field. What problems does this create?: "The greater the difference between two thought styles, the more inhibited will be the communication of ideas."[18] The members of the two collectives will look upon one another as if they were speaking two dialects of the same language, with key words defined differently and apparently similar reasoning leading to different conclusions. Very likely each will accuse the other of wrongheadedness in their assumptions and definitions, but rarely of reasoning incorrectly. This bears directly on one of the more controversial facets of Kuhn's theory of science: the incommensurability of theories across paradigms.

13
A New Image for Science

Had the expression existed in 1962, it would be fair to say that Thomas Kuhn's *The Structure of Scientific Revolutions* (*SSR*) 'went viral' across the academic world. Its two main theses were that the then-prevailing methodology of historians of science misrepresented the nature of science, and that nonlogical factors play a fundamental role in scientific reasoning. Its reception was followed by a dramatic growth in science studies as an academic field and the redirection of the history, philosophy, and sociology of science, but whether it was cause or symptom is open to debate.

SSR is a monograph, at 173 pages barely longer than Fleck's *Genesis*, which was little known in American academic circles until its translation into English in 1979. Both books are about how scientific theories change and both propose overlapping explanations of how science works. Kuhn was told about the German language edition of *Genesis* and its main ideas, especially the ideas of thought styles and thought collectives, after he had already started working on what became *SSR*. Earlier, he had had numerous conversations with Michael Polanyi about Polanyi's view that nonlogical "tacit knowledge" played a fundamental role in scientific reasoning. Arguably, many of the pieces out of which Kuhn's theory of science was constructed were not original to Kuhn. There is a clear ideational "arc" from Whewell, Peirce, and Duhem through Mannheim, Hessen, Fleck, Merton, Bernal, Polanyi, the later Wittgenstein, Quine, and Hanson to Kuhn. What Kuhn accomplished was a selective integration of earlier ideas into a distinctive theory of science.

Kuhn argued two connected theses in *SSR*. The first is that, if reconceived methodologically, the history of science could give us a better understanding of how scientific knowledge is produced; the second, that scientific theory change is not incremental and continuous, but discontinuous, produced by "revolutionary" breaks with earlier theories. At the opening of *SSR*, Kuhn announces the need for a new approach to the history of science, a "historiographic revolution" that will have subversive consequences for our understanding of science:[1] "History, if viewed as a repository for more than anecdote or chronology, could produce a decisive transformation in the image of science by which we are now possessed."[2] Kuhn's use of the word "possessed" is telling, signaling the polemical character of the criticism to come. The image by which we were possessed (in 1962) had been drawn by historians of science and by scientists themselves from "the study of finished scientific achievements," as recorded in science classics and in textbooks, whose goal is to insinuate that image into teaching the next generation of scientists how to practice its trade. A

> concept of science drawn from [such sources] is no more likely to fit the enterprise that produced them than an image of a national culture drawn from a tourist brochure or a language text. This essay attempts to show that we have been misled by [historians of science and scientists] in fundamental ways.[3]

The prevailing understanding of science derives from selective presentations of static, after-the-fact *justifications* of the completed products of scientific reasoning. It ignores completely the dynamic process out of which those completed products were produced. This is, of course, precisely the process of *discovery* that had been bracketed by the logical positivists as irrelevant to the truth of theories once formulated. Recall that the logic of justification is intrinsically ahistorical. In a deductive argument, the conclusion is already implicit in the premises that are taken to be true.

Ironically, the subject of the history of science itself becomes unhistorical if its object is an image of science drawn from science texts. The account of science in these texts

> often seemed to imply that the content of science is uniquely exemplified by the observations, laws, and theories described in their pages . . . that scientific methods are simply the ones illustrated by the manipulative techniques used in gathering textbook data together with the logical operations employed when relating those data to the textbook's theoretical generalizations,[4]

and these are invariably presented as established truths. This is profoundly misleading.

Historians of science have reinforced an image, or concept, of science in which scientific development is a "piecemeal process" by which facts, theories, and methods "have been added, singly and in combination, to the ever-growing stockpile that constitutes scientific technique and knowledge."[5] (Recall John Herschel's Baconian theory of science, for example.) The image of science that follows is that scientific knowledge is the result of the rigorous application of logical reasoning to objective facts about nature derived from objective observation and experimentation. The task of the historian of science becomes one of establishing 'who did what, when' in a linear progression culminating in the truths codified in the textbooks from which the object of their studies is drawn.

Such a methodology is an instance of what the historian Herbert Butterfield called a "Whig" history of science. A Whig history is one written backward from the present as if the present were the goal at which the past had been aiming. In a Whig history of science, past science is judged right or wrong relative to present science taken as true (or at least on the way to the truth). It tells a story about science that selectively identifies just those people, ideas, and theories from which the present body of knowledge can be seen as deriving. Other people, ideas, and theories are ignored or dismissed as wrong because they did not contribute to what we now claim to know.

"In recent years," Kuhn wrote, "a few historians of science have been finding it more and more difficult to fulfill the functions that the concept of [science as continuous] development-by-accumulation assigns to them." Instead, they have been led by their studies to suspect that "science does not develop by the accumulation of individual discoveries and inventions." These "few historians" find "growing difficulties in distinguishing the 'scientific' component of past observation and belief from what their predecessors had readily labeled 'error' and 'superstition.'" They feel increasingly certain that "those once current views of nature" dismissed as wrong by science now—among them Aristotelian physics, Ptolemaic astronomy, the phlogiston theory of combustion, and the caloric theory of heat—were, in their own contexts, "neither less scientific nor the product of human idiosyncrasy than those current today."[6]

If these discarded views are stigmatized as "myths," then the new history of science reveals that "myths can be produced by the same methods and held for the same sorts of reasons that now lead to scientific knowledge."[7] And the same is true for "errors." Recall that Galileo's reasoning in reaching conclusions that we now consider wrong—for example, about comets, the cause of the tides, and heat having weight—was the same as his reasoning in reaching those conclusions we now consider right. So, too, Huygens' reasoning in defending his spherical wave theory of light against Newton's corpuscular theory, Lavoisier's in defending the caloric theory of heat against the theory of heat as motion, Bohr's theory of a virtual radiation field against the reality of photons, etc.

To borrow one of the hallmark terms from Kuhn's theory of science, to which we turn next, the "historiographic revolution" that Kuhn was calling for amounted to a "paradigm" change in the methodology of historians of science. The Whig paradigm must be abandoned in favor of studying past science as if we did not know how the future would judge it. Would all reasonable people with the relevant knowledge in 1632 have sided with Galileo against Tycho Brahe? Recall that Galileo himself rejected Kepler's version of Copernicus's theory, a version that we now consider to be right, in favor of a "pure" Copernicanism that we now consider wrong. A case can be made— and has, by Christopher Graney—that in 1632 Brahe's planetary theory was more "scientific" than Galileo's.[8] Under the new paradigm, a historian of astronomy would treat Brahe on a par with Copernicus, Kepler, and Galileo; not ignore his theory as a dead end; and credit him only as the source of the data that Kepler used in arriving at his theory.

Kuhn aimed at showing in SSR that a new approach to the history of science produces a very different "image" of science from the one by which we were "possessed" in 1962. One of the most important differences is that it reveals that the process of discovery can no more be separated from the process of justification than experimental facts can be separated from theory: the two are mutually implicating. Criteria of justification are already present in the methodology of the discovery process, and commitments made in the discovery process become criteria of justification. Discovery and justification must both come under the purview of the historian of science.

This is the bridge to Kuhn's second thesis. The new approach to the history of science reveals the "insufficiency of methodological directives, by themselves, to dictate a unique and substantive conclusion to many sorts of scientific questions." Scientific knowledge is not *solely* the product of logical reasoning applied to facts. It follows that science is not an archaeology-like activity, uncovering a unique hidden reality; it is ineluctably interpretive, hence pluralistic, reflecting contingent judgments and assumptions. And growth in scientific knowledge is not incrementally cumulative as traditionally conceived. Nevertheless, scientific knowledge remains objective and progressive for Kuhn, and it has the world as its object. Science is a matter of solving problems about the behavior of nature, and it is "nature itself" that is the basis of scientific problems.[9] Kuhn refers to "immense difficulties often encountered in developing points of contact between a theory and nature."[10] There are "seldom many areas in which a scientific theory . . . can be directly compared with nature," but there are some.[11]

Kuhn cites as an example of such a comparison the prediction by the general theory of relativity of the peculiar precession of the perihelion of Mercury's orbit. More broadly, one of the tasks of what he calls "normal science" is to compare facts with predictions so as to "bring nature and theory into closer and closer agreement."[12] Instruments invented for this purpose that he cites are special telescopes to measure stellar parallax (proving the Earth's motion), the Atwood machine for measuring the local gravitational force, Foucault's apparatus for measuring the speed of light in water, and the apparatus used to detect the neutrino. As a consequence of his objectivism and realism, Kuhn dissociated himself from the inference drawn by others from *SSR* that scientific knowledge was socially constructed. After *SSR*, some historians of science adopted an externalist paradigm for their discipline, seeing science as a product of its social time and place, a product of culture, hence open to social and political influences and criticism. Kuhn himself, however, remained an internalist, focusing on science itself. Thus, he urged the study of the relationship of Galileo's views to those of his "teachers, contemporaries and immediate successors," but not the relationship of Galileo's views to their sociocultural context, as in an externalist approach to the history of science.[13]

An externalist approach to *SSR* itself would put that book and its ideas into the context of the anti-establishmentarianism of the 1960s and the incorporation of science and technology into the political, social, and commercial establishment. The 1960s and '70s were a time of waves of mass anti-establishment protests: the civil rights movement, the anti–Vietnam War movement, the environmental movement, the anti–nuclear power movement, the consumer rights movement, and of course the hippies. Twenty million Americans participated in the first Earth Day, held April 22, 1970. Millions marched in countless Vietnam War protest demonstrations. Over four hundred thousand people attended the "Woodstock" rock concert in August 1969. Tens of thousands of people protested nuclear power. Environmental activists looked to Rachel Carson's *Silent Spring*, published the same year as *SSR*, as their inspiration, and consumer rights activists to Ralph Nader's *Unsafe at Any Speed*, published in 1966. New forms of rock and folk-rock music were perceived both by supporters and critics as revolutionary and subversive and were often openly political, as popular music rarely had been other than in wartime.

Concurrently, science and technology were enjoying a higher social status than ever before. After WWII, respect and support for science by political, educational, and commercial institutions and organizations was unprecedented in American history. For the first time in peacetime, the federal government allocated large sums of money in support of basic as well as applied research, military and nonmilitary. Most of this money was distributed through universities, also unprecedented in American history, and it had a dramatic impact on the teaching of science and mathematics not only at the college level, but also at the secondary and even primary levels (especially after Sputnik in 1957). To a degree, federal support for science reflected acceptance of *Science: The New Frontier*, a report written near the end of the war by Vannevar Bush, wartime Director of the Office of Scientific Research and Development, at the request of President Roosevelt. Bush's report argued that

America's future prosperity and national security would be dependent on the degree that new scientific knowledge became the basis for new technologies.

By the 1960s, science and engineering were firmly embedded in the sociopolitical establishment. As such, they became targets of antiestablishment activists. Scientists and engineers, it was claimed, had been co-opted by the military and by multinational corporations. They were responsible for the weapons being deployed in the Vietnam War and the Cold War, and they enabled corporations to produce products that were unsafe for the public and/or unsafe for the environment. At the height of their social and political status, science and engineering were being stigmatized as morally corrupt. In the best-selling *The Making of a Counter-Culture* Theodore Rozsak asked, rhetorically, how long it would be before people hired by pyromaniacs to develop new incendiaries recognize that they are morally responsible for the fires that are then set? In Francis Ford Coppola's film "The Conversation," lauded as one of the best films of the decade, a brilliant electrical engineer who denies any responsibility for what his clients do with his work is duped into becoming an accomplice to murder, forced to confront that he is a (metaphorical) prostitute, selling his skills to whoever will pay for them.

This was the social context in which *SSR* became a major intellectual "event." Kuhn's theory of how science worked was subversive, perceived by many as lending legitimacy to a sociopolitical and intellectual critique of science and of the knowledge scientists produced.

Scientific inquiry into any phenomenon begins, according to Kuhn, in a conceptually amorphous state. There are lots of data and "all of the facts that could possibly pertain to the development of a given science are likely to seem equally relevant."[14] Absent a widely agreed-upon theory or overarching conceptual framework within which to explain these data, there is "continual competition between a number of distinct views of nature," each roughly fitting available data but entailing an "incommensurable way of seeing the world and practicing science in it."[15] At some point a consensus crystallizes around one of these theories, emphasizing "only some special part of the too sizable and inchoate pool of information," and becomes a "paradigm." (Fleck's thought style and thought collective, match Kuhn's paradigm and consensus, respectively.)

In the fifth century BCE, for example, the Pythagorean paradigm emerged for explaining centuries of accumulated astronomical observations. The Earth was stationary at the center of the cosmos while the moon, sun, planets, and stars moved around it in circular orbits at uniform speeds. In spite of an acknowledged conflict with observational data, this paradigm was effectively universally adopted by all Western astronomers for the next nineteen hundred years. The canonical articulation of that paradigm for thirteen hundred of those years was the planetary theory of Claudius Ptolemaeus, a second-century Graeco-Roman Egyptian. In one of his books—*The Syntaxis*, though universally known as *The Almagest*, or "great" book, a name given to it by a tenth-century Muslim astronomer—Ptolemy worked out a detailed, purely mathematical model of the Pythagorean paradigm that allowed prediction of each planet's changing position in the night sky, within the limits of accuracy of naked eye observation. This became the standard text and reference work for all Roman, Muslim, and Christian astronomers until the sixteenth century.

In the sixteenth century, Nicolaus Copernicus proposed a new planetary theory, following the *Almagest*'s format, but intended by Copernicus to be physically real (as Ptolemy's theory was not, at least not by Ptolemy). Copernicus' *On the Revolutions of the Celestial Orbs* was published in 1543, the year of his death. In Copernicus' theory, which was based on the same data used by Ptolemy and thus a reinterpretation of that data, the sun is at the center of the universe, and the Earth as well as its moon and the other planets orbit the sun in circular orbits moving at uniform speeds; the stars are all fixed. Fifty years after Copernicus' theory was published, Tycho Brahe proposed a third interpretation of the available data. Brahe's theory matched all the available data at least as well as Copernicus' theory did and required fewer assumptions. Nevertheless, over the next century, as modified by Kepler and then incorporated into Newton's mathematical theory of universal gravitation, Copernicus' theory became a new astronomical paradigm.

Experimental inquiry into static electricity began in the seventeenth century, and Kuhn describes the many theories of what electricity was and how it behaved that were rampant in the eighteenth century. Newly invented instruments for generating, conducting, storing, and discharging static electricity produced a large mass of observational data that resisted unification into a single widely accepted theory. Two classes of theories stood out, one defining electricity as a fluid, and the other as an "effluvium." On the fluid side, there were one fluid and two fluid theories, each attempting to explain attraction and repulsion. The invention at the turn of the nineteenth century of the "voltaic pile" (our battery) by Alessandro Volta based on experiments by Luigi Galvani provided a source of continuous current electricity and new kinds of experiments and applications of electricity. In the process, the one fluid theory matured into the paradigm within which all subsequent electrical research was conducted.

Newtonian mechanics was a paradigm, as was Newton's corpuscular theory of light, succeeded as a paradigm by the wave theory of light in the nineteenth century and that by the quantum theory of light. Lavoisier's oxygen theory of combustion precipitated a paradigm change in chemistry when it merged with Dalton's atomic theory of matter and adoption of the principle of fixed proportions in chemical combinations (reactions). The cell theory of life, the germ theory of disease, the Darwin/Wallace theory of evolution, the gene theory of heredity, Maxwell's electromagnetic wave theory, the relativity and quantum theories, and plate tectonics are all instances of paradigms. These were not simply theories widely judged to be true; they were perceived as defining fertile conceptual frameworks within which all research into relevant phenomena was to be conducted. Kuhn called research conducted within a paradigm "normal science."[16]

Normal science is what the overwhelming majority of scientists do at any given time. The paradigm defines the kind of research that needs to be done in a field: what the problems are, how they are to be formulated, what kinds of experiments or further theory development need to be done, and most importantly, the criteria to be met for "correct" solutions to problems. Through education, the paradigm and its requirements for further articulation are implanted as norms in the minds of aspiring young scientists, who "accept theories on the authority of teacher and text, not because of evidence."[17] (Fleck had said the same thing.) Citing Wittgenstein on "family resemblance" in ordinary language, Kuhn claimed that a paradigm determines normal science "without the intervention of discoverable rules";[18] in Polanyi's terms, tacitly.

In the course of normal science, it is inevitable that "anomalies" arise. Experiments produce data that the paradigm theory cannot explain, or that are inconsistent with predictions made by the theory. Anomalies are not secrets; everyone knows about them. Typically, they are tolerated or can be accommodated by modifying the relevant theory. Against Karl Popper's theory of science, anomalies are not "falsifying experiences." Like Mercury's anomalous orbit they can be tolerated or ignored, or like the apparently non-quantized energies of electrons emitted from the nuclei of radioactive atoms, the theory can be modified, as by Pauli's invention of the neutrino. Sometimes, however, anomalies are perceived as threats to the truth of the paradigm, as with the apparent failure of Newton's theory of gravity to predict the orbits of the moon and Uranus correctly. The theory must parry these threats, as Newton's did, or face replacement.

At some point, a point for which no rule can be given in advance, anomalies cause a growing feeling that a "crisis" exists, a crisis of confidence in the normal science that scientists are working in and thus in the correctness of the paradigm that defined normal science. The crisis is resolved by adopting a new paradigm free of the accumulated anomalies and promising a new, richer normal science: new experiments, new discoveries, new explanations, new theories. But a crisis is not a necessary condition for the emergence of a new paradigm: "Often a new paradigm emerges, at least in embryo, before a crisis has developed far or been explicitly recognized."[19] This would seem to conflict with Kuhn's claim that, crisis notwithstanding, a paradigm is not rejected until a new paradigm is available. The history of science supports paradigm change in the absence of a crisis, nevertheless, when a new paradigm is adopted a new normal science takes shape. Inevitably, new anomalies accumulate, another crisis occurs, and sooner or later a change to yet another paradigm will occur. Paradigm change, however, is not simply a matter of dropping one theory and adopting another. A paradigm is simultaneously a theory and the conceptual framework within which that theory is embedded, requiring the adoption of new definitions, assumptions, principles, and invariants.

Emile Durkheim had argued that the social context within which an individual lives, the prevailing values and relationships, exerts a kind of social "force" on the individual, shaping his/her behavior, typically unconsciously, but sometimes consciously.[20] Paradigms, and Fleck's thought collectives, exert an analogous "force" on the thinking and practice of individual scientists. To challenge a paradigm is to challenge normal science as well, and thus the practice of science by the overwhelming majority of the scientists in that field. Doing so courts marginalization by the community.

Hugh Everett challenged the dominant Copenhagen interpretation of quantum mechanics and paid a heavy personal price. Halton Arp, a respected astronomer with impeccable professional credentials, was marginalized for challenging the Big Bang paradigm and the interpretation of the red shift that plays a central role in it. Arp argued that the red shift of light from galaxies and quasars did not always reflect distance from the Earth. He used precisely the same data that all other astrophysicists used, but he interpreted them against the grain of the prevailing paradigm. To accept his interpretation would require abandoning Big Bang cosmology, and the prevailing consensus was against doing that. The paleoanthropologist Milford Wolpoff was marginalized for challenging the prevailing 'out of Africa' paradigm that Homo sapiens evolved only in East Africa and then migrated out of Africa to the rest of the world. Wolpoff argued that homo erectus migrated out of Africa and that homo sapiens emerged in various locations as types of Homo erectus. (Recent human fossil discoveries are piling up anomalies in the 'out of Africa' paradigm and Wolpoff's theory may yet prevail.) The physicist Thomas Gold made a career out of ignoring prevailing paradigms, and his theories, often later confirmed, were repeatedly derided and ridiculed.

Gold argued a feedback theory of how the ear works that was dismissed for half a century but is now orthodox. His theory of the role of magnetic fields in generating solar flares and of their impact on the Earth via a magnetosphere that encircles the Earth is similarly orthodox today. His prediction in 1951 that certain electromagnetic waves detected by radio telescopes had an extra-galactic origin was derided. So was his prediction that the newly discovered pulsars were rapidly spinning neutron stars whose magnetic fields caused them to emit periodic pulses of electromagnetic energy the way a lighthouse beacon emits pulses of visible light. (His paper on this was refused for presentation at the first conference on pulsars.) Also derided was his prediction, well in advance of Armstrong's first footstep, that the surface of the moon was coated to a depth of several inches in dust.

Because of the "force" they exert on normal scientific practice, paradigms have a tenacious hold on most scientists' thinking. Changing paradigms means changing normal science, which can be career threatening for many scientists. Obviously, paradigms *do* change and theories *do* evolve, but on what grounds? Kuhn claimed that changing paradigms requires a jump from one conceptual framework to another and that these jumps cannot be justified on logical or on empirical grounds alone. Logically, as Duhem and Quine had shown, the prevailing paradigm can always be modified to accommodate new data. Empirically, new data can never define just one new theory to explain them. It follows that the abandonment of one paradigm for another necessarily involves nonlogical, and non-empirical, factors. Because of this, paradigm change is a discontinuous process, but it is not an irrational process: "Because scientists are reasonable men, one or another argument will ultimately persuade many of them" of the value of the new paradigm, but no one argument will persuade everyone.[21] Kuhn cites Max Planck's comment in his autobiography that a new theory sometimes triumphs only when its opponents die out![22]

Kuhn called the "transition to a new paradigm . . . a scientific revolution."[23] Because paradigm change is revolutionary, scientific knowledge does not grow in a linear progress toward one uniquely correct description of nature. The notion that scientific progress is cumulative is "entangled with a dominant epistemology that takes knowledge to be a construction placed directly upon raw sense data by the mind."[24] This epistemology must be discarded, proven wrong by the history of science.

The history of every branch of science is a history of sequential revolutions. Kuhn compared this process to evolutionary biology and to changes of styles in art. In evolutionary biology, new life forms emerge out of a process that has a discontinuous aspect to it: random mutations and natural selection by an environment that is changing independently of the mutation process. There is no progress toward a final, perfect life form. Similarly, there are aesthetic revolutions in the history of art, from Renaissance-style classicism to impressionism to a sequence of post-impressionist styles, including Fauvism, cubism, expressionism, abstract expression, "pure" abstraction, surrealism, pop art, op art, conceptual art, neorealism, etc., but there is no linear progress toward one correct style.

The term "paradigm" is, as critics immediately pointed out, highly ambiguous, used in *SSR* as an accepted model or pattern, a theory, a conceptual framework, and even more broadly as a "coherent tradition for scientific research."[25] Kuhn later gave the word up as a name for his underlying idea. The history of science reveals that the practice of science is best understood as the articulation by a community of scientists of the implications of a cluster of correlated concepts and/or a specific theory to which they are all committed. The object of this work-defining commitment is a "paradigm." Kuhn claimed that the history of science validates a two-tier structure for science: paradigms and what he called "normal science," how science is practiced under a specific paradigm.

The instances mentioned earlier—including the atomic theory of matter, the cell theory of life, evolutionary biology, quantum theory, and plate tectonics—suggest that this is a useful and valuable way of organizing scientific practice, one clearly supported by the history of science. But in the opening pages of *SSR*, Kuhn had described two mutually exclusive methodologies that historians of science employ, two paradigms for the history of science (and the externalist methodology is a third). Kuhn chose one of these and claimed that that one leads to the "correct" image of science, as a practice in which the everyday work of the scientist is determined by an overarching paradigm.

It may be the case, however, that the two-tier structure for science that he finds in the history of science is valid and valuable only in the way that any conventional taxonomy is, the criterion of value being usefulness. It is one thing to claim that science is best understood as normal science clustering under a paradigm the way species cluster under a genus, with everyday scientific activity clustering under normal science as varieties cluster under a species. It is another thing altogether to claim, as Kuhn seems to do, that his two-tier structure for science is real, not merely a useful taxonomy; that paradigms act as a sort of epistemological causal agent, generating normal science as an effect; and that his theory of science is the true one.

So much for paradigms that, however ambiguous the word may be, do make a lot of sense when you look at the history of any scientific discipline.

The emergence of a paradigm in a new field of research may be exaggerated, but it isn't wrong. Kuhn characterized the normal science that paradigms determine in a way that made normal science sound like mid-level, uncreative "mop-up work,"[26] mere "puzzle solving," rarely aiming to "produce major novelty."[27] Understandably, scientists, almost all of whom spend their careers doing normal science, objected. Kuhn said that normal science involved working out the implications of a paradigm without challenging it, but the history of science reveals much paradigm challenging after adoption. And theories typically change dramatically within a paradigm. We have seen that Copernicus' theory bears little resemblance to the articulated Copernican paradigm. The Darwin/Wallace theory bears little resemblance to evolutionary biology today. George Gamow's cosmology of the late 1940s, mockingly dubbed Big Bang, was indebted to earlier speculations by Georges Lemaître and Carl von Weiszacker. It bears very little resemblance to what has evolved into the Big Bang paradigm today after theories of nucleosynthesis, a series of inflation theories, and quantization of the universe's initial appearance.

All of these original theories are recognizable within their evolved paradigms, but barely, and mostly honorifically. Normal science, in short, encompasses a form of paradigm evolution, not just articulation of what is implicit in the original theory whose "triumph" made it a paradigm. The notion of anomalies is at least as vague as paradigms, and it is problematic that there is no rule for when a crisis emerges out of anomalies. So is the absence of a rule for a new paradigm appearing to resolve the crisis; it seems to drop down out of the sky. Still, Kuhn drew five provocative conclusions from his study of the history of science, in addition to revealing that our image of science is a function of how we do history of science.

First, that paradigm change and theory change, along with paradigm and theory formation, are dependent on nonlogical factors in addition to their dependence on logical reasoning and data. That is why new paradigms and new theories seem to drop out of the sky, like a deus ex machina at the end of a Greek play. Kuhn describes their adoption as a "conversion experience" that "cannot be justified by proof."[28]

Second, the discovery process in science is a matter of interpretation, not an "archaeological" revelation of what was already there. Lavoisier did not uncover a previously hidden something that he called "oxygen." He interpreted the same data that Joseph Priestley got from the same experiments Priestley had done, but he interpreted them as showing that combustion was a process of combining with something *in* ordinary air not as releasing something *to* the air.

Third, that because of the nonlogical factors involved—new interpretations resting on new assumptions—paradigm change is discontinuous and thus justifiably called "revolutionary," as opposed to the prevailing "image" of scientific knowledge as developing linearly and cumulatively.

Fourth, because paradigm change is a matter of new interpretations and new assumptions, successive paradigms are mutually incommensurable.

Fifth, because of this incommensurability, when paradigms change, reality changes.

That the move from data to an explanatory theory or paradigm has a nonlogical component to it was hardly original with Kuhn. Robert Hooke already had proclaimed the explanatory worthlessness of "rude heaps of unpolished material." We have seen this same claim over and over again in nineteenth- and twentieth-century philosophies of science, for example, in Duhem, Hertz, and Poincaré. Peirce's "abductive" reasoning, like Whewell's earlier re-definition of "induction," is a name for a nonlogically arrived at idea that has important explanatory power. Abductive reasoning was central to Hanson's *Patterns of Discovery*. Ten years earlier, Michael Polanyi, an eminent physical chemist turned philosopher of science, had argued in *Personal Knowledge* (1958) and in *The Tacit Dimension* (1966) that "tacit knowledge," a kind of nondiscursive knowing how to do something, along with personal judgments and commitments, played an inextricable role in scientific reasoning. Polanyi turned from physical chemistry to philosophy of science because he came to believe that an understanding of the nature of scientific knowledge was crucial for the increasingly important role in public policy decisions that science would be playing after WW II. His first book on this subject was *Science, Faith and Society*, published in 1946.

Kuhn wrote that "scientific fact and theory are not categorically separable except perhaps within a single tradition of normal-scientific practice."[29] This echoes the logical positivists' recognition in the 1930s that theory statements could not be deduced from observation data. Theory-neutral observations are a myth, and theory statements themselves incorporate nonlogical interpretations of observations. Kuhn sees a further consequence for normal science of revolutionary paradigm change: "The normal-scientific tradition that emerges from a scientific revolution is not only incompatible, but often actually incommensurable with that which has gone before."[30]

The claim of the incommensurability of successive paradigms[31] along with the claim that the world changed when paradigms changed were highly controversial among scientists who read, or heard about, *SSR*. Some took Kuhn as claiming that scientists supporting different paradigms literally could not communicate with one another. Yes, the "differences between successive paradigms are both necessary and irreconcilable,"[32] hence incommensurable, insofar as fundamental terms in each are defined differently, or are added or dropped. In order to communicate, each side would need to keep that clearly in mind in order to converse meaningfully about their respective theories/paradigms. In relativity theory, for example, space, time, matter, motion, and energy are redefined vis-à-vis Newtonian mechanics. A conversation between Einstein and Newton in which each used these terms only as they themselves defined them would find them talking past one another, as Einstein and Bohr did for decades in their arguments about the completeness of quantum mechanics.

Furthermore, because there is an ontology correlated with Newtonian mechanics and a different ontology correlated with relativistic mechanics, what Newton *meant* by reality is no longer real in relativity theory. What is real in Maxwellian electromagnetic wave theory is different from what is real in its successor theory, quantum electrodynamics. Although science refers to, and typically claims to describe, objects independent of the mind, what it actually does is describe these objects as scientists conclude that they are relative to a particular paradigm or theory. At the turn of the nineteenth century, geologists described the Earth as at or approaching equilibrium, and continents were fixed since they originally formed. In the 1960s, geologists described the Earth as far from equilibrium, kept there by heat generated by radioactive decay and a core whose temperature was about the same as the temperature of the surface of our sun. This heat generates convection currents that force molten mantle material up through ocean floor vents, keeping the continents in constant motion.

What the Earth is *for science*, what the scientific reality called "the Earth" is—and this is the only reality we can *know*—indeed did change when the plate tectonics paradigm emerged. What is "out there" independent of the mind and of our reasoning about it is only describable under contingent assumptions. (Recall Goodman.) Kuhn was taken by some to have claimed that what was "out there" changed when paradigms changed, as if the Earth had been stationary before Copernicus and only began moving when scientists agreed it did, and that the continents did not move until plate tectonics became the latest geological paradigm. This is an abuse of what Kuhn wrote: paradigm change causes scientists to "see the world of their research engagement differently . . . the world does not change with change of paradigm, [but] the scientist afterwards works in a different world."[33] About this he was quite right.

SSR marshaled a powerful argument in support of the Giants–sophist view of knowledge and of being against the view of the Gods–rationalists. Whether intended or not, *SSR* was read by critics and admirers alike as undermining the objectivity and even the rationality of scientific knowledge. For Kuhn's critics, this was a heresy to be denounced and combated. For many admirers, it was like adding an atomic bomb to their arsenal of weapons for denouncing the privileged status accorded science and its knowledge claims. Arguably, *SSR* was the trigger of what soon became the Science Wars. Kuhn himself, however, straddled both sides of the knowledge problem in science. He remained committed both to a realist, objective, and progressive view of scientific knowledge *and* to the ideas he developed in *SSR* that many took as contradicting that view.

14

The Opening Phase of the Science Wars

In Wallace Stevens' poem cited earlier, the audience asks the man with a "blue" guitar to play for them "things as they are." The man with the blue guitar explains that he cannot play things as they are because his guitar is a blue guitar, it can only play things-as-played-on-a-blue-guitar. The audience insists, 'No, we want you to play things "exactly as they are" *in spite of* playing them on a blue guitar.' The guitarist explains again; the audience persists. The audience would have persisted even if the guitarist added Jacobson's line that no one can play things as they are because all of our instruments 'lack the strings to pluck the tunes we hear.'

This captures an important truth about modern science. Scientists *know* that each of them is playing a blue guitar, that the mind with which they reason is (metaphorically) a blue guitar. At the same time, they write and talk as if they *believed* that in reasoning scientifically the mind was playing things "exactly" as they are, through some combination of method, experiment, and logic. In *SSR*, Kuhn challenged the grounds of that belief. Theory acceptance and replacement cannot be explained solely in terms of method, experiment, and logical reasoning. This calls into question the objectivity of scientific knowledge along with the prevailing conception of the rationality of science and its convergence onto the truth about reality. For many in the philosophical as well as the scientific communities, that was too much to give up. One of the earliest responses to *The Structure of Scientific Revolutions* (*SSR*; other than book reviews) was by the Harvard philosopher Israel Scheffler. His *Science and Subjectivity* (1967) was an impassioned defense of the objectivity of scientific knowledge against what he saw as an emerging and dangerous subjectivist critique of it.

Scheffler defended the image of science that Kuhn aimed at replacing. Against Kuhn, Polanyi, and Hanson, Scheffler argued that science was a "systematic public enterprise, controlled by logic and by empirical fact, whose purpose is to formulate the truth about the natural world."[1] More importantly for society, scientific reasoning and the accumulating body of scientific knowledge embody a moral principle that is a bulwark against authoritarianism, coercion of belief, and tribalism. That principle is the control of the conclusions reached by "objective" reasoning, reasoning that employs impersonal, independent, and communal criteria of reasoning correctly. The alternative to objective reason is some version of 'might makes right' and the collapse of hope for the rational conduct of human affairs. That scientific knowledge rested on subjective and intersubjective conventions to which logic and facts are subordinated, as Polanyi, Hanson, and Kuhn argued, has destructive moral, social, and political consequences.

In defense of the image of science that Kuhn and others wanted to replace, Scheffler defended a realist theory of scientific knowledge and a correspondence, as opposed to a coherentist, conception of truth. He sided with an updated version of Schlick's original version of logical positivism, linking theories to reality, against Neurath's revised version. According to Neurath, scientific knowledge is about experience, and truth is a function of the logical coherence of statements about experience with one another, not a function of a correspondence between scientific propositions/theories and the world "out there." Scheffler adopted the form of the correspondence theory of truth proposed by the Polish-born logician Alfred Tarski. According to Tarski, the sentence "Snow is white" is true if and only if snow really is white. Scheffler cites Quine's version of this: to say that the sentence "Snow is white is true" is to *say* that snow really is white. This implies that science is inescapably ontological in its claims, in what it says, not merely epistemological: "There is no way of staying within the circle of [scientific] statements [as in a coherence criterion of truth], for in the very process of deciding which of these [statements] to confirm, we are deciding how to refer to, and describe things, quite generally."[2]

It follows that the "import" of scientific statements is "inexorably referential." It is intrinsic to scientific knowledge claims that they refer to a reality beyond experience. The problem is how to test the correctness of these claims. Scheffler acknowledges that method alone cannot do the job and the test cannot be a direct comparison with facts, because there is no denying the arguments of philosophers of science that facts "are generally suspect." Nevertheless, he claimed, theory acceptance and replacement is an objective process. Scientific theories are necessarily referential in their *import*, that is, they necessarily *refer* to entities and their relationships that are said to be "out there" beyond the mind. The choice of one theory over another rests "on the referential values which statements have for us at a given time, that is, on the inclinations we have, at that time, to affirm these statements as true."[3] This is consistent with Tarski's definition of truth.

However passionate and well-intentioned, it is hard not to see that at the end of *Science and Subjectivity*, Scheffler arrived at a position very close to Kuhn's on the objectivity of scientific knowledge. If theory choice rests on "referential values" and "inclinations . . . to affirm . . . statements as true" *at a given time*, then asserting referential claims as simply true must be understood as a rhetorical ploy. Epistemologically, it is very hard to see how time and context-dependent "values" and "inclinations" can be other than subjective, yet for Scheffler, they are central to a theory's references to reality, and thus to its truth.

Paul Feyerabend's response to *SSR* was the polar opposite of Scheffler's. For Scheffler, the problem with *SSR* was that Kuhn went too far in claiming that the history of science validated a new image for science, one that challenged the objectivity of scientific knowledge. For Feyerabend, the problem with Kuhn was that *SSR* did not go far enough. Scheffler's response was to protect the rationality of science and the objectivity of scientific reasoning. Feyerabend's response was to debunk rationality altogether as the controlling feature of scientific reasoning, putting in its place what in *Against Method* (1975) he called "epistemological anarchism."[4]

Feyerabend's position in *Against Method* was that scientific method was a myth. Not only was there was no such thing as one, correct method for reasoning scientifically—we have already seen this in the diverse methods proposed by Bacon, Galileo, Descartes, and Newton—but also there was nothing methodical whatsoever about reasoning from empirical evidence to theory. The process of discovery is logically anarchic. No method, no set of rules, nor rational standards for reasoning about experience could possibly result in the richness and the complexity of scientific practice and of scientific theories. Unlike Kuhn, Feyerabend saw the pre-paradigm period as a source of richness, a methodological "Wild West" in which anything goes so long as it solves a problem and/or results in an experimentally confirmed theory.[5] The more new theories proposed, the better, because the more empirical content covered, the more different ideas, and the more theories to test against empirical evidence and against one another. The important point is to take all theories seriously, to give them all a respectful hearing however improbable they may seem. Even a rejected theory can suggest new ideas, challenge old ideas, and be useful as a means of testing rival theories.

It is from such methodological "anarchy" and "theoretical pluralism" that science progresses. But progress here does not mean getting closer to a correct account of reality. Feyerabend had started out as an empirical realist, influenced by Karl Popper, but having abandoned that position, he moved to something like Mach's phenomenalism. Scientific theories are systematic responses to experience and that's all. Abandoning ontological claims has, for Feyerabend, profound implications for the social status of science and for the nature of scientific knowledge claims. Science has no right to a privileged position with respect to knowledge and truth. This is very strong talk for a philosopher of science, and Feyerabend went even further in his next book, *Science in a Free Society* (1978). Reflecting the antiestablishment mood of the 1960s and '70s and reflecting also his involvement with the Berkeley Free Speech movement, Feyerabend warned that science had become a threat to democracy and thus to society. To protect both, society needed to assert democratic control over science and scientific experts in order to allow an equal voice to all traditions. In *Farewell to Reason* (1987), he argued for epistemological relativism as opposed to singling out one theory as alone true, or one form of thought, or one social tradition and its values as superior to all others. Having dispatched method in *Against Method*, in *Farewell to Reason* he dispatched reason and rationality as ambiguous ideas only, names for rhetorical devices wielded by those in power.

In 1965, at an international conference on the philosophy of science held in London, Feyerabend met and became friends with Imre Lakatos, an émigré Hungarian physicist-philosopher. Lakatos' response to *SSR* was like Scheffler's, not like Feyerabend's. Nevertheless, Lakatos and Feyerabend planned a joint book: half Feyerabend's anti-rational theory of science, and half Lakatos' defense of science as rational and of scientific knowledge as objective. Lakatos was killed in an automobile accident in 1974, so the book that appeared, *Against Method*, was all Feyerabend. Before the accident, however, Lakatos had worked out his own theory of science, especially in *The Methodology of Scientific Research Programs*.

Lakatos received a Ph.D. in physics from the University of Cambridge, but he also studied with Karl Popper at the London School of Economics, as Feyerabend too had done some years earlier. Like Feyerabend, Lakatos rejected Popper's philosophy of science, but he remained loyal to science as producing knowledge of reality, to the privileged position of rationality in human thought, and to the rationality of science. Lakatos agreed with Kuhn that the history of science played a key role in understanding how science works and in explaining the growth of scientific knowledge. His use of the term "growth" in the title of the paper that he read at the London conference— "Falsification and the Growth of Scientific Knowledge"[6]—is telling. For Kuhn (and for Feyerabend) theories change, but they do not grow in the sense of growing *toward* anything. Kuhn's analogy was between theory change and biological evolution: individuals and populations change but not in a way that allows ranking life forms, synchronically or diachronically, as inferior or superior relative to one another, only relative to their changing environments. For Lakatos, on the other hand, scientific knowledge does grow: it grows toward better accounts of reality.

The history of science was important for Lakatos, but in a way different than for Kuhn. Whewell was the first philosopher of science to argue that the history of science was the key to the philosophy of science. According to Whewell, laying out the chronology of a science revealed that it was Fundamental Ideas that allowed scientific knowledge to be organized deductively. Duhem, too, attributed an important role for the history of science in understanding the nature and the growth of scientific knowledge. Unlike Whewell, however, for whom the history of science served an epistemological function only, Duhem saw the history of science as revealing a trajectory of sequential ideas aimed at a final, true account of reality. For Kuhn, the history of science studied in the "right" way revealed the need for a major revision in our image of how science produces knowledge.

These are three very different conceptions of what the history of science has to teach us about science, and Lakatos added a fourth. Unlike Kuhn, Lakatos had no problem with a Whig version of the history of modern science: a selective chronology that threads its way from Galileo to the present by way of just those theories out of which present theories grew. Lakatos had applied his approach to history to mathematics in *Proofs and Refutations: The Logic of Mathematical Discovery* (1976). The goal was not to discover the foundations of mathematics, but to expose the history of proposing foundations in order to understand how mathematical knowledge grew. With respect to understanding how scientific knowledge grew, the historical selection process in Lakatos' hands becomes a test of the correctness of a new theory, a test for which the history of science serves as a database.

Can a new theory successfully retrodict the explanatory content of the theory that preceded it and that had been judged correct in its time? Bohr called this his 'correspondence principle': Maxwell's theory, for example, can be retrieved from quantum theory as a limiting case, and Newtonian mechanics can be retrieved from relativistic mechanics as a limiting case, when velocities are low relative to the speed of light and masses are relatively small. In a way that would be scandalous for a historian, Lakatos called for what he called "rational reconstruction" of earlier theories against which a new theory is tested. Ignore the historical record and the actual chronology of events. Reconstruct just the logic of the earlier theory as if it had been arrived at logically. The new theory is to be tested against this rationally reconstructed version of the theory it is meant to replace. To pass, it must subsume the earlier theory.

As presented in *The Methodology of Scientific Research Programs*, science for Lakatos is organized around research programs, not Kuhnian paradigms, though in the end the two seem similar. Research programs are made up of a "hard core" of assumptions and theories that researchers are strongly committed to and unwilling to give up, though core theories can evolve. Newtonian mechanics, for example, evolved dramatically in the course of the eighteenth and nineteenth centuries in terms of its mathematical sophistication and the scope of its problem-solving abilities, but it was still Newtonian mechanics. Surrounding the hard core is a changing "protective belt" of subordinate theories and assumptions that are inspired by the core and change readily in response to new data, new ideas, and alternative theories cognate with the core. (This sounds very much like normal science.)

Science grows when research programs change. A research program can be "progressive" or "degenerative." A progressive research program is one whose core generates testable theories and novel predictions. These keep researchers stimulated to pursue experimental confirmation or disconfirmation of these predictions, and to formulate new theories and new explanations for new phenomena or new data. (This sounds like a version of a vibrant Kuhnian normal science.) A degenerative research program fails to do this. Science grows when a degenerative research program is replaced by a progressive research program. That the new program is, according to Lakatos, *objectively* better than the one it replaces restores the rationality to science that Kuhn had taken away. The history of science is a history of successively superior research programs, a series of "revolutions" if you like, but now each new program is objectively better than its predecessor in explaining the world "out there" by means of reasoning about empirical experience.

But *are* there objective criteria for determining that a previously fertile research program has become terminally degenerative and that a proposed new research program will in time prove to be progressive? Lakatos' philosophy of science stands or falls on his claim that he has provided a logical cum methodological account of theory change in place of Kuhn's account in which logical reasoning and methodology are not *enough* to explain scientific theory change and acceptance. Lakatos did not identify objective rules for identifying research programs as degenerative or progressive, except retroactively. And note that Kuhn did not claim either that the scientific enterprise and scientific reasoning were irrational, or that methodology was irrelevant to theorizing, or even that scientific knowledge was subjective rather than objective. His claim was that the history of science reveals that what we *mean* by the objectivity of scientific knowledge and by the rationality of scientific reasoning differs from the prevailing notions of objectivity and rationality. Given that clarification, was Lakatos any more successful than Scheffler in refuting Kuhn's claim that nonlogical and non-methodological factors play an important role in the production and growth of scientific knowledge? Not if there are no objective rules for choosing between competing research programs in real time, that is, in the context of a career choice that must be made *now* by an individual scientist.

Before moving on, more needs to be said about Karl Popper. Popper's reputation is that of one of the most important philosophers of science of the twentieth century. His philosophy of science—initially presented in *The Logic of Scientific Discovery* (1934; English translation, 1959)—is the one best known to, and most strongly approved of, by working scientists. After World War II, Popper relocated from New Zealand, where he had gone before the German annexation of his native Austria, to England. At the University of London and then the London School of Economics he became an influential teacher and a prominent public intellectual. Feyerabend and Lakatos were only two of the gifted thinkers who came to study with him, formally or informally; Joseph Agassi, who remained loyal to Popper, objectivity, and rationality, was another.

As a young man, Popper had been an active Marxist, but he abandoned Marxism and became a severe critic of its underlying theory. He studied psychoanalysis for several years and later became a severe critic of Freud's theories. Through the Vienna Circle at the University of Vienna, he became very familiar with, and a severe critic of, logical positivism as well as of empiricism generally. He disapproved of the Copenhagen interpretation of quantum mechanics and of both the early and the later Wittgenstein. What he approved of was his own theory of science, presented in *The Logic of Scientific Discovery* and then refined in *Conjectures and Refutations* (1963) and *Objective Knowledge* (1972). Popper began by addressing the question, "What is science?" What is it that demarcates scientific ideas and theories from nonscientific and pseudoscientific ideas and theories? Popper's answer was the possibility of falsification. If a theory does not make predictions that can be tested against experience, typically by experiment, then it is not scientific. But there is more to the falsifiability criterion than this.

According to Popper, not only must a scientific theory be falsifiable by experiment; it also must deliberately open itself to this possibility. A theory is scientific if it identifies the tests that will prove it to be false or—recalling the Fallacy of Affirming the Consequent—show only that it *might* be true; a *scientist* is one who is prepared to accept a false outcome and move on without inventing ad hoc modifications in order to rescue a theory from falsification. On these grounds, Popper denounced Marxist "scientific materialism" and Freudian psychoanalysis, among other claimants to the rubric "science," as in fact pseudoscience. They did not open themselves to falsification, and their adherents were dogmatically committed to the respective theories, willing to explain away any adverse political events or clinical outcomes. (He accepted Darwin's theory of evolution as scientific even though it seems to fail the falsifiability test!)

The mismatch between Popper's theory and the actual practice of science, however, as well its mismatch with the entire history of modern science, is glaring. First of all, logic notwithstanding, scientists routinely treat positive experimental outcomes as confirming the truth of a theory, not just its possible truth. Recall the citation from Huygens' *Treatise on Light*, in which he acknowledges the Fallacy of Affirming the Consequent but goes on to say that when an experimental outcome matches a theory's prediction, it almost compels us to conclude that the theory is true. Second, scientists routinely respond to experimental falsification, not by abandoning a theory but by modifying it in an attempt to rescue it. Recall the Bohr and Pauli modifications proposed to rescue quantum mechanics from experimental evidence that the energies of electrons expelled from the nucleus during radioactive decay were not quantized.

Another rescue strategy that scientists use to dismiss experimental falsification is to challenge the validity of the disconfirming experiment, as with the claim that faster-than-light neutrinos had been detected, contrary to the special theory of relativity. Or, they may attribute a failure to confirm to limitations of the experimental apparatus, as in the failure of the original Laser Interferometry Gravitational Observatory (LIGO) to detect gravitational waves as predicted by the general theory of relativity. Lakatos argued against Popper that confirmation of the predictions by Leverrier and Adams of a planet beyond Uranus was *not* an example of a crucial experiment. Had the astronomer Galle *not* seen the planet where Newton's theory predicted it would be, that could be attributed to the telescope's limitations or to Leverrier's calculations; it would not have led to abandoning Newton's theory. When it was shown in 1919 that the general theory of relativity had as a logical consequence an expanding universe—ten years before that was announced by the astronomer Edwin Hubble—Einstein added a term to his gravitational equation, the cosmological constant, to eliminate that consequence/prediction.

Popper's falsification criterion is also made problematic by the Duhem–Quine argument that no negative outcome can decisively disprove a theory because a theory always has multiple components to it. A negative outcome only entails that at least one of these is false. It would be irrational to abandon a theory in the face of disconfirming evidence without first trying to rescue it by modifying one or more of the components and testing it again, and perhaps again.

Popper was aware of these problems with falsification and the gulf between his theory and actual scientific practice. In *Conjectures and Refutations* (1963) he attempted to deal with them, but the criticism only intensified. Popper was not an empiricist in the traditional sense, in which knowledge about nature begins with uninterpreted "facts." Like almost every twentieth-century philosopher of science, Popper agreed that there were no uninterpreted facts of experience, but he was also a realist about science. Theories were attempts at a true account of the world "out there." In *Conjectures*, he adopted Tarski's correspondence theory of truth with its ontological implications as noted by Quine, although he had dismissed it in *The Logic of Scientific Discovery*. Science begins with problems to be solved, not with empirical observations. Thinking about how to solve these problems leads to bold conjectures that make novel predictions, and to pursuing experiments that would refute those conjectures. Popper backed off his earlier simple model of falsification, acknowledging that experimental refutation of a theory is complex, involving "free decisions" to accept selected "basic statements" as relevant to falsification. He also acknowledged that scientific theories are always embedded in a "background" of selective assumptions. As with Scheffler's "values" and "inclinations," these concessions introduce nonlogical, subjective elements to scientific reasoning.

To accommodate the empirical fact that scientists *do* interpret successful prediction as confirmation of a theory's truth, Popper allowed predictive success to "corroborate" a theory. The more tests a theory passed, the more widely the tests were replicated, the more reasonable it was to conclude that the theory was true. (Huygens had said exactly the same thing.) Popper adapted to science a Renaissance approach to painting and to writing history, "verisimilitude." Popper's version of this was that scientific theories strove for truth-*likeness*, not to be literally true to nature. The process of repeated conjecture and refutation, he claimed, increased science's verisimilitude as theorizing is subjected to something analogous to natural selection in biological evolution. Scientific theories get better and better in the sense of becoming better adapted to the whole of our ever-widening and ever-changing experience, but there is no criterion for claiming that they come closer and closer to the way things really are. Theory replacement is said to be objective because we can measure the balance of truth content and falsity content of successive theories, but again no criterion is offered for measuring ontological truth content, only epistemological criteria. No criterion is given for measuring this balance, and the mismatch of Popper's theory with the history and practice of science seems irreparable.

Meanwhile, as Feyerabend saw, *SSR* straddled a divide, equivocating between a qualified version of the traditional image of science as producing knowledge that was objective, realistic, and progressive, and an outright rejection of that image. In its wake, many philosophers, natural scientists, social scientists, historians, and sociologists came down on one side of that divide *or* the other. Either science deserved a privileged status epistemologically, and, because of that a privileged status socially, as sole source of truths about the world, or it did not. As Scheffler observed, the stakes were high in this choice, and he must have felt that his fears for science and the rational conduct of human affairs were fully justified by Feyerabend's books.

Most philosophers and natural scientists are strongly opposed to the relativism that Feyerabend came to espouse: the claim that there are no neutral/objective standards for evaluating rival judgments of *any* kind as right or wrong, true or false, or better or worse (though pragmatic criteria or persuasive arguments can be made). Strong relativism makes science just another spinner of tales on a par with religion and literature. Not surprisingly, then, a community of philosophers of science emerged devoted to defending the objectivity, rationality, and progressive character of scientific knowledge. Concurrently, a rival community of philosophers, historians, and social scientists emerged, devoted to stronger criticisms of science, objectivity, rationality, and progress than they found in *SSR*. It was the intensifying rivalry between these two communities that in the 1980s erupted as the "Science Wars," an externalization of the Gods–Giants conflict that had been internal to modern science from its beginning.

Overlapping the publication of *SSR* and the response to it, a critique centered in France grew of the modernist conceptions of objectivity, rationality, knowledge, and truth. The leading figures in this movement included Michel Foucault, Jacques Derrida, Gilles de Leuze, Jean Baudrillard and Jean Lyotard, Luce Iriggaray and Julia Kristeva, Claude Levi-Strauss, Pierre Bourdieu, and Jacques Lacan. The revaluations of rationality proposed by these people achieved even greater "virality" and generated greater controversy, in European intellectual circles, and later worldwide, than *SSR* did among Anglo-American philosophers and scientists.

On one ground or another, all of the thinkers listed above rejected the claim that discursive reason, logically employed, can produce objective—in the sense of depersonalized, context-neutral—knowledge of a mind-independent reality. What we call mind is inescapably a human faculty, and none of its operations can escape the implications of the mind's historical and cultural embodiment. Truly, the mind is a "blue guitar." It follows that the entire Enlightenment project—inspired by the rise of modern science—of basing progress in human well-being on reason, is vacated. Applied reason—technology and from the mid-nineteenth-century, techno-science—may have made us more prosperous and more comfortable in the physical world, but it not made us better people. And it has exacted a high price, including the internalization of technocratic values and control by technocratic institutions.

Modernity, conceived as the period in human history marked by the apotheosis of reason, is thus a delusion. Jean Lyotard's most popular book, though it did not express his own philosophy,[7] was *The Postmodern Condition* (1979; 1984 in English), and "postmodern" stuck as a name for a broad, amorphous movement, however inaccurately it came to be applied to individuals, eventually coming to be used derisively. As Lyotard saw it, the distinguishing characteristic of the postmodern condition is the collapse of belief in "metanarratives": overarching conceptual and valuational foundations for the specific narratives that validate our lives. Marxism is a metanarrative, for example, as is Christianity, and so is the Enlightenment's promotion of reason-rationality-truth to which the enterprise of modern science was central as the route to secular salvation. Science uses this metanarrative as the foundation on which to rest its claim to being the sole source of truth in society. Loss of belief in in this metanarrative leads to loss of belief in a universal, objective foundation for reason, knowledge, and values. Without such a foundation, knowledge, including scientific knowledge, collapses into belief and opinion. (In *Simulacra and Simulation* Baudrillard explored the implications of this, recognizing that what we call simulations, models, and theories are created by us in the absence of a truth or reality with which they can be compared.)

Kuhn began *SSR* by claiming that the history of science would subvert the prevailing image of science by which we were possessed, *if* the history of science were practiced as he proposed it should be. Foucault claimed that the history of ideas as *he* practiced it undermined the image of *modernity* by which we are possessed. His first book, based on his doctoral dissertation, was titled *Madness and Civilization: A History of Insanity in the Age of Reason* (1961). It was followed by *The Birth of the Clinic* (1967), *Discipline and Punish* (1975), and *The Will to Knowledge* (1976, published in America as *The History of Sexuality*). In each of these books, Foucault used history to expose how truth is produced, not discovered, in the course of changing conceptions of insanity, healthcare, incarceration, and sexual norms. These changing conceptions reflect changing power relationships in society. History for Foucault reveals the pervasiveness of relational power structures in society, how they underlie thought, judgment, action, and institutions. As the structures change, so do ideas and values. Everyone manipulates and is manipulated, consciously and unconsciously, and their behavior always reflects prevailing power relationships.

As a relational structure, power is a social phenomenon analogous to language. Power relations typically are internalized by individuals who are unconscious of the presence of society *within* their self-consciousness and of the contingency and historicity of these relations, as in Marx's notion of ideology. The relationships are accepted and responded to as if they were objective facts of the matter. The upshot of all this is that the empirical no less than the social world is the result of forms of social manipulation, not the product of independent-of-us realities represented more or less accurately in our minds.

In parallel with his historical studies of particular ideas, Foucault published two more general, theoretical works that were very influential: *The Order of Things: An Archaeology of the Human Sciences* (1966) and *The Archaeology of Knowledge* (1969). In these books, he deployed methods that he called "archaeological" and "genealogical" to "excavate" from history the previously hidden factors underlying the production of the concepts, the values, and the forms of knowledge that were operative at different times. Once excavated, these were revealed to be unconscious and contingent, not at all objectively given in experience, and not at all necessary or universal.

In *The Order of Things* he applied his archaeological method to changing forms of knowledge, "epistemes," and to the tacit rules for acquiring what was accepted as knowledge that are operative in a given social setting at a particular time. Foucault exposed the fundamental role played in our thinking by ordering rules, unconscious classification schemes:

> On the one hand, the history of science traces the progress of [scientific] discovery . . . it describes the processes and products of the scientific consciousness. But, on the other hand, it tries to restore what eluded that consciousness, the influences that affected it, the implicit philosophies that were subjacent to it . . . it describes the unconscious of science.

But this unconscious

> is always the negative of science . . . What I would like to do . . . is to reveal a *positive unconscious* of knowledge: a level that eludes the consciousness of the scientist and yet is part of scientific discourse, instead of disputing its validity and seeking to diminish its scientific nature.[8]

Foucault's analysis exposes the activity of Idols in the mind that, unlike Bacon's Idols, are illustrative of the "negative" unconscious of science, altogether elude the consciousness of the scientist and so cannot be overcome, certainly not by a formulaic method.

In *The Archaeology of Knowledge*, Foucault focused on discourse, the vehicle of discursive reasoning. The book is a theory of discourse and of the embeddedness of power relationships within discourse. The title of this book is somewhat misleading. Foucault's subject is power relationships, not forms of knowledge, and he slips from his earlier "archaeological" method to a "genealogical" method adapted from his close study of Nietzsche:

> It was Nietzsche who specified the power relation as the general focus . . . of philosophical discourse . . . but he managed to think power without confining himself within a political theory to do so . . . If I wanted to be pretentious, I would give the "genealogy of morals" as a general title of what I am doing.[9]

The genealogical method, too, is profoundly subversive of concepts and values.

What Foucault saw in Nietzsche was an insight that "genealogy" was not a search for origins in order to discover what things really are, as one might use a dictionary to discover the "real" meaning of a word by tracing its etymology. Instead, a genealogical method can disclose that there is no such thing as what things really are, as there is no dictionary to tell us what words "really" mean, only how they are used. Things are constituted by how we speak/think them. For example,

> genealogical analysis shows that the concept of liberty is an 'invention of the ruling classes' and not fundamental to man's nature or at the root of his attachment to being and truth. What is found at the historical beginning of things is not the inviolable identity of their origin; it is the dissension of other things. It is disparity.[10]

("Disparity" will recur in Derrida's "difference.")

It is in exposing our *changing* conceptualizations of social and empirical experience as contingently changing structures of differences that Foucault's work echoes themes in Ferdinand de Saussure's linguistics and Claude Levi-Strauss' cultural anthropology. Universal foundations for our thinking about experience do not exist. Thinking and action are constrained by contingent internalized relational structures, as Levi-Straus described in *The Elementary Structures of Kinship Relations* (1949), *Triste Tropiques* (1955), and *The Savage Mind* (1962). Especially in the latter work Levi-Strauss contrasted the so-called savage mind against the modern Western mind to the great disadvantage of the latter, largely because of its denial of the role played by social relations in determining our identity and our consciousness. So, too, the very idea of a correspondence between our thinking and an independently existing reality is simply that: an idea. And it's a bad idea because it leads us to believe that we can think outside our thinking, that we can escape who and what we are as thinkers by some combination of logical and methodological reasoning. This is subversive of modernity's apotheosis of reason and even of Kuhn's divide-straddling image of science, let alone the image he claimed to have subverted, but Jacques Derrida pushed the subversion further.

Derrida was as much a phenomenon as a philosopher. He erupted onto the French intellectual scene in 1967 with the publication of three books: *Writing and Difference*, *Voice and Phenomenon* (English title, *Speech and Phenomenon*, 1973), and especially *Of Grammatology* (English, 1976). The first of these contained a lecture he had given the previous year at Johns Hopkins University at an international conference on structuralism, "Structure, Symmetry and Play in the Discourse of the Human Sciences." Derrida's paper called the premises of structuralism into such sharp question that when the conference proceedings were published the title of the volume was changed to reflect his critique. *Writing and Difference* also contained an essay, "Cogito and the History of Madness," sharply critical of Foucault's book on that subject—and Foucault's understanding of Descartes—as a result of which, even as their ideas swept through European "high" culture, their personal relationship became hostile. (A reconciliation was effected in 1981. It has been suggested that Foucault wrote *The Order of Things* and *The Archaeology of Knowledge* to refute Derrida's criticism of his earlier studies by providing a theoretical and methodological context for them.)

Hostile personal relationship or not, Foucault's and Derrida's ideas overlap and supplement one another. Their combined impact amounted to a devastating critique of the dominant forms of Western thinking in philosophy, natural and social science, literary theory, art, and architecture. Though he was a professor of philosophy, Derrida often referred to himself as a historian, but not in any way recognizable to "professional" historians. Derrida was a historian in the same sense that Foucault was. Derrida used a historical/genealogical method similar to Foucault's as a means of exposing hidden assumptions and value judgments as well as the inescapable uncertainty, ambiguity, and open-endedness of all discourse, spoken or written, and thus also of all reasoning.

Looking back at the history of Western philosophy since the pre-Socratics, Derrida exposed the role of an uncritically accepted metaphysical presupposition: organizing philosophical thinking around mutually exclusive binary pairs. Mainstream Western philosophy has simply incorporated as if they were facts the superiority of reality to appearance, justice to injustice, true to false, good to evil, changeless being to endlessly changing becoming, substance metaphysics to process metaphysics, philosophers to sophists, Gods to Giants. One expression of this implicit ordering in philosophical thinking is the Platonic-Aristotelian dismissal of the sophists generally, and of Heraclitus, Protagoras, and Gorgias specifically, from the community of "real" philosophers. But one can step back and ask, what are the grounds for these rankings? Why binary oppositions? Why oppositions? What happens if we invert the order, or treat the pairs as mutually implicating?

Derrida is associated most closely with the term "deconstruction." He introduced this term in *Of Grammatology*, a work that inverted the traditional subordination of writing to speech. "Deconstruction" is a name for how Derrida goes about excavating what is hidden in discourse. It is not analysis, because "analysis" connotes reducing something complex to preexisting simples out of which it is composed, as a molecule can be analyzed into its component atoms. In fact, deconstruction is not a method at all because it is not formulaic: there is no set of rules to follow to produce its results. It does not seem philosophical because philosophizing in the Western tradition is preeminently a matter of offering arguments for claims made and deconstruction does not offer arguments.

So much for what deconstruction is not, but what is it? Like Kuhn with "paradigm," Derrida came to regret having introduced the term, especially after it became, for some twenty years, ubiquitous, almost always used in ways that Derrida considered a misuse. For a time, Derrida tried to clarify what deconstruction was, but then he gave up trying because the lesson of deconstruction is that no word has a correct meaning. In the end, deconstruction is just what Derrida does in his reading of texts; it is Derrida's name for the work he does in uncovering in the text what "eludes consciousness." In reading a text, Derrida employs an extended form of Ferdinand de Saussure's theory of language as a system of arbitrary (written or spoken) signs.[11]

According to de Saussure, a language is a system of signs in which the linguistic "value" of the signs is wholly a matter of their relations to other signs *within* that language system. Linguistic value is wholly *internal* to the language system. It is the differences between one sign and all others in that language that determines its value: the value of "cat" is determined in the first instance by its differences from "hat," "bat," "rat," "sat," etc. Second, the value of "cat" (we have not yet come to its meaning) is determined by what de Saussure called its "syntagmatic" relations to other signs: which other signs are permitted to precede or succeed it in writing or speech. Linguistic signs are signifiers; they point to a signified, but the relation between signifier and signified is, like the signs themselves, wholly arbitrary and internal to the language. There is no intrinsic connection whatsoever between "cat" and anything outside the language in which that sound exists. (There is a similarity between linguistic difference here and the "disparity" between concepts and even forms of knowledge in Foucault, as noted earlier.)

The *meaning* of a sign such as "cat," what it is that the sign signifies, is a matter of associative relations that are not linguistic and thus not within de Saussure's purview as a linguist, other than to note that meaning is use context dependent. But where de Saussure (or at least the posthumously published student lecture notes titled *Course of General Linguistics* to which his name is attached as author) leaves off, Derrida begins. Associative relations are completely open-ended. There is no way, within written or spoken discourse, either for an author or a reader to delimit the associative relations of words (including "deconstruction") and thus fix the meaning of a text. We do not know, and cannot be conscious of, the influences on why we use the words we do when we write or speak. Words, phrases, similes, metaphors, metonymies, synecdoches, neologisms, irony, sarcasm, and hyperbole make reading, or listening, inescapably interpretative. Neither the author nor the reader can say *this* is what a given text means and nothing else. And Derrida reminds us that interpretation is necessarily open-ended, not merely pluralistic. It is not possible to specify in advance which interpretations are to be judged valid and which not.

De Saussure may have believed that context delimits the meaning of a word, but deconstruction reveals that context vastly *amplifies* possibilities of meaning. Strict constructionists claim that the meaning of the U.S. Constitution, or of laws, is determined by the intent of the authors, but for deconstruction conscious intent, even if it can be determined, cannot limit what words mean. In Plato's dialogue *Phaedrus*, Socrates explains that he prefers speech to writing because in speaking he can clarify misunderstandings by active verbal exchanges with individual interlocutors.[12] The written text, on the other hand, is static, and without clarification appropriate to each reader misunderstanding is guaranteed.

For Derrida, Socrates was just wrong about this. Deconstruction reveals that speaking, with or without active exchanges between speakers and listeners, is no less vulnerable to open-ended interpretation than writing, even in the presence of the author of a text. In the end, each reading of a text is a misreading because it blocks out other possible readings. It's a bit like the collapse of the wave function in quantum mechanics. Before a measurement is made, the value of what is being measured ranges over an infinite set of possibilities; once made, there is only one value. When you read, only one of a vast number of possible readings is realized. And this is just as true of "reading" the book of nature, of formulating scientific theories, as it is of reading novels or poems.

No discourse and no form of reasoning escape the combined Foucault–Derrida critique. The discourse of science, natural or social, and even the discourses of mathematics and of logic, which pride themselves on being strictly denotative, are exposed as open-ended interpretations. The search for foundations, for that which will once and for all anchor truth and knowledge claims, is quixotic. As Foucault said about the search for the essences of our concepts, ideas, and language, essences are fantasies, the imposition of closure onto the open-endedness of our thinking. The search for essences is a search for that which transcends experience for the sake of rescuing experience from terminal contingency.

To the extent that communities of scientists, mathematicians, and logicians form a consensus around one form of doing science or mathematics or logic as opposed to others, these are politicized discourses for postmodernist critics because they involve power relationships. Flecks's thought collectives with their associated "approved" thought styles, Kuhn's paradigms and their derivative normal sciences, and Lakatos' research programs are all—protests from scientists, mathematicians, and logicians notwithstanding—exposed by these critics as concealed ideologies. They are concealed by tacitly privileging forms of reasoning and by concealing their character as interpretations. The extraordinary popularity of Foucault's and Derrida's ideas in the United States, primarily from the 1970s, after their books were translated into English, made the outbreak of the Science Wars inevitable.

15

Taking Sides

The battle formations begin to take shape.

On one side, the allies of the Giants, who according to the philosopher from Elea want to "drag all things down out of the heavens and the invisible realm . . . and maintain strenuously that that alone *is* [has being and thus is real] which allows for some touching and embracing." The critics of science, too, want to "drag" science down out of the place it has won in the epistemological heaven of truth, rationality, and objective knowledge of the real. They reject the ability of reason to grasp a reality independent of our experience of it, reject the ability of reason to produce objective knowledge as opposed to empirically justified opinions and beliefs.

On the other side, are the allies of the Gods for whom "the reality with which true knowledge is concerned, [is] a reality . . . intangible but utterly real, apprehensible only by the intellect," independent of the body and its empirical experience. For them, the mind, "through its own instrumentality," is able to rise above opinion, belief, and mere hypotheses to objective knowledge, in principle to objective knowledge of the real.

In 1964, just two years after *The Structure of Scientific Revolutions* (SSR) appeared and well ahead of open attacks on the objectivity of scientific knowledge, Jerome Ravetz, an American philosopher of science and mathematician whose academic career was spent almost entirely at the University of Leeds, published *Scientific Knowledge and Its Social Problems* (SKSP). Like Kuhn, Ravetz' goal was to subvert the prevailing image of science, in his case, its image as a value-neutral, objective knowledge–producing, purely intellectual enterprise. Against this image, Ravetz argued that science was a craft and that scientists possessed know-how, not what is commonly understood as knowledge. If Ravetz is right about this, and science is know-how, then science is a form of engineering (which Dewey had argued half a century earlier). Science may be an abstract, theoretical branch of engineering, different from the more familiar concrete, practical branch, but it is engineering nonetheless. And if scientists produce a distinctive kind of know-how rather than knowledge, then the intersection of epistemology and ontology is not an issue. No one challenges the reality of the products of know-how: computer chips, cell phones, cable-stayed bridges, electric automobiles. And with evaporation of the need to justify ontological claims, the need for an other-than-pragmatic theory of truth evaporates, too.

Ravetz' claim upends the deeply entrenched subordination of know-how to knowledge in Western intellectual history, ranking the abstract, the theoretical, the universal, the necessary, and the certain above the concrete, the practical, the particular, the contingent, and the probable. (Derrida explored this same subordination.) Furthermore, if science is a craft, then the practice of science and its products, like the practice of engineering and its products, are not at all value neutral. Like all artifacts, theories must incorporate value judgments deriving from the social context in which they are produced and are to be used. Science thus has an *intrinsically* moral character, and scientists share with society responsibility for the uses, and for any societal problems, caused by applications of the knowledge they produce. (Friedrich Duerrenmatt's play "The Physicists" explores this very issue, while the film "The Conversation," referred to earlier, is about an engineer who denies this shared responsibility. Both have tragic endings.) To paraphrase Dewey, science as know-how is *of* the society in which it is practiced, not dropped down into it from a Gods-like transcendental realm.

For Ravetz, as for Kuhn and Polanyi, the target was the image of science as a body of completely impersonal, value-free, context-independent, and therefore objective knowledge. This was precisely what the founders of modern science sought and why they claimed that an impersonal method and logical reasoning about value-neutral facts were the pillars of their knowledge claims. Science, for them, simply described the world as it was, which made it amoral and apolitical. This image dissolves if we agree with Ravetz that science is a craft. Because it is not possible to draw a line separating social influences on the conditions of the practice of science from social influences on the products of that practice, Ravetz' argument opened the way to the claim that scientific knowledge was not objective because it was "socially constructed," a term that would become fighting words in the 1980s. In effect, *SKSP* anticipated what were to become the Science Wars.

Ravetz's ideas, along with those of Kuhn, Polanyi, Hanson, Scheffler, and many others, were assimilated into the academic Science, Technology and Society (STS) programs that, beginning at Cornell University, proliferated from the 1970s on. The study of the relationships between science and technology and between both and society brought together faculty from across the humanities and social sciences, as well as a number of scientists and engineers. Initially, the focus was on studying technologies in their social contexts, exposing the influence of political and commercial interests on invention, on the transformation of inventions into innovations, on their introduction into society, and on their subsequent evolution. The history, philosophy, politics, and sociology of *technology* became recognized sub-disciplines of traditional history, philosophy, political science, and sociology.[1]

The dissemination in the 1970s and 1980s of works by Foucault and Derrida, Lyotard, Kristeva and Irigaray, Jurgen Habermas, and Hans Gadamer, among many others and not forgetting Feyerabend, influenced the evolving field of *science* studies. Criticism of science became more aggressive, embedded in a critique of rationality generally. It became political, and often hostile. If there are no universal foundations for reasoning and discourse, then there is no such thing as objective, universal knowledge (or values). And there can be no such thing as a single, correct account of reality. There can be little doubt that for many on the "left," as it came to be called, there was a desire to pull science down from the pedestal on which society had placed it and from which it enjoyed a privileged position in society as a unique source of disinterested knowledge. It was now clear, to its critics at least, that there was no such thing as disinterested knowledge. Historical, philosophical, and sociological studies of science were said to have exposed science as just one form of human discourse among many.

Nelson Goodman, for example, argued in *Ways of Worldmaking* (1978) that science was a distinctive description of empirical experience, but not a unique one. Art, music, and literature produce equally authentic descriptions of experience that are no less valid responses to the human lifeworld than the scientific response. (Recall Nietzsche's perspectivalism.) The history of science, according to Goodman, is a history of successive *re*-descriptions of evolving empirical experience, each based on its own, conventional assumptions. The histories of art, music, and literature, too, are histories of successive redescriptions. There is no convergence in any of these histories to one uniquely correct artistic, musical, or literary description of our experience of the world, nor do all three together form a comprehensive mosaic that describes that world in a uniquely correct way. Goodman is clearly an ally of the Giants.

The Dutch-Canadian-American philosopher of science Bas van Fraassen also argued an anti-realist, anti-traditional image theory of science. In *The Scientific Image* (1980) and in *Laws and Symmetry* (1989) van Fraassen developed what he called a Constructive Realism that bracketed ontological claims based on scientific theories. All that theories can aspire to is "empirical adequacy," an explanatorily adequate account of empirical experience supported by observation. Whether observation is only sensory or includes the use of instruments, and to what extent, is an issue, but scientific explanation deals only with observables. That's as far as the realism of science goes. Of course, scientists routinely propose the existence of unobserved theoretical entities, including entities not directly observable even using instruments. Such entities do explanatory work, but they should not be taken as having an ontological status.

In 1979, Bruno Latour, a French sociologist and philosopher, and Steve Woolgar, a British sociologist, published *Laboratory Life: The Construction of Scientific Facts*. Latour had spent 1975 and 1976 in the laboratory of Roger Guillemin at the Salk Institute in Palo Alto, California. Latour had no knowledge of molecular biology when he arrived. He was in Guillemin's laboratory to observe how science is done in the same way that an anthropologist would learn about the culture of an unfamiliar people by living among them. The choice of Guillemin's laboratory was not accidental. Guillemin at the Salk Institute and Andrew Schally at Tulane University, together with their associated research teams, had for years been engaged in a highly publicized rivalry. Both teams were doing pioneering research on neurohormones and had had a bitter priority dispute over the isolation, synthesis, and determination of the molecular structure of thyrotropin-releasing hormone (TRH). For this work, Schally and Guillemin shared half of the 1977 Nobel Prize in Physiology or Medicine. (The other half went to Rosalyn Yallow for her research on radioimmunoassays and their medical applications.)

Because of the ongoing rivalry with Schally, there was an urgency to the work in Guillemin's laboratory that revealed more of the process of doing science during Latour's time there than might otherwise have been the case. Based on Latour's observations, what he and Woolgar revealed in *Laboratory Life* was the normally suppressed yet central role that nonobjective factors played in the production of scientific knowledge. Papers delivered at conferences, articles, and books ignore the details of life in a research laboratory. There was acquisition of materials and equipment, construction of experimental apparatus, determination that it is operating "correctly," calibration of instruments, and inventing and conducting experiments, including decisions about "good" data to be recorded and used, and "bad" data to be discarded for one reason or another. Echoing Derrida, Latour noted how data are selectively "inscribed" in "texts"—laboratory notebooks, graphs, images, etc.—whose "reading," for purposes of incorporation into publications that will make knowledge claims, is highly interpretive and not at all straightforward: "The negotiations as to what counts as a proof or what constitutes a good assay are no more or less disorderly than any argument between lawyers and politicians."[2]

Complementing these activities internal to the operation of Schally's laboratory were external activities, normally not included in scientific papers, that were essential to the laboratory's existence and to recognition by peers that it was producing new knowledge. Funding was a necessary condition for the lab's existence, and Latour observed firsthand how much time a laboratory head like Guillemin spent securing funding. Grant proposals are subtle texts that must be carefully crafted and targeted in order to raise the probability of success. The selective description of results already achieved and of results that could be achieved with needed funding must be matched to the known interests of specific funders and of the people they are likely to ask to review the proposals submitted to them. This in turn influences the kinds of experiments that are done in order to generate the data that will support claims made in the grant proposal.

Funding is a necessary but by no means a sufficient condition for a research laboratory to produce what is accepted as new knowledge. Latour and Woolgar documented the time spent keeping in touch with what was going on in other molecular biology research laboratories (especially Schally's!), in selectively sharing accounts of their ongoing research with colleagues at other institutions who they have identified as likely to be "allies" in the knowledge claims Guillemin was going to be making, and in tailoring conference papers and articles to audiences that would be open to those claims.

For Latour, but not Woolgar, what Guillemin was doing outside the laboratory suggested the application to science of Actor Network Theory, borrowed from sociology. On this view, knowledge production is a matter of deliberately creating a network of people whose knowledge interests are mutually supportive and will be reinforced by accepting new knowledge claims that fit those interests. (Recall the role Fleck assigned to thought collectives and thought styles.) Actor Network Theory highlights the role of a scientific community in validating knowledge claims as opposed to the traditional view that such claims are validated by nature. In his 1987 book *Science in Action*, Latour argued that the role of nature in validating scientific knowledge claims was negligible, adding fuel to the by-then open science wars. For his part, in his *Science, the Very Idea* (1988), Woolgar argued for a strongly relativist view of scientific knowledge.

For both Woolgar and Latour, *Laboratory Life* made a strong case for the claim that scientific knowledge was socially constructed, and their book reinforced an emerging movement committed to that claim. Their claim was that their methodology, imported from social science, revealed that in science "knowledge" is a name for a consensus among the relevant community of scientists that some theory was "true" and therefore qualified as knowledge rather than opinion or belief. Logical reasoning, consistency with experimental data, and agreed-upon (conventional) assumptions all played a role in forming this consensus, but they were only necessary conditions for a knowledge claim, not sufficient.

Note that Latour and Woolgar's methodology, together with the assumptions underlying their interpretation of the data they collected with it, clearly led to a *different* understanding of science from the traditional one. That is not at all the same as claiming that it led to a *correct* understanding of science. Recall that at the outset of *SSR*, Kuhn had claimed that his new methodology for studying the history of science would produce a very different "image" of science from the prevailing one, and he was certainly right about that. But his new, still internalist, methodology for studying science was not the only alternative to the traditional one. The externalist methodology also leads to a different image of science from the traditional one, and different from Kuhn's image as well; Lakatos' "rational reconstruction" methodology leads to yet another different image of science. The lesson would seem to be that, as with scientific knowledge itself, method alone cannot generate a uniquely definitive "image" of science.

Eight years after *Laboratory Life* was published, for example, the science writer Natalie Angier published *Natural Obsession* (1988). In it she presented an account of a year she had spent as an observer in the molecular biology laboratories of Robert Weinberg at Columbia University and of Michael Wilger at Cold Spring Harbor. In many respects, Angier's observations overlapped those of Latour and Woolgar, but as she did not share their interpretive assumptions, she was not led to challenge the objectivity of scientific knowledge. In 2009, Carole and Edmund Rifkind released the documentary video *Naturally Obsessed: The Making of a Scientist*. This was the result of three years they had spent following a group of graduate students pursuing their doctorates in Larry Taylor's Columbia University molecular biology laboratory. Once again the discovery process in science is seen to be "messy" and highly nonlogical, but it does not lead a viewer to conclude that scientific knowledge is not objective and not validated by nature.

In the 1970s, a group of faculty at the University of Edinburgh emerged as an important center of the social construction of scientific knowledge (SCSK) movement, especially of what came to be called the "strong" SCSK program. Advocates of the strong program argued that scientific theories dismissed as false should be studied on equal terms with theories accepted as true. We have already noted in the case of Galileo, for example, that scientists' reasoning to conclusions later judged wrong had the same form as for those later judged right. Kuhn had made this a central feature of his call for a new approach to the history of science. Historical studies of science were now being complemented by sociological studies, and these reinforced the claim that reasoning, and even reasoning about the same evidence, was not what differentiated "good" science from "bad' science. According to SCSK, the differentiation came from the scientific community itself, not from nature, given that we have no access to nature independent of our reasoning about it.

Barry Barnes, David Bloor, Harry Collins, and David MacKenzie were particularly influential members of the Edinburgh school of SCSK. (MacKenzie's work focused on the social construction of military technologies and on statistical science.) The titles of their books convey a sense of the nature and the growing assertiveness of the SCSK movement. Barnes published *Scientific Knowledge and Sociological Theory* (1974), followed by *Natural Order: Historical Studies of Scientific Culture* (1979), coauthored with the American historian and sociologist of science Steven Shapin. Shapin would go on to coauthor with Simon Schaffer the highly influential *Leviathan and the Air Pump: Hobbes, Boyle and the Experimental Life* (1985) and on his own *A Social History of Truth* (1994).

In *Leviathan*, Shapin and Schaffer reconstructed Hobbes' arguments against the experimental method as a means of producing knowledge of nature. In particular, Hobbes challenged the interpretation that Boyle and Hooke placed on the outcomes of their experiments using an air pump of their own design. The Shapin–Schaffer reconstruction reveals not only the strength of Hobbes' case against the experimental method, but also the political dimensions of the controversy and its eventual resolution in favor of the experimental method (echoing Foucault on the ubiquity of power relationships in society.) They conclude, also echoing Foucault, "As we come to recognize the conventional and artificial status of our forms of knowing, we put ourselves in a position to realize that it is ourselves and not reality that is responsible for what we know."[3]

Meanwhile, David Bloor published *Knowledge and Social Imagery* (1976) in which he proposed a methodology for SCSK: treat all scientific theories as effects whose causes are to be identified; treat all theories, whether judged right or wrong by scientists, impartially and symmetrically; apply SCSK to itself, that is, to the results of its own studies of science. In 1985 Harry Collins published *Changing Order* in which he applied Bloor's methodology to three case studies in contemporary science: introducing a new type of laser into a research laboratory; the claim by University of Virginia physicist Joseph Weber that he had detected gravitational waves; and research into extrasensory perception (ESP). The laser was eventually judged by it users to be working correctly and its results reliable. Weber's claim was dismissed by his colleagues in spite of his reputation as an accomplished experimental physicist. The ESP research was dismissed by the scientific community as pseudoscience. Predictably, Collins was able to show the operation of nonlogical and nonobjective factors in reaching each of these judgments. After *Changing Order*,

Collins focused on gravitational wave research from Weber to the construction and operation of the Laser Interferometry Gravitational Observatory (LIGO). *Gravity's Shadow: The Search for Gravitational Waves* was published in 2004, eleven years before an upgrade to LIGO that had been made in 2002 finally detected such waves. What emerges is a contrast between the persistence of support for LIGO in the face of some fifteen years of no results and the rapid dismissal of Weber's claim.

Karin Knorr Cetina is an Austrian anthropologist turned sociologist who became interested in the sociology of science in the 1970s, coauthoring *Determinants and Control of Scientific Development* (1975). The titles of her books announce their commitment to SCSK: *The Manufacture of Knowledge: An Essay on the Constructivist and Contextual Nature of Science* (1981) and *Epistemic Cultures: How the Sciences Make Knowledge* (1999). The first of these books is an anthropological study of laboratories in the spirit of *Laboratory Life*. It pays greater attention than *Laboratory Life* did to the epistemological and ontological issues raised by the practice of experimental science, and it was an important contribution to the emerging SCSK movement. The second book compared the "cultures" of particle physics and molecular biology, showing how very different they were in spite of apparent similarities.

Andrew Pickering is a British physicist who turned to the philosophy, history, and sociology of science, receiving a second Ph.D. from the Science Studies Unit at the University of Edinburgh. In *Constructing Quarks: A Sociological Analysis of Particle Physics* (1984) Pickering gave a detailed, highly technical history of experimental and theoretical particle physics from the early post–WW II period through the early 1980s. His focus was the development of the quark theory of elementary particles and the electroweak theory that unified the electromagnetic force and the so-called weak force in nuclear physics. Pickering argued that his history, with its account of reciprocal connections among theory, experimental design, and interpretation of experimental outcomes, validated the claim that scientific knowledge was socially constructed.

Constructing Quarks was of particular importance to the SCSK movement because, unlike anthropological studies such as *Laboratory Life* and *Manufacturing Knowledge*, its author fully understood from the inside, as it were, the science he was using as a case study. Pickering wrote that the

> preponderance of mathematics in particle physics is no more hard to explain than the fondness of ethnic groups for their native language . . . On the view advocated [here], there is no obligation upon anyone framing a view of the world to take account of what twentieth century science has to say.[4]

While other books by advocates of SCSK could be dismissed by scientists, *Constructing Quarks* could not. It played a significant role in bringing the simmering science wars into the open, provoking a lengthy joint review by the American physicist turned historian of physics Sylvan Schweber and the Canadian historian and sociologist of science Yves Gingras. It is telling that their review was coauthored by one author who was highly respected by physicists as well as by historians and philosophers of science, and another highly respected as a sociologist. Even more telling was that they chose to publish this review in the journal *Social Studies of Science*, thereby confronting SCSK on its home ground.[5]

Schweber acknowledged the technical competence of Pickering's history, but he and Gingras rejected Pickering's argument that this history supported the claim that scientific knowledge was socially constructed. They rejected Pickering's claim on the grounds that his interpretation of particle physics *presumed* the validity of social construction and selectively included events in the history of particle physics to support that conclusion. In addition, Schweber and Gingras claimed that Pickering assumed on the part of particle physicists the primacy of theory to data, the necessity of what were contingent judgments, and that the particle physics community (rather than nature) played the determining role in validating knowledge claims.

In a later book, *The Mangle of Practice: Time, Agency and Science* (1995)—not to be confused with his 2008 book titled *The Mangle in Practice*—Pickering reintroduced a role for nature in the construction of scientific knowledge claims, and with it a kind of realism. In constructing theories, scientists encounter "resistances" that constrain, but do not determine, the form a theory can take if it is to work, that is, if it is to explain and predict. (William James had said something very similar, also for ontological reasons.) At the same time, theories incorporate "accommodations" to these resistances that are objective and universal in the sense of being the same for everyone, but it is the scientific community that decides which accommodations to make, which preserves a role for the scientific community in the "construction" of scientific knowledge. (Recall Poincaré's conventions and Fleck's "active associations.") Pickering characterized scientific practice as having a "temporal structure" in the form of a "dialectic of resistance and accommodation," a dialectic played out over time between our reasoning and the world that exists "out there" independent of our reasoning about it.[6]

The preceding pages present an extremely brief sketch of the challenge to the production and status of scientific knowledge from the "left," so to speak, from allies of the Giants. In a word, the conclusion shared by the SCSK community and its sympathizers was that science is a *belief* system, not at all a body of knowledge. In their contemporary context—aggressively critical of entrenched social, political, and educational institutions and values—even these few examples give a sense of the provocation that led to the Science Wars. The response to the provocation from the "right" comprised a variety of defenses of the objectivity of scientific knowledge, but also defenses of traditional institutions and values against the demand for leveling them in the name of eliminating elitism.

Common to the responses from the "right" was an insistence on the compelling realism of science as well as on the objectivity and universality of scientific knowledge. Existentially at least, it is impossible to deny that there is a world of objects outside the mind (including the body in which the mind finds itself), and that the properties and behaviors of these objects are independent of our awareness of and reasoning about them. Scientists produce causal-theoretical accounts of the relationships between the properties and behaviors of objects outside the mind. These accounts are said to be objective and universal because scientific truth claims are based on logical reasoning about our empirical experience and are confirmed by that experience.

Using instruments as simple as the barometer, telescope, and microscope and as complex as the Large Hadron Collider, gene sequencing machines, and neutrino telescopes, scientists have discovered objects "out there" that are beyond our sensory experience, but which science shows can be causally linked to sensory experience. It seems extremely implausible that atoms, viruses, DNA, continental plates, and even the Higgs boson and black holes do not exist "out there," but only in the minds of the scientific community. Or that space, time, matter, and energy do not really have the properties deduced from the general and special theories of relativity.

The defense of science ultimately comes down to "It works" and in ways that literature, art, and religion do not. Given that it works, that its success makes the most sense if we accept that it is an account of reality, how can what science studies have revealed about science be accommodated to a defense of science as objective knowledge of the world, not a form of know-how, not opinion or belief, and not a self-promoting ideology? Scheffler and Lakatos were early neo-objectivists, philosophers who defended the objectivity of scientific knowledge in the wake of *SSR*. What follows are synopses of theories of science proposed by a selection of other such "neo-objectivists" responding to the post-*SSR* critiques of reason and objectivity.

Carl Hempel and Willard van Orman Quine have been referred to earlier. Both were important philosophers of science in the 1950s, pre-*SSR*, and both were active and influential after *SSR* as well, in support of science as producing objective knowledge. Hempel's realist and objectivist philosophy of science centered on identifying scientific explanation with a "covering law," or nomological-deductive model, of inference. There was nothing psychological and thus subjective about how an explanation explains or about judging what makes one explanation "good" and another "bad." To explain a phenomenon is just to deduce it from a true, universal, covering law together with relevant empirical data. Hempel also identified prediction with deductive inference. The word "prediction" becomes synonymous for him with "conclusion of a deductive argument," even though "prediction" has different semantic connotations.

How the truth of the universal theory premise is established is pivotal. Hempel leaves it hypothetical, so that, as for Whewell, scientific knowledge is deductive epistemologically, but not ontologically. The connection with the world "out there" is empirical but not logical. With the maturation of quantum mechanics and other statistical theories, Hempel found it necessary to include statistical laws as premises in his model. In that case, the conclusion could only be probably, and not necessarily, true. Calculating those probabilities required a theory for deriving quantitative degrees of confirmation from a given statistical covering law and relevant empirical information. This required support from a formal version of inductive logic, parallel to formal models of deductive logic, but like his mentor Rudolf Carnap, Hempel was unable to produce such a model.

We encountered Quine in the Quine–Duhem thesis and the underdetermination of theories by evidence thesis, a corollary of which is the indeterminacy of translation: there is no uniquely correct way to translate a text from one language to another. His 1950 essay "The Twin Dogmas of Empiricism" was widely perceived by philosophers as undermining the credibility of logical empiricism by denying the distinction between synthetic and analytic statements, and by vacating the verification criterion.[7] It was the vulnerability of logical empiricism on these points that prompted Hempel to develop his nomological-deductive philosophy of science. By 1962, when *SSR* was published, logical empiricism was already losing support, and Quine was well into the development of a philosophy of science based on a "naturalized" epistemology. Quine worked within the "analytic" philosophy tradition that had its roots in the mathematical logic of Frege and Russell, in Russell's broader philosophy of science, and in Wittgenstein's *Tractatus Logico-Philosophicus*. The analytic approach called for philosophical inquiry to begin with careful attention to language, because of the intrinsic ambiguity and imprecision of ordinary language. Ideally, philosophy should be conducted in a formal logic, free of imprecision and ambiguity. For Quine, first-order predicate logic was *the* fundamental form of logic and uniquely capable of anchoring an account of the world as it really is.

Quine's position was that there is only one body of what is justly called "knowledge" available to human beings and that is science. Scientific knowledge alone is capable of "limning the true and ultimate structure of reality."[8] Again, it is "within science itself . . . that reality is to be identified and described."[9] Science is not only the best way of learning about the world; it is the *only* way. One implication of this is that philosophy cannot be considered as standing outside science. Philosophers cannot judge the merit of scientific knowledge claims on philosophical grounds, because there are no grounds outside of science for judging *any* knowledge claims.[10] (There is a circularity here that Quine just shrugged off.) To be legitimate today, philosophy must be subject to the same epistemological criteria as science. This includes epistemology itself, which is "contained in natural science as a chapter of psychology."[11] As cognition begins with input to the brain from the senses, the issue for epistemology is not a priori reflection on the nature of knowledge, but discovering how the linguistic expression of reasoning about the world emerges from a biological process that begins with sensory stimuli. This is a question involving neuroanatomy, neurophysiology, psychology, and linguistics, not epistemology as philosophers have practiced it since Plato and Aristotle.

In Quine's view, the "best" science at a given time is all that we can say that we know about the world, *at that time*. Scientific theories are always corrigible and fallible, subject to changing experience. As theories change, what we can say that we know about the world also changes. Coordinately, Quine's conception of truth is that truth is simply an assertion. To say that snow is white is equivalent to saying, 'It is true that snow is white,' and that we should use "true" in this way when talking about snow and its color. If experience proves that we are wrong, then we accept that we are wrong, and we adapt what we say about snow accordingly, as we adapted what we said about swans after the discovery of black swans. Quine's philosophy of science is that science is empirical, realist, and objective in that its assertions are controlled by our experience of the way the world is, not as in SCSK by the consensus view of scientists.

Donald T. Campbell was an eminent social scientist before he turned to the philosophy of science. As a social scientist, his central concern was methodology, and this served as a bridge to the philosophy of science in the context of the reaction to *SSR*. In defense of the realism and objectivity of science, Campbell proposed what he called "evolutionary epistemology," which overlaps Quine's naturalistic epistemology. The human brain evolved within the world and is thus *of* the world. Consistent with Dewey, Campbell argued that our reasoning is intrinsically coupled to, and therefore intrinsically about, the world, not some internal representation of the world. If the latter were the case—and Richard Rorty, in *Philosophy and the Mirror of Nature* showed that that was the dominant view in modern philosophy—then proving that our internal representation corresponds to the way the world is "out there" becomes an intractable problem. (Kant famously solved it by making the internal representation a product of the mind, uncoupling it from the way the world *really* is.)

Given its evolutionary origin, what else could the mind reason about but the world, as that world is present to the mind in empirical experience? Campbell took the analogy with the evolutionary thesis a step further. Quine's naturalized epistemology made the question 'How do we know?' a matter of neuropsychology. Campbell was interested in cognition itself, in how theories were made, not as Quine was in the neurophysiological processes underlying cognition. He applied an analog of biological evolution to the reasoning of scientists. Scientific theories emerge from data via a process of blind variation and selective retention. Campbell agreed with Feyerabend in one point only: there is no algorithm for going from data to theory. Feyerabend called the process of arriving at a theory from data "anything goes"; Campbell called it "blind variation." But for Campbell, the world then steps in and, contra SCSK (and Feyerabend), it is the world that controls the selective retention of the ideas that the mind of the scientist had "blindly" thrown up. The theories that improve our "fitness" to our empirical experience are the ones we call "true." Fitness is the criterion of truth, not correspondence.

Like Quine, Campbell was an empiricist, realist, and objectivist about science. He was a passionate opponent of SCSK, calling its proponents "ontological nihilists." If our knowledge claims about the world are socially constructed, then we lose all cognitive contact with the reality within which our bodies, and therefore our minds, evolved. At the same time, science was, for Campbell, forever conjectural, fallible, and corrigible. Because our knowledge claims are dependent on constantly changing experience, they are always contingent, never necessary. One of his favorite expressions was "cousin to the amoeba, how can we be sure?" The answer is that we cannot, but what we claim to *know* about the world is about the world in which we function and is at least a partially correct account of it.

Campbell regularly used a mapping metaphor in describing the relationship between theories and the world. (Phillip Kitcher also used this metaphor in his philosophy of science, for example, in *Science, Truth and Democracy*). A map on a scale of one to one is not a map at all; it *is* what it was supposed to be a map of. Maps are always selective and symbolic representations of something, but the "truth" of a map nevertheless is assessed objectively: either it correctly represents its object or it does not. A hydrologic map of a state, for example, never *corresponds* to its water resources, nor does it picture them (hence the map's symbolic nature), but it either identifies them and their relationships correctly, or it doesn't. The same applies to a road map, a topographical map, etc. Maps clearly are social constructs, but the judgment of the truth/correctness of a map is objective because the correlation with its object is. What remains open is whether the correctness of a conceptual map is judged by its correspondence with experience or its correspondence with the real world. Either way, that judgment is objective, not a social construct.

This suggests that the notion of the social construction of scientific knowledge has been confused from the beginning, as the Canadian philosopher of science Ian Hacking argued in *The Social Construction of What?* (2000). Prisons are socially constructed, but there is a real difference between being on one side of the bars or the other, and that difference is objectively determinable. At the same time, who goes to jail, for how long, and what the conditions in the jail are, are all contingent and socially constructed. That scientific theories are socially constructed does not entail that there is no real, objectively determinable difference between a true theory and a false theory. The entire process of discovery may be socially constructed, the conditions of the practice of science may be socially constructed, but, contra Latour, either empirical experience or the world plays a determinate role in the justification of scientific knowledge claims.

Mary Hesse, a leading British philosopher of science, and later Ronald Giere, argued that scientific theories should be understood as models. In *Models and Analogies in Science* (1963), Hesse argued that any satisfactory description of scientific practice had to make reference to the use by scientists of models and analogies because it is not possible to connect experimental outcomes to a theory without them. (Recall Duhem's opposition to models and explanation because they were surreptitiously ontological.) In a 1958 paper she had written,

> To interpret an experiment directly in theoretical terms so that it can be a test of the theory is always to say more than the corresponding phenomenal statements would say . . . the phenomenal description of an experiment has a relationship to the scientific theoretical description [that] is similar to [the relationship] between Holingshed and Shakespeare . . .[12]

Once again we see the assertion that data do not speak for themselves. An explanation, Hesse wrote, always contains a metaphorical re-description of what is to be explained, and models and analogies are the metaphors employed in science.[13]

A model's value, for Hesse, is the theory's value: making predictions that are experimentally confirmed. Giere employed the mapping metaphor and the model metaphor in his philosophy of science. The mapping metaphor supported Giere's perspectivalism, or as he called it, "perspectivism." Theories, like maps, are by their very nature selective, hence partial, experience-bound perspectives on the reality behind experience. Models add another dimension to Giere's theory of science. Theories are models in the same sense that a computer simulation is a model. A computer simulation can produce results that parallel the results produced by "the real thing" without being that thing and without producing them the way the real thing does. A simulation of a nuclear weapons test or of the merger of two black holes or of the Earth's atmosphere can produce the same data that the real thing produces, but not the way the real thing produces it. What a model/simulation models are relationships that are conjectured to be the same as the causal relationships among the hypothesized elements of the real thing. The test of a theory as a model is the ability of the model to correctly reproduce known data, but more importantly, to produce new data from its "operation." This can take the form of drawing logical consequences from a theory—as in the case of deducing $E = mc^2$ from the special theory of relativity—or from running a computer program and analyzing its output. A "good" computer simulation of the Earth's atmosphere should produce a picture of the atmosphere at some time in the future. The term "prediction" typically is used in both for both theories and models, but as already noted, neither theories nor models make predictions; they have logical consequences, and both are validated (in part at least) by experimental confirmation of those consequences. What's the difference? As Duhem warned, "prediction" is intrinsically ontological, while "logical consequence" and "outcome" are not.

Hilary Putnam, one of the leading American philosophers of the second half of the twentieth century, proposed a defense of the objectivity and realism of science that he called Internal Realism. In *The Many Faces of Realism* (1987) he wrote that "the mind and the world jointly make up the mind and the world."[14] Putnam agreed that scientific accounts of experience necessarily rest on assumptions that are conventions, but he rejected the relativism of SCSK. Internal realism accepts the tacit realism of our everyday experience of a world of ordinary objects and adds the reality of scientific objects on the same grounds. There are multiple possible conceptual frameworks, "versions," he calls them, that can produce a scientific account of experience. Different versions produce worlds with different objects, and while there is no objective basis or foundation for judging one version better than another in advance of experience, it is the world "out there" that determines which versions work. Each version produces outcomes and, contra SCSK, it is the world that selects which versions work and which do not, not the scientific community.

There are echoes in this book of Putnam's of Dewey on the mutual determination of mind and world, as well as of Fleck's active and passive associations: the active associations being our choice of definitions, assumptions, and rules of reasoning, and the passive being the logical consequences of those choices. In *The Shaky Game* (1986), the American physicist Arthur Fine proposed an empirical realism that he called the Natural Ontological Attitude, which attempted to finesse the problem of justifying realism. Let's agree to accept the reality of scientific objects on the same tacit grounds that we accept the reality of the world of ordinary experience. We don't ask for a proof that the objects of ordinary experience really are "out there," and neither should we do so for scientific objects. If accepting the existence of these objects and reasoning about them collectively and objectively anticipates new experiences (recall Hertz on this), it is appropriate to call them real as we call people, trees, and cars real, without demanding proof that they correspond to objects independent of experience. Fine acknowledges that scientists in fact do treat scientific objects as if they were real in the sense of being "out there," but that is a fact about scientific practice.

A group of philosophers of science co-located at one time at Stanford University—among them Susan Cartwright and Peter Galison—made the experimental dimension of scientific practice rather than theory construction and justification central to their accounts of science. Cartwright, in *How the Laws of Science Lie* (1983) and *Nature's Capacities and Their Measurement* (1989), and Galison, in *How Experiments End* and *Image and Logic: A Material Culture of Microphysics*, describe how instruments and experiments are used to produce the data that are incorporated into theories. That scientists treat scientific objects and their causal relations realistically is a fact about the practice of science, and that's that.

The American philosopher of science Wesley Salmon was an empiricist, an unabashed objectivist about scientific reasoning, and a realist about scientific theories and their objects. At the same time, statistics and probability played a central role in his account of theory construction, such that scientific knowledge claims could never be certain. He presented his ideas in a large number of books and papers, especially (for our purposes) in *Scientific Explanation and the Causal Structure of the World* (1984), *Causality and Explanation* (1998), and the posthumous *Reality and Rationality* (2005). Salmon admired Hume and also Hans Reichenbach's application of probability theory to science. Scientific reasoning was inductive only, as for Hume, and not deductive as for Popper. And scientific explanation required identifying causal mechanisms at work in nature, but understood probabilistically, as for Reichenbach, not deterministically as in the Principle of Sufficient Reason.

The number of different defenses of the realism and objectivity of scientific knowledge claims by itself tells us that none of them made a compelling case. The effort reflects a challenge from the "left," especially after *SSR*, Foucault, Derrida, and SCSK. Kuhn himself lined up on the side of realism and objectivity, but *SSR* had long since taken on a life of its own.

16
The Science Wars Go Public

In the 1980s, criticism of science became openly hostile, together with criticism of Western culture generally, both denounced as sexist, racist, colonialist, homophobic, elitist, imperialist, and capitalist. The critics shared an anti-foundationalism—denying the possibility of objective criteria for knowledge, truth, and values—reinforced by the works of Foucault and Derrida among others, entailing epistemological and valuational relativism. Science and Western social institutions were attacked as embodying imperialistic ideologies that exploited the rhetoric of knowledge, truth, and reason to legitimate their social power. When it came, the reaction against this criticism was broader than a defense of science only. In 1985, Alain Renaut and Luc Ferry began what would become a sustained and widening attack on postmodernism and on the ideas of Foucault and Derrida in particular. They saw in these ideas a profound threat to liberal democracy and individual human rights. If everything is socially constructed, including the self, the way is opened to the primacy of society over the individual and the legitimation of totalitarian regimes, notably fascism and communism. If all forms of reasoning and valuing, all claims to knowledge, are ideological, and there are no objective grounds on which to rank them, then the way is opened to a pervasive 'might makes right' mentality in society, as Scheffler had warned. It is also a self-defeating argument in that the feminist critique of science, for example, is no less an ideology than the patriarchal science it denounces as ideological, and on what grounds can it be judged superior to it?

Renault and Ferry pointed to the influence of Nietzsche and Heidegger on Foucault and Derrida, emphasizing the distrust of democracy by both of those philosophers. Renaut's *Era of the Individual* traced the history of Western individualism through the modern era. He made a distinction between the social atomism form of individualism—each person an encapsulated self, like an Leibnizian monad—and respect for the autonomous subject in a social setting. Postmodernism, he argued, undermined that respect because it denied that there was, or could be, such a thing as an autonomous subject. This was, he argued, a much greater political threat to society than the claim that scientific knowledge was a social construct, though the two threats were connected. They were both products of the postmodern dismissal of the modernist conceptions of knowledge, truth, reason, and values.

In 1987, University of Chicago professor Alan Bloom published *The Closing of the American Mind: How Higher Education Has Failed Democracy and Impoverished the Souls of Today's Students.*" To the surprise of the author no less than the publisher, this passionate and witty rant became a runaway national bestseller, number one on the New York Times nonfiction bestseller list for four straight months. To Bloom, universities have surrendered traditional standards and values—in particular, the so-called Great Books of the Western World—to the facile, politically driven, culture-leveling agendas of activist liberal faculty. Consistent with Foucault's genealogical method and Derrida's deconstructionism, these academics teach that there are no foundations on which to pronounce one form of culture superior to others, or one cultural product superior to others. The traditional educational curriculum for them is a form of propaganda aimed at perpetuating the white, male, heterosexual, capitalist control of society. In the name of an anti-elitist egalitarian populism—reminiscent of Kurt Vonnegut's terrifying short story "Harrison Bergeron"—anything goes for these culture activists in literature, poetry, music, and art. Let all voices be heard without prejudice.

Bloom was appalled by this. His book was an attempt to discredit postmodernism-based cultural agendas of all sorts because accepting them meant that there was no such thing as excellence and that critical opinion was solely a matter of personal preference. The Rolling Stones (he was particularly disgusted by Mick Jagger) are on a par with Bach and Beethoven; comic books with Dante; science fiction with Shakespeare; "Father Knows Best" with "Death of a Salesman"; so-called outsider art with Michelangelo and Raphael. Not only were students being robbed of learning the best products of Western culture from the Greeks to the present, but democracy itself was reduced to demagoguery. The most successful politicians would become those who spoke to the lowest common denominator of the self-interests of the electorate. (Saul Bellow, Bloom's friend and University of Chicago colleague, made a Bloom-like character the hero of his novel *Ravelstein*.)

In the same year that *The Closing of the American Mind* was published, the literary theorist E. D. Hirsch also published a bestseller, though not on Bloom's scale. *Cultural Literacy: What Every American Should Know* was a less acerbic criticism than Bloom's of the American educational system, covering K–12 as well as college. Hirsch's academic reputation came from his reader-based theory of texts, one that was 180 degrees from Derrida's deconstruction. There was, Hirsch argued, a level of textual understanding that was objective and the same for everyone who read a text "correctly." At the same time, each reader brought a subjective level of interpretation to a text, whether they understood the text correctly or not. Ideally, the reader both understands a text *and* reads it interpretively. *Cultural Literacy* was consistent with Hirsch's literary theory, but claimed that there were things that everyone should know in order to be considered educated, whatever else they may or may not know; Hirsch identified what these were in *Cultural Literacy*, provoking considerable backlash from the same liberal intellectual community that Bloom had pilloried. Not surprisingly, Hirsch was active in the uniform core curriculum movement that was anathema to the postmodernists. Although Hirsch's criticism of the educational system supplemented Bloom's, Bloom was an intellectual elitist and Hirsch an educational populist.

The defenders of the objectivity of scientific knowledge against social construction of scientific knowledge (SCSK) were primarily philosophers, who wrote for intellectuals, hence with low public visibility. The bestsellers of Bloom and Hirsch, on the other hand, were written for a broad public audience and gave high visibility to the consequences of postmodernism for traditional conceptions of truth, knowledge, reason, and values. As a result, the public was aware of a culture war before it was aware of a war against science. That changed in 1994 with the publication by Paul Gross and Norman Levitt of *Higher Superstition: The Academic Left and Its Quarrels with Science* (*HS*).

Gross was a marine biologist at the University of West Virginia, and Levitt a mathematician at Rutgers University. *HS* is a polemic against the full spectrum of academic critics of science. By contrast, the Nobel Laureate in physics Steven Weinberg's *Dreams of a Final Theory* (1992) is an impassioned affirmation of (natural) science as objective knowledge of reality, but he only devoted half a dozen pages of that book to critics of science.[1] In those pages he singled out *Laboratory Life*, *Constructing Quarks*, Feyerabend, the feminist Donna Harding, and Theodore Rozsak's *Where the Wasteland Ends*.

Gross and Levitt's target in *HS* is "muddleheadedness," which, they say in the book's opening sentence, "has always been the sovereign force in human affairs, a force far more potent than malevolence or nobility."[2] Muddleheadness is a threat to society because it "lubricates our hurtful impulses and ties our best intentions in knots." *HS* is a crusade against muddle-headedness, in particular the form of muddle-headedness to be found in the writings of what, "with great misgiving," Gross and Levitt call "the academic left." While this is a diverse group with diverse intellectual, social, and political agendas, with respect to science there is, they say, "a noteworthy uniformity of tone, and that tone is unambiguously hostile. To put it bluntly, the academic left dislikes science."[3]

Gross and Levitt acknowledge that there are valid grounds for criticizing the *uses* to which scientific knowledge has been and is being put by society. That is not their concern in *HS*. What is their concern is the "open hostility to the *actual content* of scientific knowledge and toward the assumption . . . that scientific knowledge is reasonably reliable and rests on a sound methodology." Their "greatest hope" is for *HS* to "stimulate awareness and debate" about an issue of importance to society at large, because it is "not without historical precedent that incoherent or simply incomprehensible opinions have had great and pernicious social effect."[4]

So much for the opening of the book. In the final chapter, Gross and Levitt proclaim—they offer no argument—that science "is above all a reality-driven enterprise . . . Reality is the unrelenting angel with whom scientists have agreed to wrestle."[5] To conduct a "serious investigation of the interplay of cultural and social factors with the workings of scientific research in a given field" requires "patience, subtlety, erudition and a knowledge of human nature," in addition to an "intimate appreciation" from the ground up of the field being studied.[6] Clearly, none of the critics of science whose muddle-headedness was mocked in *HS* meet these qualifications.

In between the first and last chapters, Gross and Levitt took on a cross section of cultural and epistemological critics of science, noting the influence on all of Foucault and Derrida, including: the sociologist and political activist Stanley Aronowitz and the culture critic Andrew Ross, together founding editors of the postmodernist journal *Social Text* (of which more will be presented shortly); Latour, Shapin, and Schaffer; feminist critics of science Donna Haraway and Sandra Harding, Evelyn Fox Keller, Katherine Hayles, and Helen Longino; and the "radical environmentalists" Carolyn Merchant, Steven Best, and Jeremy Rifkin.

Gross and Levitt cite passage after passage from these and other authors, arguing their incoherence, vacuity, incomprehensibility, self-contradiction, and manifest ignorance of science and mathematics, though statements about science and mathematics are routinely used by these critics to make their postmodern points. Here is one example, from an essay by Derrida:

> The Einsteinian constant is not a constant, not a center. It is the very concept of variability—it is, finally, the concept of the game. In other words, it is not the concept of some thing—some center from which an observer could master the field— but the very concept of the game.[7]

Gross and Levitt are particularly harsh on Derrida. They deal with the charge by Steven Best, echoing Foucault, that scientists systematically exploit the connotations of "truth" to retain their power and privilege in society,[8] but to argue as Derrida and his followers do that all discourses are on a par and all of them resist definitive meaning is over the top for them.

Gross and Levitt saw Nietzsche's perspectivalism as a common denominator of all of the critics of science. Sandra Harding, for example, had written that science "is just one of the many ideologies that propel society and it should be treated as such," that is, just one of many competing or cooperative ideologies. Science "is not only sexist, but also racist, classist, and culturally coercive."[9] Among the feminist critics of science, Helen Longino comes closest to acknowledging the objectivity of scientific knowledge, if it is the product of a mosaic of multiple perspectives, but then concludes that the judgment of objectivity in such a case would come from us, not from nature.

In the end, all there are for the culture critics are socially/culturally/historically shaped responses by individuals and groups to their life experience, and among these responses, science, literature, art, religion, myth, and politics. And all these in all their forms are to be treated with equal respect. Gross and Levitt were not convinced. *HS* was particularly well received by the natural science community, but also by many outside that community and outside academe as well. Admirers saw it as having exposed postmodern attacks on science as ill-informed, nonsensical, and narrowly political. A year after the book appeared, Gross and Levitt organized a conference at the New York Academy of Science at which some forty academics across a wide range of academic disciplines critiqued the "academic left" critics of science, rejecting SCSK and reaffirming the objectivity of scientific knowledge. The presentations by these academics were then edited by Gross and Levitt into *The Flight from Reason and Science* (1997). By that time, *HS* had inadvertently instigated an intellectual scandal that, in retrospect, was the peak of the Science Wars.

Alan Sokal, a respected theoretical physicist at New York University, read *HS* and decided on an "experiment" to test the intellectual integrity of the "academic left" critics of science. After reading the works of a number of the authors discussed in *HS* in order to pick up their jargon, Sokal wrote a paper that parodied postmodern science studies. The title of the paper was "Transgressing the Boundaries: Towards a Transformative Hermeneutics of Quantum Gravity."[10] On the face of it, this was a scholarly paper by a prominent physicist arguing in favor of the social construction of scientific knowledge. In fact, the physics in the paper was utter nonsense, and Sokal's argument was that quantum gravity was a social construct, an invention of the physics community. (Which in fact it was, but no more so than dark matter and dark energy are, or Dirac's positron and Pauli's neutrino were. What happens *after* conceptual invention is what matters in science.)

From reading *HS*, Sokal knew of the postmodern science studies journal *Social Text* and its editors, Stanley Aronowitz and Andrew Ross. He also knew that Ross was planning a special issue of *Social Text* with the title "Science Wars" to be published in the spring/summer 1996 issue. Sokal sent his paper to Ross, who accepted it for publication without further review, assuming that the physics was correct because of Sokal's reputation. (At that time, *Social Text*'s editors chose the papers to publish without outside review.)

Meanwhile, Sokal arranged with the editor of another journal, *Lingua Franca*, to reveal his hoax immediately after the "Science Wars" issue of *Social Text* was mailed to subscribers.[11] The revelation provoked a national uproar among intellectuals. In the main, scientists—but also nonscientist opponents of postmodernism, especially cultural and political conservatives—responded to the hoax by heaping ridicule on social studies of science generally and on the postmodern "academic left" in particular. Publication of Sokal's paper was taken as proof that these critics of science were utterly ignorant of science and intellectually dishonest to boot. Sokal's paper was published without review solely because it confirmed the views of the postmodern critics of sciences. To opponents, this was perceived as confirming that criticism of science was a means of promoting political agendas.

Of course, the Sokal hoax proved nothing at all about the validity of post-modern criticism of science, including the claim that scientific knowledge was socially constructed. At most it proved that Ross was too trusting an editor (Aronowitz was not involved in the publication decision), but the passionate debate that erupted in the wake of the revelation of the hoax ignored nuances. It was fueled by the mutual hostility that had been building steadily since *SSR* between proponents of the traditional view of scientific knowledge as objective and realist and opponents of that view. The science war really *was* a war, albeit one fought with words, sometimes with arguments.

From the science studies side, Sokal was denounced by many as having played an intellectual dirty trick on Ross. The hoax was unethical and proved nothing more than Sokal's malice, and that of the natural science community as a whole, toward those who were, finally, revealing the true nature of science and challenging its status in society. Interestingly, there were those in the scientific community who supported Ross and criticized Sokal as having acted intellectually dishonestly. Mara Beller, a respected Israeli historian and philosopher of physics, author of *Quantum Dialogues*, published an article in *Physics Today* citing passages from the writings of Bohr and Heisenberg that sound just as much like gibberish as anything in Sokal's paper.[12] David Mermin, a Cornell University physicist and columnist for *Physics Today*, criticized Sokal's "experiment" while his colleague Kurt Gottfried defended Sokal.

In 1997, while the Sokal hoax debate was still raging, *The Flight from Reason and Science* was published, containing the papers presented at the pre-hoax New York Academy of Science conference of the same name. In 1998, Noretta Koertge, an American historian and philosopher of science, edited a volume of post-hoax essays titled *A House Built on Sand*. The 'house built on sand' "refers to interdisciplinary endeavors called Science, Technology and Society studies (STS) or Science and Culture Studies" *as co-opted by postmodernists*.[13] Koertge identifies "percepts" that are "widely shared" by this group, among them, that the "quest for scientific knowledge is quixotic," that science is 'politics by other means,' that science has no privileged claim to knowledge, that one society's science is no better than another's, and that science is terminally ideological and can only be reformed from the outside.

The book's contributors include prominent philosophers, physicists, life scientists, philosophers and historians of science, and an engineer. There is an introductory chapter by Sokal, "What the *Social Text* Affair Does and Does Not Prove," that is moderate in its tone. So is Phillip Kitcher's "A Plea for Science Studies," a defense of studies of the social, historical, and cultural influences on scientific practice and reasoning, *not*, however, based on the postmodernist agenda. "Something has gone badly wrong in contemporary science studies," he wrote. That something is the polarization of perspectives on science—the two sides of the "divide" discussed earlier—that in principle have a great deal in common and together could result in a richer understanding of science than either one alone provides. This requires getting beyond the Science Wars and an 'us versus them' mentality.[14]

By the time *A House Built on Sand* appeared, the passion had gone out of the controversy sparked by Sokal's hoax, and the temperature of the open Science Wars was declining from a boil to a simmer, in part because of the turn against Derrida, Foucault, and postmodernism generally, but the wars were far from over. In 2001 the physicist Allan Franklin, in *Are There Really Neutrinos?*, could still write, "The attitude of many humanists toward science has changed from indifference to distrust and even hostility. Science is under attack from many directions, and each of them denies that science provides us with knowledge." [15] Meanwhile, science studies expanded from the SCSK "strong program" into studies of scientific practice consistent with some form of objectivism and realism. The emergence of philosophy of biology, for example the work of Michael Ruse, alongside philosophy of physics contributed to this. So did historical and philosophical studies that focused on experiments and instruments rather than on the construction of theories, among them works cited earlier by Peter Galison and Susan Cartwright, and books by Lorraine

Daston, among them *Biographies of Scientific Objects* (2000), *Objectivity* (2010, with Peter Galison), and *Against Nature* (2019).

The Science Wars did not fade away altogether, however, not in academe and not in society at large. The cultural and political agenda–driven critics of science had with some justification been lumped together as the "academic left" by Gross and Levitt. Feminist, Marxist, and "progressive" liberal criticism of science as patriarchal, racist, elitist, and co-opted by establishment institutions and values have continued. Already in the 1980s, however, awareness of the vulnerability of science's claim to being a privileged source of truth had been picked up by those on the political "right," and it had emboldened the so-called religious right.

If, as some academics had concluded, science is a belief system, not a body of logically arrived-at truths, if science is one story among the many others that humans tell about their experience of the world, and if it is just one perspective on the human condition, then why are we only teaching the science story to children and young adults? And why are we teaching it as the one *true* story? Pressure to incorporate "creation science" into the high school biology curriculum intensified. So did the pressure to eliminate, or at least sharply limit, the teaching of evolution, making it clear to students that evolution was a hypothesis and as such unproven, not at all a fact. Through contracts with publishers of high school biology texts, the departments of education of a number of U.S. states—among them, Alabama, Louisiana, Tennessee, Kansas, Kentucky, New Mexico, Ohio, and South Carolina—had set limits on how much of the theory of evolution was to be included in those texts and how it was to be presented. Attempts to impose teaching "creation science" alongside evolution provoked legal challenges that invariably resulted in rulings that creationism was a religious teaching, not a scientific hypothesis, and thus could not be included in a public-school science class.

The religious right shifted tactics, pressing to include intelligent design— that life phenomena themselves provided empirical evidence for deliberate design no less than for evolution—on the grounds that intelligent design was a scientific hypothesis even if creation science were not. In *Darwin's Black Box: The Biochemical Challenge to Evolution* (1996), the Lehigh University biochemist Michael Behe argued that cellular biochemistry was simply too complex to have evolved spontaneously, driven by random mutations, from the ground up, as required by Darwinian evolution. The relationships among a cell's chemical reactions, he claimed, had the character of a system of interdependent elements that could only have been introduced by a designer with a vision of the end result to be achieved. In *The Edge of Evolution* (2006), and in *Darwin Devolves* (2019), Behe argued that there are narrow limits to the role that natural selection and random mutation can play in explaining the evolution of life forms, and play effectively no role at all in explaining the origin of life.

Behe was the star witness in *Kitzmiller v. Dover (PA) Area School District (2005)*, the first federal trial over a mandate—by the Board of the Dover Area School District—to teach creation science and intelligent design in high school biology classes. This mandate was challenged by the American Civil Liberties Union, among other plaintiffs, as a violation of the establishment clause of the U.S. Constitution. Under cross-examination, Behe made concessions on points of science that weakened the case for the defense, and Judge John. E. Jones III ruled that both creation science and intelligent design were religious beliefs and as such could not be taught in science classes in a publicly funded school.

Three years later Behe again was a star witness, in *Association of Christian Schools International (ACSI) v. Roman Stearns, et al.* This was a suit brought by the ACSI against the University of California for not accepting for college credit high school courses at ACSI member schools that used creation science / intelligent design texts. Judge S. James Otero ruled in favor of the University of California's right to set its own academic standards. The ACSI appealed to the Ninth Federal District Court, which in 2010 upheld Judge Otero's ruling. Later that year the U.S. Supreme Court let that decision stand by refusing to review it.

The challenge to science from the religious "right" is a "front" in the Science Wars because, however ironically, it is consistent with attacks by the typically anti-religious "left" on science's claim to produce objective knowledge of the way the world really is. If, as Cardinal Bellarmine told Galileo in 1616, we bracket ontology and adopt a pragmatic epistemological stance such that science produces corrigible interpretations of experience that are justified by explanatory power contingent upon conventional assumptions, there is no conflict between science and religion. Religions can claim to be bodies of truths about reality based on revelation, and people can respond to such claims in any way they like. (Duhem's version of this, for example, allowed his physics to be compatible with his faith as a religious Catholic; similarly for van Fraassen.) Religious belief, or any level of disbelief, would be as consistent with accepting the truth of the products of science as they are with accepting the "truth" of the products of engineering.

Stepping back for a synoptic view of the situation, we see that the Science Wars were fought on four overlapping fronts. There was the political front, attacking science as co-opted by a corrupt establishment; the cultural front, an intellectual-led broadside against the modernist-Enlightenment valuation of objectivity, reason, and rationality and its standard bearer, modern science; the social studies of science front, attacking the claim of natural science to be objectively true knowledge of reality; and the religious front, attacking the scientific community as gatekeepers of what is or is not truth.

The Structure of Scientific Revolutions played no role at all in opening the social or the cultural fronts. It did bring to the attention of a large number of readers problems with the prevailing image of science. The rapidity with which *SSR* spread through the academic community, across all disciplines, and among intellectuals generally suggests an already existing openness to displacing the "image of science by which we are possessed today." *SSR* became the leading edge of a re-imaging of science, drawing from new studies of the history of science itself, from new social studies of the practice of science, and from new interest in the writings of Fleck, Polanyi, and Hanson, among others. The upshot was a broad awareness that science's claim to objective knowledge of reality was problematic on logical grounds, and that scientists had been aware of this from the outset of modern science. Theories are inescapably conjectural because they are dependent on assumptions that are contingent, corrigible, fallible, probable, and historical. Theories continually change, adapting to new data, new experiments, new instruments, new analytical tools, and new assumptions, including new assumptions about "old" data, new definitions, and new conceptual "measures" to apply to experience. Episodically, established theories are replaced by newer theories in a process that is neither wholly logical nor wholly objective.

Logically, then, and consistent with the history of science, theories can at most claim to be interpretations of empirical experience, justifiable on contingent grounds, not uniquely correct "pictures" of reality. While few if any scientists would assert that a particular theory at a given time was definitively true, the overwhelming consensus of natural scientists at least, as reflected in their rhetoric, is that science does produce a progressively more correct picture of reality. Ignoring the clear evidence of the history of science, scientists balk at the characterization of theories as "merely" probable, transient interpretations of experience justified not by correspondence with reality but by some combination of explanatory power, predictive success, and technological applications.

Social studies of technology had grown alongside, actually in advance of, social studies of science. Historians and sociologists applied an externalist methodology to studying technological innovation and engineering, especially the relationship of engineering to science. This revealed the centrality of social context to the exploitation of technical know-how, including selective application of scientific knowledge. This did not, however, lead to "technology wars," nor did claims of the social construction of technologies (SCT) lead to any ontological crises, challenging the reality of the products of techno- logical knowledge. Why was that? All artifacts are contingent and socially constructed in the sense of being the result of a highly selective, values-based mutual interaction between know-how and the social context in which it is applied and adopted (or not). But no one in the SCT community denied the reality of automobiles, the electric utility network, nuclear power plants, com- munication satellites, computer chips, the Internet, or the "cloud." Why did the social studies of *science* provoke a "war," while the social studies of *tech- nology* did not?

Know-how is intrinsically ontological, but it is a common-sense ontology. It involves action on the world as we experience it through our senses, and its truth is determined solely by our sense experience of the outcomes of that ac- tion. Know-how and its products are as real as the world of common-sense ex- perience is taken to be. Knowledge in the Platonic-Rationalist tradition is also intrinsically ontological, but it dismisses common-sense ontology as an illu- sion, replacing it with one or another version of an unsensed reality that can only be grasped by the prepared mind. Know-how, reflecting its dependence on a constantly changing, Hercalitean ontology, is itself constantly changing and as a result is particular and contextual. Knowledge in the Parmenidean- Rationalist tradition is context independent and timeless because it has as its object an unexperienced ontology that is universal, is unchanging, and cannot be experienced by the body.

As the history of modern science and the history of the philosophy of modern science reveal, modern science is epistemologically Heraclitean while retaining a Parmenidean-Rationalist ontology. This was and remains a deliberate choice on the part of scientists and most philosophers of science. There is nothing internal to scientific theories that mandates it. The philoso- phies of science of Mach, Hertz, Poincaré, and van Fraassen show that nothing changes in science by not making this choice. Resistance to not making it, however, is pervasive, as reflected in the rhetoric of science and the many phi- losophies of science that struggle to make science objective knowledge of a reality beyond the mind. It is this choice, and the inability to justify it logically, that provoked all four fronts of the Science Wars. As only a pragmatically jus- tified account of experience, theologians would no more be threatened by sci- ence than by entrepreneurs and engineers, and by bridges and cell phones. So, too, without the claim by scientists of objective knowledge of reality, the social construction of knowledge is no more radical than the social construction of technologies. Without scientists allying themselves with the timeless view of Being of the Gods, the ideas of Foucault, Derrida, and company are a more or less convincing critique of the various forms taken by a perennial Western intellectual "pathology"—privileging the universal, necessary, and certain over the particular, contingent, and probable.

It is at the turn taken by science at the intersection of epistemology and ontology that scientific knowledge claims become problematic. Feyerabend put it this way:

> Why are so many people dissatisfied with what they can see and feel? Why do they look for surprises behind events? Why . . . do they take it for granted that this hidden world is more solid, more trustworthy, more 'real' than the world from which they started?[16]

There is, after all, no logical bridge from science's empirical-experimental epistemology to an unexperienced ontology, yet taking the ontological turn is a deeply embedded feature of scientific practice and an inescapable feature of scientific rhetoric.

Allow science the scope of revealing "all the Choir of Heaven and Furniture of the earth," to hijack a phrase of Berkeley's, and how can it not be the case that scientific objects, from DNA to quantum energy fields, are real? There is, of course, the historic fact that science's ontology changes all the time, sometimes modestly, sometimes drastically. It is also relevant that our ordinary sense-based experience of the world, dismissed out of hand by science as completely wrong vis-à-vis reality, also "works." The world that we see, hear, smell, taste, and feel is a product of our nervous system. That world does not exist "out there," and yet it is sufficiently well correlated with what it is that we experience—perhaps an independently existing world causing our experience—for us to function effectively. If that were not the case, our survival would be more than a miracle; it would be impossible. And the same is true for all living organisms, each sensing a world often radically different from that of other organisms, yet able to function in a way that works for those life forms. The idea that all are experiencing the same world is just that, an idea. It would seem to follow that science, too, is best construed as a functionally justified mapping onto experience without adding 'and it is a uniquely correct mapping of reality.'

What is it about science that makes it hard for scientists to say that science is *a* source of truth, but not the only source? Engineers would never insist that the design of some artifact was the only correct design. The engineering design process has resisted every attempt to reduce it to value-free science. It is pervaded by explicit willfulness and contingent value judgments, among them, market niche, size, weight, appearance, materials, marketability, reliability, cost, performance, legal constraints, manufacturability, maintenance, and time to market. Depending on the weights assigned to these factors, which is where non-rational willfulness enters into the design process, multiple designs may be judged equally "good."

Scientists do not leave it at that, with theories being "designs" for dealing with experience in a specific, scientific way, subject to specifically scientific constraints? Given those constraints, in which empirical experience plays a pivotal role, multiple theory-designs that work are possible, for example, that a collision with an extraterrestrial object caused the extinction of the dinosaurs, or massive volcanic eruptions did, or some combination of the two. As a matter of practice, however, scientists do not leave it at that, and perhaps cannot, because one of the constraints on explanation imposed by the scientific community is that only one explanation at a time can be true. That is a corollary of realism. If theories aim at correspondence with the world "out there," then only one can be true, or at least partially true. That is precisely the point Einstein was making in his 1905 theory of light explaining the photoelectric effect. There is "a profound formal difference" between continuous wave theories and discontinuous particle theories. No explanation of what is real can employ mutually exclusive theories.

What, in the end, do scientists *know*? Consider what scientists claimed to know in 1900 with what their predecessors had claimed to know about the same subjects in 1800. Quite reasonably, scientists in 1900 could say that no one in 1800 really *knew* anything scientific at all. Now consider what scientists in 2000 claimed to know—in physics, chemistry, biology, geology, and psychology—with what scientists in 1900 had confidently claimed to know about these subjects. Respectfully or not, one must say that scientists in 1900 *knew* nothing at all. And what is the probability that scientists in 2100 will say the same thing about what scientists claim to know today? By induction, very high; so do scientists *know* anything today?

An understanding of the contingent nature of scientific knowledge is not an intellectual nicety. It bears directly on the very practical matter of science-related public policy issues, among them the COVID-19 pandemic, climate change, and applications of artificial intelligence, robotics, genetic engineering, and stem cell research. Misunderstanding the nature of scientific knowledge precludes an effective role for science in formulating public policies. And the public, abetted by scientists, *does* misunderstand the natures of science and of scientific knowledge. Wittingly or unwittingly, in their representations to the public, scientists suppress the evolving nature of scientific knowledge. Science is represented as being "archaeological" rather than "interpretational," as uncovering unchanging truths about reality rather than as continually refining empirically justified interpretations of experience. Science is represented as *knowledge* of the world, objective knowledge, the result of reason applied to, tested against, and confirmed by the world itself. The public is assured that personal, social, political, and cultural values play no role in the production and justification of scientific knowledge. The "proof" that science produces true knowledge of the world is that its knowledge claims rest solely on the logic of its explanations, on experimental confirmation, and on prediction of new phenomena, including new technologies. But scientists have always known that these criteria do not and cannot prove the correspondence of a theory with reality.

That theories are true because they correspond with reality is a misrepresentation of science, useful perhaps in gaining for science the status it enjoys in society, and public funding of research, but with harmful social consequences. As became evident in the political struggle in the United States over a national policy with respect to global warming and again with the COVID-19 pandemic, when scientific theories are revised even as policy options are being discussed, people conclude that the scientists do not yet know, that they are guessing, and that this guessing reflects a hidden social or political agenda. Significant portions of the public interpret theory change as a sign that the changed theory was false, so why accept the new one as true? This is a direct consequence of ignorance of the conjectural, corrigible, and evolving nature of scientific knowledge, subject as it is to available information. And this ignorance is exploited by those with a political agenda that would benefit from thwarting or limiting scientific input into a policy decision.

In a lecture at Edinburgh University in 1929, John Dewey asked, "Are there in existence the ideas and the knowledge that permit experimental method to be effectively used in social interests and affairs?"[17] For Dewey, "experimental method" meant the problem-solving methodology employed in science and engineering in which reasoning is reflexively shaped by its consequences in a kind of cognitive feedback loop. This methodology, in which pragmatically validated scientific knowledge—which for Dewey as for Ravetz is really a form of know-how—is coupled to science-based engineering, has proven its practical problem-solving power and fertility. It underlies all of the society-transforming technological innovations of the nineteenth and early twentieth centuries. What Dewey was asking, then, was, do we know enough about this methodology to apply it to the formulation of more effective public policies?

If Dewey's "we" refers to the general public today, the simple answer is 'no.' The public does not know enough about the manner of production and the truth status of scientific knowledge to do this, though we do know enough about engineering's methodology. Public policies are organized around action-limiting and action-promoting value judgments. They are by definition political. Scientific knowledge is promoted as value neutral both in its production and certainly in its application. It follows that *as scientists*, scientists have no expertise in making the social/political value judgments that enter into deciding how their knowledge should be used. Their expertise extends to identifying the foreseeable consequences of one policy or another, but this is qualified by the nature of theory change: new information can lead to theory change and new consequences. Seeing theory change as a flaw rather than as a virtue opens science-related public policies to the charge of being flawed.

The public needs to understand the intrinsically conjectural, contingent, and corrigible nature of scientific knowledge. Scientists produce the best experience-validated accounts of experience available to us at a given time, accounts that are routinely updated in response to new experiences. Science does not give us certain knowledge on which to base action, only reason-and-experience-based probabilities derived from an empirical-experimental methodology that is permanently open to new evidence and new ideas. If global atmospheric and ocean circulation models and their predictions change as new instruments are deployed and new data are acquired, that reflects the *strength* of science to be the best experience-validated account of experience available that that time. Given the political nature of the public policy formulation process, however, this is challenging. Public policies entail financial and legislative commitments. The expense involved and the social impact of policy-associated legislation are justified by their being right. Policy *change* is perceived as the result of a mistake having been made, which can be suicidal for a politician who supported the policy and who voted for the relevant legislation. To the extent that the public perceives science as producing once-and-for-all truths, unanimous agreement among scientists is expected. Change in the science underlying a policy implies that policy was wrong, that wrong things were done, at great public expense, because the science was wrong. The scientists didn't know what they claimed to know, and by listening to them we were misled, so what do we do now?

Everyone understands that technologies evolve, that perfectly functional computers and cell phones, for example, are regularly replaced by new models that are perceived to be better. If, as Dewey and Ravetz argued, science were perceived as a highly sophisticated form of know-how, it too would be expected to evolve. Newer models of the COVID-19 virus, or of global climate change, would be accepted as improved, hence better, models, even if their predictions contradicted predictions by the older models, and certainly if they extended those predictions. Disagreement among scientists would be expected in the same way that we expect differences in the forms in which new technologies are implemented. Science—in the form of the consensus of the scientific community at a given time—would be perceived as providing us with the most reasonable basis on which to formulate science-relevant public policies at that time. The policy formulation process would need to incorporate a mechanism for monitoring the evolution of the science knowledge base of a policy and for adjusting the policy appropriately.

The Science Wars did not result in a better understanding of the nature of scientific knowledge and how it is produced. No understanding emerged that science is intrinsically conflicted, siding with the Gods *and* the Giants, with the archaeological *and* the interpretational views of scientific knowledge. The critics on the "left" reduced science to a belief system with a political agenda of its own, terminally weakening its ability to play an effective role in public policies. Supporters on the "right" defended the objectivity of scientific knowledge and reality as the object of scientific theories, reinforcing an expectation by the public that cannot be met that science produces once-and-for-all knowledge. It is not beyond the public's comprehension to understand that science is functionally interpretational, but promotionally archaeological, and that independent of its internal conflict as both archaeological and interpretational, science remains the best *instrumental* understanding we have of human experience and the most reasonable means we have for affecting future experience.

Notes

Introduction

1. In Slater, 67. Russell addressed this question in *Our Knowledge of the External World* (1928).
2. In Jean Baudrillard, 7.
3. Gaukroger, 11.

Chapter 1

1. Plato, *Sophist*, 241b.
2. Ibid., 246a.
3. Ibid.
4. Ibid.
5. Plato, *Phaedrus*, 185e.
6. Plato, *Thaeatetus*, 185e; also Phaedo, 66a, and *Sophist*, 250a.
7. Plato, *Meno*, 98a; also *Republic*, 477e and 529b.
8. Plato, *Thaeatetus*, 186c.
9. Plato, *Republic*, 534b.
10. Plato, *Phaedo*, 66a.
11. Plato, *Republic*, 510a–511a.
12. Aristotle, *Posterior Analytics*, 88b30.
13. Ibid., 89a1.
14. Aristotle, *Nicomachaean Ethics*, 1139b20.
15. Aristotle, *Metaphysics*, 997b5.
16. Aristotle, *Posterior Analytics*, 71b and 71a40.
17. Ibid., 100a7.
18. Aristotle, *Nicomachaean Ethics*, 1139b28 and 1140b; also *On the Soul*, III, 5.
19. Aristotle, *Posterior Analytics*, 99b–100b.
20. Aristotle, *Metaphysics*, VI, 1; *Posterior Analytics*, 71b15 and 75a40.
21. See especially Untersteiner.
22. Plato, *Cratylus*, 386c.
23. Ibid., 386d–e.
24. Ibid., 439a–440b; also *Phaedrus*, 265ff and 506e.
25. Gorgias, *Helen*; also Untersteiner on Gorgias.
26. Plato, *Phaedrus*, 247c–e.
27. Bernstein, 8.

Chapter 2

1. See, for example, Edgerton on perspective drawing and science, Brown on cartography, and Ramelli on machine drawings. On mathematics and astrology in Renaissance natural philosophy, see McGrath.

2. Bacon, *Advancement of Learning*, in Ellis and Spedding's translation, 87.

3. Bacon, *New Organon*, 34.

4. Bacon, *Great Instauration*, in Ellis and Spedding, 252.

5. Bacon, *New Organon*, 48ff.

6. Ibid., 42.

7. Bacon, *Great Instauration*, loc. cit., 254.

8. Barbara J. Shapiro, 24.

9. Ibid., 24.

10. Bacon, *Great Instauration*, loc. cit., 256.

11. Barbara J. Shapiro, 25.

12. Mazur, *Enlightening Symbols*.

13. Descartes, *Rules for the Direction of the Mind*, Rule II.

14. Ibid., Rules XII and II.

15. Ibid., Rule V.

16. Descartes, *Discourse on Method*, VI, 120ff.

17. Descartes, *Rules*, Rule I.

18. Ibid., Rule IV.

19. Ibid., Rule VIII.

20. Ibid., Rule III.

21. Ibid., Rule IV.

22. Ibid., Rule II.

23. Ibid., Rule XI.

24. Descartes, *Meditations*, II.

25. Ibid., III.

26. Descartes, *Rules,* Rule IV.

27. Lennon, 24.

28. Spinoza, *Ethics*, Part II, Proposition 7.

29. Descartes, *Rules*, IX.

30. Ibid., Rule XI.

31. Descartes, *Discourse on Method*, I, 85.

32. See, for example, Klaas van Berkel, *Isaac Beekman on Matter and Motion*.

33. Descartes, *Discourse*, V, 115.

34. Descartes, *Rules* XI and XIV.

35. Ibid., Rule XI.

36. Descartes, *Discourse*, VI, 119/120.

37. Ibid., 109.

Chapter 3

1. In his *The Optical Part of Astronomy* (1604) and then in his *Dioptrics* (1611) after learning of the spyglass' invention.

2. Galileo, *Sidereal Messenger*, 102.

3. Galileo, *Assayer*, 13.

4. Galileo, *Dialogue*, 328.

5. Ibid., 463.

6. Galileo, *Assayer*, 12.

7. Galileo, *Dialogue*, 52.
8. Galileo, *Assayer*, 19.
9. Ibid., 1.
10. Ibid., Preface.
11. Ibid., 6.
12. Galileo, *Dialogue*, 53.
13. Ibid., 51 and 417.
14. Ibid., 403.
15. Ibid., 58.
16. Ibid., 103.
17. Galileo, *Discourses*, 41.
18. Ibid., 3.
19. Ibid., 1.
20. Ibid., 6.
21. Ibid., 18.
22. Ibid., 254.
23. See, for example, Cartwright.
24. Galileo, *Discourses*, 178.
25. Heilbron, 201.
26. See Graney.

Chapter 4

1. See, especially, Leshem.
2. Sara Dry, *The Newton Papers*.
3. Betty Jo Teeter Dobbs in Leshem, 1.
4. Newton, *Principia*, 943.
5. Newton, *Opticks*, Query 28.
6. Westfall, 239.
7. Newton, in Allen E. Shapiro, "Newton's Experimental Philosophy."
8. Newton, *Principia*, 408ff.
9. See Barbara J. Shapiro for a full discussion of this subject.
10. Newton, *Opticks*, xxii–xxiii.
11. See note 4.
12. See note 5.
13. Newton, *Principia*, 943–944.
14. Newton, *Opticks*, Query 30.
15. Ibid., Query 4.
16. Ibid., Query 1.
17. See note 15.
18. Ibid., Query 29.
19. Ibid., Query 31.
20. Ibid.
21. Huygens, *Treatise*, vi.
22. Huygens, *Cosmotheoros*, in Buchwald and Feingold, 53.
23. Ibid.

24. Leibniz, *Leibniz-Clarke*, Leibniz' first letter.
25. See note 19.
26. Hertz, 2.
27. See Losey for a detailed account.

Chapter 5

1. In Popkin, 8.
2. Dewey, *Quest*, 12.
3. This dispute in its full social context is the subject of Shapin and Schaffer.
4. Spinoza, Part II, Proposition 7.
5. Ibid., Propositions 29 and 32.
6. See, for example, Lolordo.
7. Locke, 10–11.
8. Huygens, *Treatise*, vi.
9. Barbara J. Shapiro, 60.
10. Berkeley, *The Principles of Human Knowledge* and *Three Dialogues Between Hylas and Philonous*.
11. Hume, *Treatise*, 14.
12. Ibid., 23.
13. Ibid., 67ff.
14. Kant, *Prolegomena*, 10.
15. Kant, *Critique*, 22–23.
16. Ibid., Introduction to first edition, 41.
17. Ibid., Introduction to second edition, 41.
18. Ibid., 43.
19. Hume, 305–306.
20. Aristotle, *On the Soul*, 433b11–12.

Chapter 6

1. d'Alembert, 16.
2. Ibid., 122.
3. Ibid., 143ff.
4. Ibid., 8–9.
5. Ibid., 24.
6. On Physiocracy, see Fox-Genovese.
7. On Destutt de Tracy, see Kennedy; on de Biran, see Huxley.
8. There is a very large literature on taxonomies and reality, especially the reality of species. See Frank E. Zakhos, *The Species Concept in Biology*, and Jan Saap, *The New Foundations of Evolution*, on the conventionality of the species concept, also Theodosius Dobzhansky's *Genetics and the Origin of Species*. See Ernst Mayr's *The Growth of Biological Thought* and *Systematics and the Origin of Species* for a defense of the reality of species.
9. Laplace, 4.

Chapter 7

1. Einstein, *Geometry and Experience*. Full text available online.
2. See, for example, Graham Priest et al., *The Law of Non-Contradiction*, and also his *Introduction to Non-Classical Logic*.
3. Kurt Godel, "On Formally Undecidable Propositions of *Principia Mathematica* and Related Systems," 1931, *Monatshefte fur Mathematik*. Several English translations. The one approved by Godel himself was by Jean van Heijenoort in *From Frege to Godel*, 1966. For an explication, see E. Nagel and J. Newman, *Godel's Proof*; Alan Turing, "On Computable Numbers with an Application to the Entschedungsprobla." Turing wrote this paper during a stay at Princeton, where he worked with Alonzo Church.
4. See, for example, Paul Benacerraff and Hilary Putnam, Editors, *Philosophy of Mathematics: Selected Readings*; Roger Penrose, *The Road to Reality: A Complete Guide to the Laws of the Universe*; Mark Steiner, *The Application of Mathematics as a Philosophical Problem*.
5. d'Alembert, 93.

Chapter 8

1. Galileo, *Discourses*, 178.
2. "Bolzano's Logic," *The Stanford Encyclopedia of Philosophy*.
3. Muelders, 76, 139, 149, 151.
4. Ibid., 131.
5. Hertz, 2.
6. Duhem, 223.
7. Duhem, in Ariew and Barker, 66.
8. Whewell, Charles Sanders Peirce, and William James also believed that, ultimately, an experience-based epistemology would intersect ontology, giving us knowledge of the world as it is.
9. Mach, *Science of Mechanics*, 577. See also, Paul Pojman, "Ernst Mach," The online *Stanford Encyclopedia of Philosophy*.
10. Hanson, 6.
11. In Toulmin, 70. See https://plato.stanford.edu/archives/spr2019/entries/ernst-mach/.

Chapter 9

1. Aristotle, *Nicomachaean Ethics*, Book X.
2. Descartes, *Meditations*, II.
3. Spinoza, *Ethics*, V, 42.
4. Kant, *Groundwork for the Metaphysics of Morals*.
5. Hegel, *Phenomenology of the Spirit*.
6. Vico, *On the Most Ancient Wisdom of the Italians: Unearthed from the Origins of the Latin Language*. On Vico, see Donald J. Verene, *Vico's Science of Imagination*.
7. See, for example, the entry "Friedrich Schlegel" in the online *Stanford Encyclopedia of Philosophy*.
8. Nietzsche, *Beyond Good and Evil*, I, 6, in *The philosophy of Nietzsche*, p. 386.

9. Nietzsche, *Gay Science*, in Nietzsche, 301.
10. Nietzsche, *The Anti-Christ*, in Nietzsche, 2.
11. Nietzsche, *Gay Science*, in Nietzsche, 107.
12. Nietzsche, *Genealogy of Morals*, III, paragraph 12 in Nietzsche, 744–745.
13. See note 8.
14. For a concise overview of process philosophy, see Nicholas Rescher, *Process Philosophy*. Alfred North Whitehead's *Process and Reality* is a fully developed process metaphysics, very different from Bergson's in being compatible with modern science.
15. Bergson, *Introduction to Metaphysics*, 36.
16. Ibid., 41.
17. Ibid., 22. See also Milic Capek, "The Myth of Frozen Passage: The Status of Becoming in the Physical World," in *Boston Studies in the Philosophy of Science*, II, pp. 441–464; also Capek's "Time in Relativity Theory: Arguments for a Philosophy of Becoming," in J. T. Fraser, *The Voice of Time*, pp. 434–454.
18. Bergson, 5.
19. James, *Principles*, 628.
20. Proust, *Swann's Way*, "Overture," 34.
21. Bergson, 24.
22. Bergson, 83.

Chapter 10

1. Poincaré, *Science and Hypothesis*, 24.
2. Ibid., 211.
3. Poincaré, *The Value of Science*, 14.
4. Poincaré, *Science and Method*, xxiii.
5. Meyerson, 19.
6. Bridgman, vii.
7. Ibid., viii.
8. Ibid., 1.
9. Ibid., 5.
10. Peirce, in *Charles Sanders Peirce: Selected Writings*, 132.
11. James, *Pragmatism and the Meaning of Truth*, 14.
12. Ibid., 11ff, and again 124ff.
13. James, *Pragmatism and the Meaning of Truth*, 172–173.
14. James, *Essays*, 9.
15. James, *Pragmatism*, 20.
16. Ibid., 100.
17. Dewey, in Boydston, *Later Works*, volume 12, p. 16.
18. Ibid.
19. Dewey, in Boydston, *Middle Works*, volume 6, p. 67.
20. Russell, *Our Knowledge of the External World*, 11.
21. Russell in Wittgenstein, *Tractatus*, 7.
22. Wittgenstein, *Tractatus*, 31.
23. Ibid.
24. Schlick, in Scheffler, 102.
25. Reichenbach, *Axiomatization*, 5.

Chapter 11

1. Einstein, in Stachel, 178.
2. Ibid., 177.
3. Ibid., 178.
4. Ibid.
5. Einstein, *Annals of Physics*, volume 336, Sept. 2013, 56–75.
6. Wigner, *Communications in Pure and Applied Mathematics*, volume 13, number 1, Feb. 1960. For a range of views on mathematics and reality see Paul Benaceraff and Hilary Putnam, Editors, *Philosophy of Mathematics: Selected Readings*; Roger Penrose, *The Road to Reality: A Complete Guide to the Laws of the Universe*; and Mark Steiner, *The Application of Mathematics as a Philosophical Problem*.
7. For example, Bohr, *Atomic Physics and Human Knowledge*, and Heisenberg, *Physics and Philosophy: The Revolution in Modern Science*.
8. Cited in Robert P. Crease, *The Great Equations*, 259.
9. Wheeler, in Halpern, 251.
10. Feynman, ibid., 252.

Chapter 12

1. Descartes, *Meditations*, II.
2. Joyce, 186.
3. Hume, *Treatise*, 192ff.
4. Jacobson, 15.
5. Hanson, 6.
6. Descartes, in *Dioptrics*, Discourse 6.
7. Fleck, 46.
8. Ibid., xxvii.
9. Ibid., 39.
10. Ibid., 142.
11. Ibid., 38.
12. Ibid., 37.
13. Ibid., 51.
14. Ibid., 95.
15. Ibid., 51.
16. Ibid., 48.
17. Ibid., 49.
18. Ibid., 109.

Chapter 13

1. Kuhn, *Structure*, 3.
2. Ibid., 1.
3. Ibid.
4. Ibid.
5. Ibid.

6. Ibid., 2.
7. Ibid.
8. See Graney.
9. Kuhn, 168–169.
10. Ibid., 30.
11. Ibid., 26.
12. Ibid.
13. Ibid., 3.
14. Ibid., 16.
15. Ibid., 4.
16. Ibid., 17.
17. Ibid., 80.
18. Ibid., 45.
19. Ibid., 86.
20. Durkheim.
21. Ibid., 158.
22. Ibid., 151.
23. Ibid., 90.
24. Ibid., 96.
25. Ibid., 80.
26. Ibid., 24.
27. Ibid., 36.
28. Ibid., 151–152.
29. Ibid., 7.
30. Ibid., 103.
31. Ibid., 147ff.
32. Ibid., 103.
33. Ibid., 111.

Chapter 14

1. Scheffler, 8.
2. Ibid., 122.
3. Ibid., 123.
4. Feyerabend, *Against Method*, 17. In a footnote on p. 1, Feyerabend explains why he prefers the term "epistemological Dadaism" to "epistemological anarchism."
5. Ibid., 295.
6. In Lakatos and Musgrave. Larry Laudan defended a progressive theory of science based on research traditions rather than on Lakatos' research programs.
7. See, for example, Jean Lyotard, *The Differend*.
8. Foucault, *The Order of Things*, xi.
9. Sheridan, 116.
10. Ibid., 118.
11. Ferdinand de Saussure, *Course in General Linguistics*.
12. Plato, *Phaedrus*, 274c.

Chapter 15

1. See especially Stephen H. Cutcliffe, *Ideas, Machines and Values: An Introduction to Science, Technology and Society Studies.*
2. Latour and Woolgar, 186.
3. Schaffer and Shapin, 344.
4. Pickering, *Constructing Quarks*, 186–187.
5. Schweber and Gingras, "Constraints on Construction."
6. Pickering, *Mangle*, 22.
7. Quine, "Twin Dogmas of Empiricism."
8. Quine, *Word and Object*, 221.
9. Quine, *Theories and Things*, 21.
10. Ibid., 67.
11. Quine, *Ontological Relativity*, 83.
12. Hesse, "Theories, Dictionaries, and Observation."
13. Hesse, "A New look at Scientific Explanation."
14. Putnam, 1987. It should be noted that Putnam often changed his philosophical commitments.

Chapter 16

1. Weinberg, 185–190, in a chapter titled "Against Philosophy." On the "distrust and even hostility" of many humanists toward science as objective knowledge, see Allan Franklin, *Are There Really Neutrinos?*, 5.
2. Gross and Levitt, *Higher Superstition*, 1.
3. Ibid., 2.
4. Ibid., 15.
5. Ibid., 234.
6. Ibid., 235.
7. Ibid., 79.
8. Ibid., 95.
9. Weinberg, 189.
10. Sokal, in *Science Text*.
11. *Lingua Franca*, May/June, 1996.
12. Beller, *Physics Today*, volume 51, 9, p. 29 (1998). The hoax was defended by the physicist Kurt Gottfried in *Physics Today*, volume 50, 1, p. 61 (1997).
13. Koertge, 3.
14. Kitcher, in Koertge, 32.
15. Franklin, 1.
16. Feyerabend, quoted in the posthumous collection *Conquest of Abundance: A Tale of Abstraction Versus the Richness of Being*, vii.
17. Dewey, *Quest*, 218.

Bibliography

Alembert, Jean d'. *Preliminary Discourse to the Encyclopedia of Diderot*. Translator Richard N. Schwab and Walter E. Rex. New York: Bobbs-Merrill, 1963.

Ariew, Roger, and Peter Barker, Editors. *Essays in the History and Philosophy of Science*. Indianapolis: Hackett, 1996.

Aristotle. *The Complete Works*. Princeton: Princeton University Press, 1984.

Bacon, Francis. *The New Organon*. New York: Bobbs-Merrill, 1960.

Bacon, Francis. *The Philosophical Works of Francis Bacon*. Translators James Spedding and Robert Leslie Ellis. London: Routledge, 1905.

Baudrillard, Jean. *Simulacra and Simulation*. Translator Sheila Farla Glaser. Ann Arbor: University of Michigan Press, 1994.

Benaceraff, Paul, and Hilary Putnam. *Philosophy of Mathematics: Selected Readings*. Cambridge: Cambridge University Press, 1984.

Bergson, Henri. *Creative Evolution*. Translator Arthur Mitchell. New York: Modern Library, 1944.

Bergson, Henri. *Introduction to Metaphysics*. Translator T. E. Hulme. Indianapolis: Hackett, 1999.

Bergson, Henri. *Time and Free Will: An Essay on the Immediate Data of Consciousness*. Translator F. L. Pogson. New York: Harper and Row, 1960.

Berkel, Klaus van. *Isaac Beeckman on Matter and Motion: Mechanics in the Making*. Baltimore: Johns Hopkins University Press, 2013.

Berkeley, George. *The Principles of Human Knowledge and Three Dialogues Between Hylas and Philonous*, New York: Penguin, 1988.

Bernstein, Richard. *Beyond Objectivity and Relativism: Science, Hermeneutics and Praxis*. Philadelphia: University of Pennsylvania Press, 1983.

Biran, Maine de. See Huxley.

Bohm, David, and Basil J. Hiley. *The Undivided Universe: An Ontological Interpretation of Quantum Mechanics*. London: Routledge, 1995.

Bohr, Niels. *Atomic Physics and Human Knowledge*. New York: Dover, 2010.

Bolzano, Bernard. In the online *Stanford Encyclopedia of Philosophy*. Editor Edgar Morscher. 2018.

Bridgman, Percy. *The Logic of Modern Physics*. New York: Macmillan, 1927.

Brown, Lloyd Arnold. *The Story of Maps*. New York: Dover, 1980.

Buchwald, Jed, and Mordechai Feingold. *Newton and the Origin of Civilization*. Princeton: Princeton University Press, 2012.

Carnap, Rudolf. *The Logical Structure of the World and Pseudoproblems in Philosophy*. Translator Rolf A. George. Lasalle, Ill.: Open Court, 2003.

Carnap, Rudolf. *The Philosophical Foundations of Physics: An Introduction to the Philosophy of Science*. Editor Martin Gardner. New York: Basic Books, 1966.

Cartwright, Nancy. *How the Laws of Science Lie*. Oxford: Clarendon, 1983.

Crease, Robert P. *The Great Equations*. New York: Norton, 2008.

Cutcliffe, Stephen H. *Ideas, Machines and Values: An Introduction to STS Studies*. Lanham, MD: Rowman and Littlefield, 2000.

Derrida, Jacques. *On Grammatology*. Translator Gayatri Chakravarti Spivak. Baltimore: Johns Hopkins University Press, 1976.

Derrida, Jacques. *Speech and Phenomena*. Translator David B. Allison. Evanston: Northwestern University Press, 1973.

Derrida, Jacques. *Writing and Difference*. Translator Alan Bass. Chicago: University of Chicago Press, 1978.

Descartes. *The Philosophical Works*. Translators Elizabeth S. Haldane and G. R. T. Ross. Cambridge: Cambridge University Press, 1973.

Dewey, John. *Experience and Nature*. New York: Dover, 1958.

Dewey, John. *The Middle Works 1899–1924*, volume 6. Editor Jo Ann Boydston. Carbondale: Southern University Press, 1978.

Dewey, John. "Logic: The Theory of Directed Inquiry." In *The Later Works of John Dewey*, volume 12. Editor Jo Anne Boydston. Carbondale: Southern Illinois University Press, 2008.

Dewey, John. *The Quest for Certainty*. Carbondale: University of Illinois Press, 1990.

Diderot, Denis. *L'Encyclopedie*. Oxford: Pergamon, 1969.

Dobbs, Betty J. Teeter. *The Foundations of Newton's Alchemy*. Cambridge: Cambridge University Press, 1975.

Dry, Sarah. *The Newton Papers*. Oxford: Oxford University Press, 2014.

Duhem, Pierre. *The Aim and Structure of Physical Theory*. Translator Philip P. Wiener. Princeton: Princeton University Press, 1991.

Durkheim, Emile. *The Rules of the Sociological Method*. Translators Sarah A. Solvay and John H. Mueller. New York: Free Press, 1966.

Edgerton, Samuel Y. *The Heritage of Giotto's Geometry: Art and Science on the Eve of the Scientific Revolution*. Ithaca: Cornell University Press, 1994.

Edgerton, Samuel Y. *The Mirror, the Window and the Telescope: How Linear Perspective Changed Our Vision of the Universe*. Ithaca: Cornell University Press, 2009.

Einstein, Albert. *Geometry and Experience*. London: Methuen, 1922.

Feyerabend, Paul. *Against Method*. London: Verso, 1975.

Feyerabend, Paul. *Conquest of Abundance*. Chicago: University of Chicago Press, 1999.

Fine, Arthur. *The Shaky Game: Einstein, Realism and the Quantum Theory*. Chicago: University of Chicago Press, 1986.

Fleck, Ludwik. *Genesis and Development of a Scientific Fact*. Translators Fred Bradley and Thaddeus J. Trenn. Chicago: University of Chicago Press, 1981.

Foucault, Michel. *The Archaeology of Knowledge*. Translator A. M. Sheridan-Smith. London: Tavistock Press, 1994.

Foucault, Michel. *The Order of Things: An Archaeology of the Human Sciences*. New York: Vintage, 1994.

Fox-Genovese, Evelyn. *The Origins of Physiocracy: Economic Revolution and Social Order in Eighteenth Century France*. Ithaca: Cornell University Press, 1976.

Fraassen, Bas Van. *The Empirical Stance: The Terry Lectures*. New Haven: Yale University Press, 2002.

Franklin, Allan. *Are There Really Neutrinos? An Evidential History*. Cambridge: Perseus Press, 2001.

Gadamer, Hans-Georg. *Reason in the Age of Science*. Translator Frederick G. Lawrence. Cambridge, Mass.: MIT Press, 1984.

Galilei, Galileo. *The Assayer*. Translator Stillman Drake. www.stanford.edu.

Galilei, Galileo. *Dialogues Concerning the Two Great World Systems*. Berkeley: University of California Press, 1967.

Galilei, Galileo. *Discourses on Two New Sciences*. Translators Henry Crew and Alfonsio de Salvio. New York: Dover, 1954.

Galilei, Galileo. *The Sidereal Messenger*. Translator Albert Van Helden. Chicago: University of Chicago Press, 1989.

Galison, Peter. *How Experiments End*. Chicago: University of Chicago Press, 1987.

Galison, Peter. *Image and Logic: A Material Culture of Microphysics*. Chicago: University of Chicago Press, 1997.

Gaukroger, Stephen. *The Emergence of a Scientific Culture: Science and the Shaping of Modernity 1210–1685*. Oxford: Clarendon, 2007.

Giere, Ronald N. *Scientific Perspectivism*. Chicago: University of Chicago Press, 2010.

Godel, Kurt. "On Formally Undecidable Propositions of *Principia Mathematica* and Related Systems." In *A Sourcebook in Mathematical Logic 1879–1931*. Editor Jean van Heijenoort. Cambridge: Harvard University Press, 1967.

Goodman, Nelson. *Ways of Worldmaking*. Indianapolis: Hackett, 1978.

Graney, Chistopher M. *Setting Aside All Authority: Giovanni Battista Riccioli and the Science Against Copernicus in the Age of Galileo*. Notre Dame: University of Notre Dame Press, 2015.

Gross, Paul R., and Norman Levitt. *Higher Superstition: The Academic Left and Its Quarrels with Science*. Baltimore: Johns Hopkins University Press, 1994.

Habermas, Jurgen. *The Theory of Communicative Action*. Translator Thomas McCarthy. Boston: Beacon Press, 1984.

Hacking, Ian. *The Social Construction of What?* Cambridge, Mass.: Harvard University Press, 1999.

Halpern, Paul. *The Quantum Labyrinth: How John Wheeler and Richard Feynman Revolutionized Time and Reality*. New York: Basic Books, 2018.

Hanson, Norwood Russell. *Patterns of Discovery*. Cambridge: Cambridge University Press, 1958.

Harding, Sandra. *The Science Question in Feminism*. Ithaca: Cornell University Press, 1986.

Harding, Sandra. *Whose Science? Whose Knowledge? Thinking from Women's Lives*. Ithaca: Cornell University Press, 1991.

Hausman, Carl. *Charles S. Peirce's Evolutionary Philosophy*. Cambridge: Cambridge University Press, 1995.

Hegel, G. F. W. *The Phenomenology of Spirit*. Translator Terry Pinkard. Cambridge: Cambridge University Press, 2019.

Heilbron, J. L. *Galileo*. New York: Oxford University Press, 2010.

Heisenberg, Werner. *Physics and Philosophy: The Revolution in Modern Science*. New York: Harper, 2007.

Helmholtz, Hermann. *On the Sensations of Tone*. Translator Alexander J. Ellis. New York: Dover, 1954.

Hempel, Carl. *Philosophy of Natural Science*. New York: Prentice-Hall, 1966.

Hertz, Heinrich. *The Principles of Mechanics Presented in a New Form*. London: Macmillan, 1894.

Hesse, Mary. *Forces and Fields: The Concept of Action at a Distance in the History of Physics*. New York: Dover, 1962.

Hesse, Mary. *Models and Analogies in Science*. Notre Dame: Notre Dame University Press, 1966.

Hesse, Mary. "A New look at Scientific Explanation." *Review of Metaphysics*, volume 17, issue 1, (September 1963), pp. 98–108.

Hesse, Mary. "Theories, Dictionaries, and Observation." *British Journal for the Philosophy of Science*, volume IX (May 1958), pp. 12–28.

Hobbes, Thomas. *Leviathan*. New York: Penguin Classics, 2017.

Hume, David. *Enquiry Concerning the Human Understanding*. Editor Eric Steinberg. Indianapolis: Hackett, 1993.

Hume, David. *Treatise on Human Nature*. New York: Oxford University Press, 2000.

Huxley, Aldous. *Themes and Variations*. New York: Harper, 1950.

Huygens, Christiaan. *Treatise on Light*. Translator Sylvanus P. Thompson. New York: Dover, 1984.

Isocrates. *Isocrates II: Antidosis*. Translator George Norlin. Cambridge, Mass.: Harvard University Press (Loeb Classical Library), 1929.

Jacobson, Howard. *The Very Model of a Man*. New York: Viking, 1992.

James, William. *Essays in Radical Empiricism*. New York: Longmans Green, 1922.

James, William. *Pragmatism and the Meaning of Truth*. Cambridge, Mass.: Harvard University Press, 1978.

James, William. *The Principles of Psychology*. New York: Dover, 1950.

Joyce, James. *A Portrait of the Artist as a Young Man*. London: Penguin Classics, 2003.

Kant, Immanuel. *Critique of Pure Reason*. Translator Norman Kemp Smith. New York: St. Martin's Press, 1965.

Kant, Immanuel. *Groundwork for the Metaphysics of Morals*. Translator Mary Gregory. Cambridge: Cambridge University Press, 2012.

Kant, Immanuel. *Prolegomena to Any Future Metaphysics*. Translator Gary Hatfield. Cambridge: Cambridge University Press, 2004.

Keller, Evelyn Fox. *Reflections on Gender and Science*. New Haven: Yale University Press, 1985.

Kennedy, Emmett. *A Philosophe in the Age of Revolution: Destutt de Tracy and the Origins of Ideology*. Philadelphia: American Philosophical Society, 1978.

Kitcher, Philip. *The Advancement of Science: Science Without Legend, Objectivity without Illusion*. New York: Oxford University Press, 1993.

Kitcher, Philip. *Science, Truth and Democracy*. New York: Oxford University Press, 2001.

Koertge, Noretta. *A House Built on Sand*, New York: Oxford University Press, 1998.

Kuhn, Thomas. *The Essential Tension*. Chicago: University of Chicago Press, 1977.

Kuhn, Thomas. *The Structure of Scientific Revolutions*. Chicago: University of Chicago, 1962.

Lakatos, Imre. *The Methodology of Scientific Research Programs*. Editors John Worrall and Gregory Currie. Cambridge: Cambridge University Press, 1970.

Lakatos, Imre, and Alan Musgrave, Editors. *Criticism and the Growth of Knowledge*, (including Lakatos' important paper "Falsification and the Methodology of Science"). Cambridge: Cambridge University Press, 1970.

Laplace, Pierre Simon Marquis de. *A Philosophical Essay on Probabilities*. Translators Frederick Wilson Truscott and Frederick Lincoln Emory. New York: Dover, 1951.

Latour, Bruno. *Science in Action: How to Follow Scientists and Engineers Through Society*. Cambridge, Mass.: Harvard University Press, 1987.

Latour, Bruno, and Steve Woolgar. *Laboratory Life: The Construction of Scientific Facts*. Princeton: Princeton University Press, 1979.

Laudan, Larry. *Science and Relativism*. Chicago: Chicago University Press, 1990.

Leibniz, Gottfried Wilhelm. *The Leibniz-Clarke Correspondence*. Editor H. G. Alexander. Manchester: Manchester University Press, 1956.

Lennon, Thomas. *The Battle of the Gods and Giants: The Legacies of Descartes and Gassendi 1655–1715*. Princeton: Princeton University Press, 2016.

Leshem, Ayval. *Newton on Mathematics and Spiritual Purity*. Dordrecht: Kluwer, 2003.

Locke, John. *An Essay Concerning Human Understanding*. Editor Roger Woolhouse. London: Penguin, 1997.

Lolordo, Antonia. *Pierre Gassendi and the Birth of Early Modern Philosophy*. Cambridge: Cambridge University Press, 2006.

Longino, Helen. *Science as Social Knowledge*. Princeton: Princeton University Press, 1990.

Losey, John. *Theories on the Scrap Heap: Science and Philosophy on the Falsification, Rejection and Replacement of Theories*. Pittsburgh: University of Pittsburgh Press, 2005.

Lyotard, Jean. *The Differend*. Minneapolis: University of Minnesota Press, 1989.

Mach, Ernst. *The Analysis of Sensations*. Translator C. M. Williams. New York: Dover, 1954.

Mach, Ernst. *Science of Mechanics: A Critical Account of Its History and Development*. Chicago: Open Court, 1988.

Mannheim, Karl. *Ideology and Utopia: An Introduction to the Sociology of Knowledge*. Translators Louis Wirth and Edward Shils. Eastford, Conn.: Martino Fine Books, 2015.

McGrath, Mary Quinlan. *Influences: Art, Optics and Astrology in the Italian Renaissance.* Chicago: University of Chicago Press, 2013.

Merchant, Carolyn. *The Death of Nature: Women, Ecology and the Scientific Revolution.* New York: HarperOne, 1980.

Merton, Robert K. *The Sociology of Science: Theoretical and Empirical Investigations.* Chicago: University of Chicago Press, 1973.

Meyerson, Emile. *Identity and Reality.* New York: Dover, 1962.

Mill, John Stuart. *System of Logic: Ratiocinative and Inductive.* London: Benediction Classics, 2011.

Muelders, Michel. *Helmholtz: From Enlightenment to Neuroscience.* Translator Lawrence Garey. Cambridge, Mass.: MIT Press, 2010.

Newton, Isaac. *Opticks.* New York: Dover, 1952.

Newton, Isaac. *The Principia: Mathematical Principles of Natural Philosophy.* Translators I. B. Cohen and Anne Whitman. Berkeley: University of California Press, 1999.

Nietzsche, Friedrich S. *The Philosophy of Nietzsche.* New York: Modern Library, 1927.

Novalis. *Philosophical Writings.* Translator and editor Margaret Mahony Stoljar. Albany: State University of New York Press, 1997.

Nye, Mary Jo. *Michael Polanyi and His Generation: Origins of the Social Construction of Science.* Chicago: University of Chicago Press, 2011.

Pais, Abraham. *Niels Bohr's Times: On Physics, Philosophy and Polity.* Oxford: Clarendon, 1991.

Peirce, Charles S. *Charles Sanders Peirce: Selected Writings.* Editor Philip P. Wiener. New York: Dover, 1958.

Penrose, Roger. *The Road to Reality: A Complete Guide to the Laws of the Universe.* New York: Vintage, 2007.

Pickering, Andrew. *Constructing Quarks: A Sociological History of Particle Physics.* Chicago: University of Chicago Press, 1984.

Pickering, Andrew. *The Mangle of Practice: Time, Agency and Science.* Chicago: University of Chicago Press, 1995.

Plato. *Plato: The Collected Dialogues.* Editors Edith Hamilton and Huntington Cairns. Princeton: Bollingen, 1964.

Poincaré, Henri. *Science and Hypothesis.* New York: Dover, 1952.

Poincaré, Henri. *Science and Method.* Translator Francis Maitland. New York: Cosimo Classics, 2007.

Poincaré, Henri. *The Value of Science.* Translator George Bruce Halstead. New York: Dover, 1958.

Pojman, Paul. "Ernst Mach." In the online *Stanford Encyclopedia of Philosophy* (Spring 2019 Edition). Editor Edward N. Zalta. 2019.

Polanyi, Michael. *Personal Knowledge: Towards a Post-Critical Philosophy.* Chicago: University of Chicago Press, 1958.

Popkin, Richard. *A History of Skepticism from Savonarola to Bayle.* New York: Oxford University Press, 2003.

Priest, Graham. *An Introduction to Non-Classical Logics.* Cambridge: Cambridge University Press, 2008.

Priest, Graham, J. C. Beall, and Bradley Armour-Garb. *The Law of Non-Contradiction.* Oxford: Clarendon, 2004.

Proust, Marcel. *Remembrance of Things Past.* Translator C. K. Scott Moncrieff. New York: Random House, 1934.

Putnam, Hilary. *The Many Faces of Realism.* Editor James Conant. Cambridge, Mass.: Harvard University Press, 1987.

Quine, Willard van Orman. *Ontological Relativity and Other Essays.* New York: Columbia University Press, 1969.

Quine, Willard van Orman. *Pursuit of Truth.* Cambridge, Mass.: Harvard University Press, 1990.

Quine, Willard van Orman. *Theories and Things*. Cambridge: Belknap Press, 1986.

Quine, Willard van Orman. "Twin Dogmas of Empiricism." *Philosophical Review*, volume 60 (1951), pp. 20–43.

Quine, Willard van Orman. *Word and Object*. Cambridge, Mass.: MIT Press, 2013.

Ramelli, Agostino. *The Various and Ingenious Machines of Agostino Ramelli*. London: Scolar Press, 1987.

Reichenbach, Hans. *Axiomatization of the Theory of Relativity*. Translator Maria Reichenbach. Berkeley: University of California Press, 1969.

Rescher, Nicholas. *Process Metaphysics*. Albany: SUNY Press, 1996.

Russell, Bertrand. *Our Knowledge of the External World*. New York: New American Library, 1960.

Salmon, Wesley. *Scientific Explanation and the Causal Structure of the World*. Princeton: Princeton University Press, 1984.

Saussure, Fredinand de. *Course in General Linguistics*. Chicago: Open Court, 1998.

Schechter, Eric. *Classical and Non-Classical Logics*. Princeton: Princeton University Press. 2005.

Scheffler, Israel. *Science and Subjectivity*. Indianapolis: Hackett, 1982.

Schweber, Sylvan, and Gingras, Yves. "Constraints on Construction." *Social Studies of Science*, volume 16 (1986), pp. 372–383.

Shapin, Steven, and Simon Schaffer. *Leviathan and the Air Pump: Hobbes, Boyle and the Experimental Life*. Princeton: Princeton University Press, 1985.

Shapiro, Alan E. *Fits, Passions and Paroxysms: Physics, Method and Chemistry and Newton's Theory of Colored Bodies*. Cambridge: Cambridge University Press, 1993.

Shapiro, Allen E. "Newton's Experimental Philosophy." *Early Science and Medicine*, volume 9 (2004), pp. 185–217.

Shapiro, Barbara J. *Probability and Certitude in Seventeenth Century England*. Ithaca: Cornell University Press, 2000.

Sheridan, Alan. *Michel Foucault: The Will to Truth*. London: Routledge, 2015.

Slater, John G. *Bertrand Russell*. Bristol: Thoemmes, 1994.

Sokal, Alan. "Transgressing the Boundaries: Towards a Transformative Hermeneutics of Quantum Gravity." *Science Text*, Spring/Summer (1996), pp. 217–252.

Spinoza, Baruch. *Spinoza: Complete Works*. Indianapolis: Hackett, 2002.

Stachel, John. *Einstein's Miraculous Year*. Princeton: Princeton University Press, 1998.

Stallo, J. B. *The Concepts and Theories of Modern Physics*. Cambridge: Belknap Press.

Steiner, Mark. *The Application of Mathematics as a Philosophical Problem*. Cambridge MA: Harvard University Press, 1998.

Toulmin, Stephen, Editor. *Physical Reality: Philosophical Essays on Twentieth Century Physics*. New York: Harper, 1970.

Turing, Alan. "On Computable Numbers with an Application to the Entschedungsprobla." *Proceedings of the London Mathematical Society*, volume s2–42 (1, 1937), pp. 230–265.

Untersteiner, Mario. *The Sophists*. Translator Katheleen Freeman. New York: Philosophical Library, 1954.

Verene, Donald J. *Vico's Science of Imagination*. Ithaca: Cornell University Press, 1991.

Vico, Giovanni. *The New Science*. Translator Dave Marsh. New York: Penguin Classics, 2000.

Vico, Giovanni. *On the Most Ancient Wisdom of the Italians: Unearthed from the Origins of the Latin Language*. Translator L. M. Palmer. Ithaca: Cornell University Press, 1988.

Weinberg, Steven. *Dreams of a Final Theory: The Search for the Fundamental Laws of Nature*. New York: Pantheon, 1993.

Westfall, Richard. *Never at Rest: A Biography of Isaac Newton*. Cambridge: Cambridge University Press, 1983.

Wittgenstein, Ludwig. *Tractatus Logico-Philosophicus*. London: Routledge and Kegan Paul, 1958.

Index

For the benefit of digital users, indexed terms that span two pages (e.g., 52–53) may, on occasion, appear on only one of those pages.

a priori knowledge, 12, 117–19, 123. *See also* self-evident truth; synthetic a priori truths
abduction, 206
    Charles Peirce on, 117, 148–49, 168–69, 206
    induction and, 117, 148–49, 206
    nature of, 148–49
Abraham, 138
absolute knowledge, 8–9, 141, 143–44
abstraction, 152, 226–27
academic left, 245, 247, 248, 250
Actor Network Theory, 230
Adelard of Bath, 21
aesthetics, 137
aether theories, 107–8
air, 98
air pump experiments, Boyle–Hooke, 65–66, 232
alchemy, 52, 70
analytic philosophy, 236–37
analytical reasoning and methodology, 142–43
analytic–synthetic distinction, 157, 159, 236–37. *See also* synthetic a priori truths
Angier, Natalie, 231
anomalies as threats to paradigms, 201–2
anti-foundationalism, 243
Antiphon, 12–13, 14
Apollonius of Perga, 24–25
Archimedes, 36–37
    Galileo and, 36–37, 41–42, 43, 44
    on mathematical physics, 24–25, 36–37, 41–42, 43, 44
    writings, 41–42, 43
Aristotle, 26, 78
    on atomism, 106
    characterizations of, 13–14, 22
    fallacies of deductive reasoning, 28–29
        (*see also* Fallacy of Affirming the Consequent)
    first principles and, 12, 22–23, 26
    on knowledge, 10–11, 12, 13
    physics and, 172
    Plato and, 10, 11–14, 22, 26
    on practical vs. theoretical wisdom, 152
    Protagoras and, 12–13
    on reality, 10, 13
    on reason, 78, 130
    on reasoning, 10–11, 12, 22–23, 28–29
    sophists and, 12–13, 222–23
    on universals and particulars, 10–12, 22–23, 120
    writings, 10, 21–23, 26–27
Arp, Halton, 202
art, 139
art history, 185, 203–4, 228
association, Hume's principles of, 93, 180
Association of Christian Schools International (ACSI) v. Roman Stearns, 251
associations, active vs. passive, 191–92
assumptions. *See also* suppositions
    induction and, 63, 74, 111, 114
astronomers, 13, 48–49, 95–96, 176–77, 199–200. *See also* Copernicus; Galileo; Kepler
astronomy, 48–49, 94–95, 101, 117–18. *See also* Big Bang; gravity; heliocentrism; Milky Way; stars
    Catholic Church and, 43, 46
    Galileo and, 37–41, 43, 46, 47–49
atom, "solar system" model of the, 168
atomic theory of matter, 69–70, 106–7, 119, 128, 168, 191–92
    Bohr and, 165–66
    Dalton and, 102, 106, 131, 200–1
    Einstein and, 163–64
    Planck and, 128, 163–64, 165–66
atomic weight, 191–92
atomism, 33, 106. *See also* logical atomism; social atomism
atomists, 106
Augustine, St., 12, 32–33

Bacon, Francis, 27–28, 30, 34–35
  deduction and, 26, 28, 29–30, 117
  Descartes and, 25–26, 30, 31–32, 34–35,
    67, 180
  experimental method of, 26–30, 31–32,
    34–35, 83, 102–4, 115–16, 119, 124–25
  Galileo and, 40, 45–46
  Hobbes and, 64–65, 67
  on Idols of the mind, 26–27, 115–16, 117,
    127–28, 185–86, 220–21
  induction and, 26–27, 28, 29–30, 34–
    35, 115–16
  intuition and, 26, 28
  Jean d'Alembert and, 86–87
  knowledge problem and, 30
  mathematics and, 25–26, 28
  on mind, 26, 27
  natural philosophy, natural philosophers,
    and, 25–27, 29–30, 82
  The New Organon, 26–28, 104
  Newton and, 53, 55
  overview and characterizations of, 25–
    26, 30
  reasoning and, 25–27, 28, 67, 115–16, 117,
    180, 185–86, 187
  William Whewell and, 116, 117, 119
Barberini, Maffeo (Pope Urban VIII), 47
Baudrillard, Jean, 1, 219
Beeckman, Isaac, 33
behavioral psychology, 184
Behe, Michael, 250–51
belief, 148. *See also specific topics*
Bellarmine, Roberto, 46, 47
Beller, Mara, 248
Bergson, Henri, 142, 143–44
  on change, 140
  empiricism and, 141, 142, 144
  on intuition, 141, 142–44
  on memory, 141, 142, 143
  metaphysics, 140–42, 144
  on rationalism, 141
  on reality, 140, 142
  on reason, 137, 140, 141–43, 180–81
  time and, 140, 141–42, 144
  writings, 140
Berkeley, George, 72–73
  critique of matter, 72–73, 93
Berlin Circle, 154, 157–58
Bernoulli, Jacob, 90
Bernstein, Richard, 19
Bible, 21, 46, 50–51, 81, 138

Big Bang, 173–74, 202, 204–5
Biran, Maine de, 93–94
black-body radiation
  Einstein and, 106, 162, 163–64
  Planck and, 161–62, 163–64
Bloom, Alan, 244–45
Bloor, David, 232–33
Boas, Franz, 180–81
Bohr, Niels, 125, 162–63, 166, 168–69, 171–
  72, 216
  complementarity principle, 171
  correspondence principle, 213
  vs. Einstein, 162–63, 165–67, 206–7
  Heisenberg and, 171
  radiation and, 163, 168, 196, 216
  Rutherford and, 128, 168
Bolyai, Janos, 109–10
Bolyai–Lobachevskian geometry, 109–10
Born, Max, 166
Boscovich, Roger, 55
botany, 98–99
Boyle, Robert, 65
  Hobbes and, 65–66, 231–32
Boyle's law, 65
Bradley, James, 94
Brahe, Tycho, 17, 116–17
  Copernicus and, 17, 48–49, 200
  Galileo and, 37, 48–49, 196–97
  Kepler and, 30, 37, 116–17, 119–20, 196–97
Breteuil, Gabrielle de, 84–85
Bridgman, Percy, 147
Buridan, Jean, 21

Cabanis, Pierre, 93–94
caloric theory of heat, 104–5, 196
  Antoine Lavoisier and, 98, 103–4, 196
Campanus of Novara, 22
Campbell, Donald T., 237–39
Cardan, Jerome, 23
Carnap, Rudolf, 153–54, 158, 236
Cartesian self
  Hume and, 92–93, 180
  unconscious and, 182
Cartesianism. *See also* Descartes, René
  Spinoza's version of, 33
Cartwright, Susan, 241, 249–50
Cassini, G. D., 95
Catholic Church, 36, 69
  Copernican heliocentrism and, 36, 46, 47
  Galileo and, 36, 43, 46–47
  skepticism and, 18, 69

Catudal, Jacques, 19
causal connection, 68, 74–75, 78, 146–47
    Hume on, 74, 76, 78
causality
    Hume's critique of, 73, 75
    Kant and, 75, 76
cause, defined, 68
cell theory of life, 131–32
certainty, 28, 73–74. *See also* deduction;
        skepticism; uncertainty principle
    definitions, 71
    empiricism and, 28, 43–44
    Kant and, 77, 78–79, 133
    Locke on, 71–72
    quest for, 64, 69, 73–74, 78–79, 133, 149
Cetina, Karin Knorr, 233
Chambers, Ephraim, 85
Charcot, Pierre-Martin, 181–82
chemistry, 106, 131, 132, 184–85, 200–1
Clairaut, Alexis, 96–97
Clarke, Samuel, 60, 61
classification, 98–99, 124
classification schemes, 86, 98–99, 220. *See
        also* taxonomies
Cogito, ergo sum. See "I think, therefore I am"
cognition, nature of, 189–90
coherence, 118–19, 159–60
colligation, 117–19
Collins, Harry, 232–33
combustion, theories of, 17, 65, 102. *See also*
        oxygen theory of combustion
comets, 41, 94–95
complementarity (physics), 171
complementary concepts and
        complementarity, 171
Composition, Newton's method of, 57, 58
Comte, Auguste, 120–22
conceptual frameworks, 240–41
Condillac, Pierre, 93–94, 186
Condorcet, Marquis de, 90–91
conservation laws, 113–14
consilience, 118–19
constructionism, social. *See* social
        constructionism
constructionism, strict, 224
constructive realism, 228–29
Copenhagen interpretation of quantum
        mechanics, 166, 172–74, 202, 215
Copernican heliocentrism, 29, 48–49, 117–
        18, 200, 204–5
    Catholic Church and, 36, 46, 47

    Galileo and, 36, 37, 39, 40, 46, 47–49
    Kepler and, 117–18, 196–97, 200
Copernicus, Nicolaus, 76, 117–18, 196–97, 200
    Galileo and, 40
    Ptolemy and, 17, 48, 200
    Tycho Brahe and, 17, 48–49, 200
corpuscular theory of light, 57, 59, 97, 101–2,
        196, 200–1. *See also* wave theory of light:
        vs. particle theory
correspondence principle, 213
correspondence rules, 156–57, 158, 192–93
Cosimo II de' Medici, Grand Duke of
        Tuscany, 38–39
cosmological constant, Einstein's, 216, 246
cosmology, 204–5. *See also* Big Bang
covering law model, 158–59, 235–36. *See also*
        nomological-deductive model
creationism and "creation science," 250–51
creative thinking, 179–80
"crisis" and paradigm change, 201–2
Critique of Pure Reason (CPR). See under
        Kant, Immanuel
Cultural Literacy (Hirsch), 244–45
cultural relativism, 89, 180–81

d'Alembert, Jean le Rond, 87, 96–97, 113
    Bacon and, 86–87
    on deduction, 87
    Diderot and, 82, 85–86
    Encyclopedie and, 82, 85–88, 121–22
    Hume and, 86–87
    on "irresistible impulse" to believe in
        external world, 87, 103, 119–20
    mathematics and, 82, 87, 96–97
    Preliminary Discourse, 86, 94, 115
    on reason, 86, 87–88
    tree of knowledge, 86–87, 98, 121–22
    writings, 82, 115
Dalton, John
    atomic theory of matter, 102, 106,
        131, 200–1
deconstruction, 223–25
deconstructionism, 244–45
deduction, 58. *See also* induction:
        deduction and
    Aristotle's fallacies of deductive
        reasoning, 28–29
    Bacon and, 26, 28, 29–30, 117
    intuition and, 12, 22–23, 26, 28, 31–33, 35
    and knowledge in the strong sense, 9, 10,
        11, 12, 22–23, 26, 32–33, 34, 108, 110

deduction (*cont.*)
  William Whewell and, 116, 117–19, 148–
    49, 213, 236
deductive-nomological model, 235–37. *See
  also* covering law model
Derrida, Jacques, 221–23, 224–25, 246
  characterizations of, 222
  criticism of, 243, 246–47, 249–50
  deconstruction and, 223, 224–25
  deconstructionism, 244–45
  Foucault and, 222, 246–47
  Kuhn and, 221–22, 223
Descartes, René, 25–26. *See also*
    Cartesian self
  Bacon and, 25–26, 30, 31–32, 34–35, 67, 180
  Discourse on Method, 30, 33
  enumeration and, 30–31, 34
  first principles and, 33–34, 35, 45
  Hobbes and, 65–66, 67, 69–70
  induction and, 30, 31, 34–35
  innate ideas and, 12, 31–34, 35, 45, 87,
    113, 179
  and knowledge in the strong sense,
    33, 34, 65
  on matter and space, 106
  natural philosophy of, 56, 64, 69–70
  Pierre Gassendi and, 19, 54–55, 69–70, 71
  rationalism and, 18–19, 35, 69–70, 148
  on reasoning, 180
  Rules for the Direction of the Mind, 30
  skepticism and, 30, 32
  Spinoza, Cartesianism, and, 33, 35, 67–68,
    130, 179
  on truth and the mind, 12
Dewey, John, 151–53, 256–57
  Charles Peirce and, 151–52
  on experience, 151–53
  know-how and, 227, 256, 257–58
  on knowledge, 149, 151, 152–53, 256–57
  pragmatism and, 151, 152
  on the quest for certainty, 64
  on science, 149, 151, 152–53, 226–27,
    256, 257–58
  on truth, 153
  William James and, 149, 151, 152, 153
DeWitt, Bryce, 173–74
dialectical reasoning, 9
Diderot, Denis, 85–86
  Encyclopedie and, 82, 85–86
  Jean d'Alembert and, 82, 85–86
  know-how and, 82, 85–86

Dirac, Paul, 170–71, 173, 174–75
discourse, 220–21, 222, 225, 228
discursive reasoning, 136, 178, 180, 218–19,
    221. *See also* reasoning
Duhem, Pierre, 124–26, 239, 240
  Catholicism, 124, 251
  Ernst Mach and, 124, 126–27
  on the goal of science, 126
  history of science and, 124, 126–27, 213
  on metaphysics, 125–27
  ontology and, 126, 239, 240
  physics and, 124, 125–27, 251
  underdetermination and, 124, 159, 236–37
  William Whewell and, 126, 213
  writings, 124
Duhem–Quine thesis, 124–25, 159,
    216, 236–37
duration. *See also* time
  experience of, 141–43, 144
  Henri Bergson on, 141–42, 144 (*see also*
    Bergson, Henri: time and)
  intuition and, 142–43, 144
Durkheim, Émile, 202

Earth Day 1970, 198
economics, 91–92
  mathematics and, 89–90
education. *See* universities
Eightfold Way (physics), 175
Einstein, Albert, 110–11, 113–14, 165, 175.
    See also general theory of relativity;
    special theory of relativity
  black-body radiation and, 162, 163–64
  vs. Bohr, 162–63, 165–67, 206–7
  cosmological constant, 216, 246
  Émile Meyerson and, 146–47
  Kant and, 141–42
  Newton and, 117–18, 206–7
  particle theory of light and, 162,
    164, 165–66
  on photoelectric effect, 164, 165–66,
    168, 255
  quantum mechanics and, 105, 165–67
  theory of Brownian motion, 128, 131
electromagnetic wave theory of light, 128,
    130–31, 161–62, 168
  Maxwell's, 101–2, 105–6, 107–8, 125–26,
    130–31, 145, 160, 161–62, 163, 207
  Pierre Duhem's criticism of, 125–26
electromagnetic waves, 161–62
  Einstein and, 161–62

frequency, 128
photoeletric effect and, 162
emergent entities, 184–85
empiricism, 70, 120. *See also under* Fallacy of
    Affirming the Consequent
    Berkeley and, 72
    conflation of certainty and, 43–44
    Galileo and, 43–44, 45–46
    Giants and, 7
    Henri Bergson and, 141, 142, 144
    Hobbes and, 65
    Hume and, 73, 93
    John Stuart Mill and, 120
    Locke and, 71–72, 73
    logical, 153, 154, 157–59, 188–89, 236–37
    Pierre Gassendi and, 69–70
    Popper and, 215, 217
    pragmatism and, 147, 151, 152
    prospective, 151, 152
    Quine and, 124–25, 159, 238
    radical, 150
    rationalism and, 45–46, 141, 142
    skepticism and, 71–72, 73, 74–75
    William James and, 149–50
Empiricus, Sextus, 25, 69
empty space, 163, 171–72
encyclopedic law, 121–22
energy, 106. *See also* mass–energy
    equivalence
    kinetic (*see mv²*)
    quantum field theory and, 106, 171–
        72, 174
Engels, Friedrich, 186
enumeration, Descartes and, 30–31, 34
enumerative induction, 119. *See also* induction
epistemological anarchism, 211
epistemological relativism, 211–12, 243. *See*
    also relativism
epistemology, 1, 21, 64, 113, 233, 237. *See also*
    certainty
    Kuhn and, 203, 204
    probabilist, 122
    Quine's naturalistic, 236–38
    "ultimate problem" in, 118
epistemology and ontology, 21, 103, 108, 123,
    210, 251, 253–54
    fusion of, 99–100
    history of science and, 126
    as independent, 112, 128, 150, 254
    intersection of, 4, 8, 33, 69, 106–7, 112,
        161, 226–27, 253–54

equilibrium, 151
ether. *See* aether theories
Euclid, 10–11, 12, 21–22, 108–10, 111
Euclidean geometry, 10–11, 12, 68–69, 77–
    78, 108–10. *See also* geometry
Euler, Leonhard, 96–97
Everett, Hugh, III, 172–73
evolution, 121, 151–52, 168, 238, 250–51
    Bergson's metaphysics and, 140
evolutionary biology, 203–5, 212
evolutionary epistemology, 237–38
evolutionary philosophy of science, 127
explanation and explanatory theories, 105–6
externalist methodology, 197–98, 204,
    231, 252–53

faith, 138
    vs. reason, 22–23
fallacies of deductive reasoning, Aristotle's,
    28–29. *See also* Fallacy of Affirming the
        Consequent
Fallacy of Affirming the Consequent
    and empirically confirming theories, 58,
        62, 111, 119, 124–25, 159, 215–16
    overview, 28–29
falsification, 212, 215–17
Fechner, Gustav, 179
Feyerabend, Paul, 211–12, 254
    Donald Campbell and, 238
    Imre Lakatos and, 212, 215
    Israel Scheffler and, 211, 212, 218
    Kuhn and, 192–93, 194, 211, 212
    Popper and, 211–12, 215
    Quine and, 237–38
    rationality and, 211–12
    relativism and, 218
    The Structure of Scientific Revolutions
        (*SSR*) and, 211, 212, 218
    writings, 211–12
Feynman, Richard, 174
Fine, Arthur, 241
first principles, 12, 134
    Aristotle and, 12, 22–23, 26
    Descartes and, 33–34, 35, 45
Flamsteed, John, 50
Fleck, Ludwik
    active and passive associations, 191–92,
        234, 241
    *Genesis and Development of a Scientific
        Fact*, 188, 192–93, 194
    Kuhn and, 192–93, 194, 212

Fleck, Ludwik (*cont.*)
　on logical positivism, 188–89
　overview, 188
　on relativism, 192
　theory of science, 188–89, 192–93, 201
　on thought collectives, 188, 189, 199, 202,
　　216, 225, 230
　on thought styles, 188–91, 192–93, 194,
　　199, 225, 230
forces, 24, 51–52
Foucault, Michel, 220–22, 243–44, 246
　Derrida and, 222, 246–47
　genealogical method, 220, 221, 222, 244
　influence of, 228, 246
　knowledge and, 223–24, 232
　Nietzsche and, 221, 243–44
　power, power relations, and, 219–20,
　　221, 232
　science and, 220–21, 225, 228, 242, 246–47
　on search for essences, 225
　writings, 219–20, 221, 222
Foucault's pendulum, 94
Fourier, Joseph, 108
　"Fourier's move," 105–6, 107–8, 145
　Laplace and, 105
　Maxwell's electromagnetic wave theory of
　　light and, 107–8, 145
　ontology, purpose of scientific theories,
　　and, 105–6
　theory of heat flow, 105–8, 115, 130–
　　31, 145
　writings, 105
Franklin, Allan, 249–50
Frege, Gottlob, 111–12
　Bertrand Russell and, 111–12, 154–55
French Revolution, 90
Freud, Sigmund, 181–82
Fundamental Ideas (Whewell), 116, 117–19,
　126, 148–49, 213

Galiani, Ferdinando, 91
Galileo Galilei, 41
　Archimedes and, 36–37, 41–42, 43, 44
　The Assayer, 40, 41–42
　astronomy and, 37–41, 43, 46, 47–49
　Bacon and, 40, 45–46
　Catholic Church and, 36, 43, 46–47
　characterizations of, 36
　Christopher Scheiner and, 39, 41
　Copernican heliocentrism and, 36, 37, 39,
　　40, 46, 47–49

*Dialogues Concerning the Two Great World
　Systems*, 41, 42, 46, 47, 48, 187–88
*Discourses and Mathematical
　Demonstrations Concerning Two New
　Sciences*, 43, 45, 62, 110–11, 118–19
　empiricism and, 43–44, 45–46
　geometry and, 41–42, 43, 44
　induction and, 36–37, 45–46
　Kepler and, 37, 38, 39, 41, 47–48, 196–97
　Kuhn and, 197–98, 266n.16
　life history, 37–38
　mathematical physics and, 23, 36, 42, 43
　method, 36–37, 40, 41–42, 43, 45–46
　natural philosophy and, 36, 37, 40, 42
　Newton compared with, 50
　overview, 36–38
　personality, 50
　telescopes and, 36, 38, 39, 40, 42–43, 47–48
　Tycho Brahe and, 37, 48–49, 196–97
Galison, Peter, 241, 249–50
Galvani, Luigi, 200
gases, kinetic theory of, 131, 163–64, 167
Gassendi, Pierre, 54–55, 69–70, 71
　Descartes and, 54–55, 69–70, 71
Gaukroger, Stephen, 3
genealogical method, 220, 221, 222, 244
general theory of relativity, Einstein's, 28–29,
　　146–47, 157, 165–66, 197–98. *See also*
　　space and time
　gravitational waves and, 16, 29, 216 (*see
　　also* gravitational waves)
　Newtonian physics and, 54–55, 117–
　　18, 206–7
genius, 179–80
geometry, 41–42, 68–69, 77–78, 110
　Descartes and, 30
　Euclidean, 10–11, 12, 68–69, 77–
　　78, 108–10
　Galileo and, 41–42, 43, 44
　Huygens and, 58
　Lobachevskian, 109–10
　non-Euclidean, 109–10, 111
　Plato and, 33, 108
germ theory of disease, 131–32, 189–90
gestalt psychology, 127–28, 184–86
Giants (Greek mythology), 14, 69
　allies of the, 12–13, 14, 23, 25, 64, 226,
　　228, 234–35
　knowledge and, 64, 74–75
　Protagoras and, 12–13, 14
　reason and, 14, 136

science and, 74–75
    sophists and, 7, 12–13, 207–8
Giants and Gods, 63
    knowledge and, 14–15, 19, 20, 77, 159–60,
        207–8, 234–35, 258
    reason and, 15, 45–46, 149
    science and, 15, 18–19, 20, 63, 159–60,
        234–35, 258
Giants–Gods battle, 7–8, 14–15, 18–19, 20,
        25, 63, 64, 69, 77, 207–8
    Kant and, 77
    metaphysics and, 140–41
    Plato and, 63, 136
    reasoning and, 15, 20, 149
    science wars and, 218
    William James and, 149
Giere, Ronald N., 239, 240
Gilbert, William, 42–43
Gingras, Yves, 234
God, 23–24, 72–73. See also Bible
    Abraham and, 138
    Adelard on, 21
    Augustine on, 12, 22–23, 32–33
    Bacon on, 28
    and Creation, 35, 50–51
    Descartes and, 32–33, 35, 56
    imagination and, 23–24, 28
    Kierkegaard and, 137–38
    Leibniz and, 60–61, 140–41
    nature of, 51, 130, 137
    Newton on, 50–52, 60–61
    Protestants, Catholic Church, and, 18
    universals and, 12, 22–23, 28, 32–33
    William Whewell on, 119
Gödel, Kurt, 112, 153–54
Gods. See also Giants and Gods
    rationalism and, 8, 12, 140–41, 207–8
Gold, Thomas, 202–3
Goodman, Nelson, 176, 177, 228
Gorgias, 12–13, 14, 135
Gravesande, Willem 's, 97
gravitational waves, detection and
        measurement of, 15–16, 29, 216, 232–33
gravity
    Descartes' vortex-based theory of, 51, 95
    general relativity theory and, 28–29 (see
        also general theory of relativity)
    Leibniz and, 60–61
    Newton's theory of, 50, 51–52, 55–56, 60–
        61, 94–97, 116, 117–18, 131–32
    quantum, 247

Greek philosophers, 3, 7, 12–13, 106. See also
    sophists; specific philosophers
Groddeck, Georg, 182
Gross, Paul R., 245–47
Grosseteste, Robert, 22
Guillemin, Roger, 229–30
Gulliver's Travels (Swift), 82

habits, defined, 148, 153
Hacking, Ian, 239
Halley, Edmund, 89–90, 95
Halley's comet, 95
Hamann, J. G., 75, 132–33, 134
Hamilton, Alexander, 92
Hanson, Norwood Russell 127–28, 185, 194,
        206, 209–10, 227–28, 252
Harding, Sandra, 247
Hariot, Thomas, 38, 39, 41
heat, 98, 103. See also caloric theory of heat
    motion theory of, 98, 103–5, 196
    nature of, 103–4
heat flow. See also thermodynamics
    Fourier's theory of, 105–8, 115, 130–
        31, 145
Heisenberg, Werner., 166, 170–71. See also
    matrix mechanics; uncertainty principle
heliocentrism, 94. See also Copernican
    heliocentrism
Helmholtz, Hermann, 122–23, 182–83
Hempel, Carl, 158–59, 235–36
    induction, deduction, and, 158–
        59, 235–37
    Kuhn and, 158, 235–37
    overview, 158, 235–36
    Rudolf Carnap and, 158, 236
Heraclitus, 7, 140
Herschel, Caroline, 94–95
Herschel, John, 119, 122
    A Preliminary Discourse on the Study of
        Natural Philosophy, 115, 116
    William Whewell and, 116, 119
Herschel, William, 94–95
Hertz, Heinrich, 62, 123, 124
Hesse, Mary, 239–40
Hessen, Boris, 187–88
Higher Superstition: The Academic
    Left and Its Quarrels with Science
    (HS), 245–48
Hilbert, David, 112–13
Hirsch, E. D., 244–45
historiographic revolution, 194–95, 196–97

Hobbes, Thomas, 66–67, 180, 232
  Descartes and, 65–66, 67, 69–70
  empiricism and, 65
  experimental method and, 64–65, 232
  Locke and, 65, 70–71
  natural philosophy and, 64–67, 70
  overview, 64–65, 67
  reasoning and, 66–67
  Robert Boyle and, 65–66, 231–32
  science and, 64
homogeneity vs. heterogeneity of
    sunlight, 53, 66
Hooke, Robert, 17, 29–30, 50, 66, 206
  Newton and, 17, 50, 66
Hooke–Boyle air pump experiments, 65–
    66, 232
House Built on Sand, A (Koertge), 249–50
Hubble, Edwin, 176–77
humanists, 24, 80
  and the problem of knowledge, 25
  science and, 24, 25, 249–50
Hume, David, 74–75, 78, 91, 121–22
  Adam Smith and, 92
  Cartesian self and, 92–93, 180
  on causal connection, 74, 76, 78
  critique of causality, 73, 75
  empiricism and, 73, 93
  on induction, 119–20, 158, 241–42
  Jean d'Alembert and, 86–87
  Kant and, 75–76, 78–79, 133
  on mind, 92–93, 180
  principles of association, 93, 180
  on reason and reasoning, 73–74, 78, 136,
    180, 241–42
  on sensation, 73, 74, 87, 183
  skepticism and, 73, 75–76, 136
  Treatise of Human Nature, 74, 75
Huygens, Christiaan, 44, 58–60
  Descartes and, 59–60
  Locke and, 70, 71–72
  Newton and, 54–55, 58–59, 70 (*see also*
    wave theory of light: vs. particle theory)
  theory confirmation and, 59–60, 72, 74–
    75, 216, 217
  Treatise on Light, 58–59, 71–72, 216
  wave theory of light, 59–60, 97, 101–
    2, 196
hyperbolic geometry. *See* Lobachevskian
    geometry
hypothetico-deductive theory, 158–59
hysteria, treatment of, 182

"I think, therefore I am" (Descartes), 32–33,
    130, 179
ideal theories, 126
ideals, 189
ideas, Hume's rules of the association of,
    93, 180
ideology, 186
Idols (Bacon), 26–27, 115–16, 117, 127–28,
    185–86, 220–21
incompleteness theorems, Gödel's,
    112, 153–54
individualism, 179–80, 188, 243–44
induction, 11, 30, 74–75, 117, 158
  abduction and, 117, 148–49, 206
  assumptions and, 63, 74, 111, 114
  Bacon and, 26–27, 28, 29–30, 34–
    35, 115–16
  Carl Hempel and, 158–59, 236
  deduction and, 10, 11, 18–19, 26, 28,
    29–30, 31, 34, 35, 36–37, 45–46, 55–
    56, 58, 87, 111, 117–19, 133, 159–60,
    236, 241–42
  definitions, 116, 117, 118–19, 148–49, 206
  Descartes and, 30, 31, 34–35
  Galileo and, 36–37, 45–46
  Hume on, 119–20, 158, 241–42
  interpretation and, 63
  John Stuart Mill and, 119
  limitations, 30, 74–75, 110
  mathematics and, 110
  Newton and, 57–58
  Rudolf Carnap and, 158, 236
  science and, 15–16, 18–19, 111, 118, 120,
    133, 241–42
  William Whewell on, 116, 117–20, 148–
    49, 168–69, 206
innate ideas, Descartes and, 12, 31–34, 35, 45,
    87, 113, 179
Instrument (Aristotle), 21–22, 26–27
instruments, scientific, 15–18, 65–66
intelligent design, 250–51
internal realism, 240–41
internalist methodology, 197–98, 231
interpretation and induction, 63
interpreting experience vs. uncovering
    reality, 63. *See also* Giants and Gods
intuition, 23–24, 31, 122, 142, 144
  Aristotle on, 12, 22–23, 26
  Bacon and, 26, 28
  deduction and, 12, 22–23, 26, 28, 31–33, 35
  Descartes and, 12, 31–33, 179

Henri Bergson on, 141, 142–44
Kepler and, 30
nature of, 31–32, 142–43
non-discursive, 142
vs. reason, 142–43
reasoning and, 12, 30
spatial and temporal, 77–78
isochronous timing. See pendulum,
    period of a
Isocrates, 152

James, William, 149–50
    Charles Peirce and, 147, 149, 151–52
    Dewey and, 149, 151, 152, 153
    empiricism and, 149–50
    on knowledge, 149, 150
    pragmatism and, 147, 149, 150
Janet, Pierre, 181–82
Joyce, James, 179–80

Kamil, Abu, 24–25
Kant, Immanuel, 12, 74–75, 77–78, 111, 118–
    19, 133, 237–38
    causality and, 75, 76
    certainty and, 77, 78–79, 133
    Critique of Pure Reason (CPR), 75, 76–
        77, 78–79
    Einstein and, 141–42
    Hume and, 75–76, 78–79, 133
    1771 inaugural lecture, 75
    J. G. Hamann and, 75, 133
    long practice of attention, 77
    rationality and, 130, 133
    reason and, 77, 111
    revolution in philosophy, 76
    skepticism and, 75–76, 133
    transcendental idealism and, 133
Kepler, Johannes, 196–97
    Copernican heliocentrism and, 117–18,
        196–97, 200
    Galileo and, 37, 38, 39, 41, 47–48, 196–97
    laws of planetary motion, 30, 116–17
    Newton and, 94, 116–18, 200
    Tycho Brahe and, 30, 37, 116–17, 119–
        20, 196–97
Khwarizmi al-Khati, al-, 24–25
Kierkegaard, Søren, 137–38, 180–81
    on "comedy of the higher lunacy,"
        64, 78–79
    familial approaches/stages to human
        existence, 137

Fear and Trembling, 138
God and, 137–38
Schopenhauer and, 138, 180–81
kinetic theory of gases, 131, 163–64, 167
Kitcher, Phillip, 238–39, 249
know-how, 81–82, 83–84, 85–87,
    106, 252–53
Denis Diderot and, 82, 85–86
Dewey and, 227, 256, 257–58
Jerome Ravetz and, 227, 256, 257–58
ontology and, 253
science and, 81, 82, 106, 148, 235, 252–53,
    256, 257–58
knowledge, 64. See also specific topics
defining, 2–3, 4, 8, 9–10, 18–19, 36, 71, 153
Dewey on, 149, 151, 152–53, 256–57
empty, 122, 152–53
Hobbes on, 65
meanings and connotations of the term, 2–
    3, 4
metaphysics and, 4
nature of, 1, 3, 9–10, 70–71, 73–74, 152–53
and the nature of knowing, 4
objectivity and, 1–2, 3, 4
vs. opinion/belief, 2–3, 4
Plato on, 7, 8, 9–10, 12–13, 26
science as the only source of, 3
terminology and related terms, 2, 4
truth and, 2–3, 4, 9
knowledge in the strong sense, 10, 32–33, 64,
    70, 73–75
causality and, 74–75
deduction and, 9, 10, 11, 12, 22–23, 26,
    32–33, 34, 108, 110
defined, 3
Descartes and, 33, 34, 65
Friedrich Schlegel on, 134–35
Gods and, 14–15
Hobbes and, 65
Hume and, 73–74, 78
Locke and, 70, 71, 72
mathematics and, 12, 33, 73–74, 108
natural philosophy and, 78
Plato and, 9, 12, 14, 108
science and, 6, 12, 19–20, 72, 88, 94,
    117, 124
skepticism and, 69, 74–75
Socrates and, 9, 14
truth and, 9
universals and, 22–23
vs. weak sense of knowledge, 3, 4, 88

knowledge in the weak sense, 32, 74–75
    defined, 3, 32
    Giants and, 23, 159–60
    inductive reasoning and, 22–23
    logical positivists and, 159–60
    science and, 4, 6, 19–20, 88
    vs. strong sense of knowledge, 3, 4, 88
knowledge problem, 30, 76
    birth of, 7
    Kant and, 75, 76
    Kuhn and, 207–8
    philosophical responses to, 116, 122 (see
        also specific philosophers)
Koertge, Noretta, 249–50
Kuhn, Thomas, 196, 197–98, 203, 209–10,
    212. See also Structure of Scientific
    Revolutions
    Carl Hempel and, 158, 235–37
    Derrida and, 221–22, 223
    epistemology and, 203, 204
    Galileo and, 197–98, 266n.16
    history of science and, 194–95, 196–97,
        200, 203–4, 205, 211, 212–13, 214, 219–
        20, 231–32
    Imre Lakatos and, 212–15
    Israel Scheffler and, 209–11
    Jerome Ravetz and, 226–28
    knowledge problem and, 207–8
    logical positivism and, 206, 210
    Ludwik Fleck and, 192–93, 194, 212
    Michael Polanyi and, 185–86, 194, 201,
        206, 209–10, 227–28
    normal science and, 197–98, 200–2, 204–
        5, 206, 214, 225
    objectivity, subjectivity, and, 207–8, 209–
        11, 214–15, 242
    on paradigm change, 196–97, 203, 204–6, 207
    paradigms and, 189–90, 192–93, 199, 200–
        1, 204–5, 206–7, 211, 223, 225
    on physics, 197–98, 200
    rationality and, 209, 214
    on scientific revolutions, 203–4
    on theory change, 194–95, 212,
        213, 214–15
    theory of science, 190–91, 192–93, 194–95,
        196–97, 199, 204
    thought styles and, 190–91, 192–93, 194

laboratories, research. See also
        Laboratory Life
    funding, 229–30

Laboratory Life (Latour and Woolgar), 229,
    230, 233, 245
Lagrange, Joseph-Louis, 105
Lakatos, Imre
    Kuhn and, 212–15
    Paul Feyerabend and, 212, 215
language. See also linguistics
    mirroring the real world, 155–56
    ordinary, 155
    and thought, 134, 135–36
Laplace, Pierre Simon Marquis de, 68–69,
    99–100, 104, 105, 170–71
    Fourier and, 105
Laser Interferometry Gravitational
    Observatory (LIGO), 70–71
Latour, Bruno, 229. See also Laboratory Life
Lavoisier, Antoine, 17, 104
    caloric theory of heat and, 98, 103–4, 196
    experiments, 17, 102
    Joseph Priestly and, 17, 102, 205
    oxygen theory of combustion, 17, 102,
        103–4, 200–1, 205
Law, John, 89–90
Leibniz, Gottfried Wilhelm, 140–41
    calculus and, 50, 60
    causal connection and, 67, 74–75
    characterizations of, 68, 140–41
    Descartes and, 35, 68
    God and, 60–61, 140–41
    natural philosophy and, 60–61, 68, 84–85
    vs. Newton, 50, 54–55, 60–61, 62–63, 97
    overview, 68
    philosophical principles, 68–69 (see also
        Principle of Sufficient Reason)
    physics and, 54–55, 60–61, 62–63, 97
    principle of sufficient reason, 146–47
    on reasoning, 67
Lennon, Thomas M., 19
Leonardo of Pisa, 24–25
Lepaute, Nicole, 95
Levi-Strauss, Claude, 216, 221–22
Levitt, Norman, 245–47
light, speed of, 125, 161–62, 164. See also
        special theory of relativity
light particles. See under Newton, Isaac
light waves. See electromagnetic wave theory
        of light; Huygens, Christiaan; wave-
        particle duality
lightprint, 168
linguistic signs and linguistic value, 223–24
linguistics, 134. See also language

comparative, 134, 180–81
    meaning and, 135–36
    philosophy and, 155–56
    Sapir–Whorf hypothesis and, 180–81
    science and, 178
Linnaeus, Carl, 98, 99
Lobachevski, Nikolai Ivanovitch, 109–10
Lobachevskian geometry, 109–10
Locke, John
    on certainty, 71–72
    empiricism and, 71–72, 73
    Essay Concerning the Human
        Understanding, 70
    Hobbes and, 65, 70–71
    Huygens and, 70, 71–72
    on knowledge, 70–71
    and knowledge in the strong sense,
        70, 71, 72
    Newton's Principia and, 70
    overview, 70
    on sensation, 70–71
logical atomism, 154–55, 156
logical empiricism, 153, 154, 157–59, 188–
    89, 236–37. See also empiricism
logical positivism, 154, 155, 157, 158–59
    Bertrand Russell and, 154, 155, 156–57
    criticism of, 159–60, 188–89, 192, 215
    Kuhn and, 206, 210
    Moritz Schlick and, 122, 153–54, 156–
        57, 210
    Rudolf Carnap and, 158
    science and, 153–54, 156–57, 158–60,
        188–89, 192, 195, 206, 210
    Vienna Circle and, 155, 157–58, 188–
        89, 215
    Wittgenstein and, 154–55, 156–57
Lucretius, 25, 69, 106
Lyotard, Jean, 218, 219, 228

Mach, Ernst, 128, 166
    on atomic theory, 128
    characterizations of, 126–27, 128
    Heisenberg and, 166, 171
    neurophysiology and, 126–28
    overview, 124, 126–27
    phenomenalism, 154, 171, 211–12
    philosophy of science and, 126–27, 128–
        29, 130–31, 137, 149–50, 253–54
    Pierre Duhem and, 124, 126–27
    psychology, the mind, and, 126–
        28, 145–46

reasoning and, 128, 130–31
    writings, 124, 127
macro level of scientific thinking, 190–91
magical natural philosophy and natural
        philosophers, 23–24
magnetic fields, 107, 203. See also
        electromagnetic wave theory of light
magnets, 34, 42–43, 107
"Man is the measure of all things"
    (Protagoras's man–measure doctrine),
    13, 14, 181
Mannheim, Karl, 186–87, 194
Many-Worlds Interpretation of Quantum
        Mechanics (MWI), 173–74
mapping, theories, and the world, 238–39
Margolis, Joseph, 19
Marx, Karl, 186
mass, 164–65
mass–energy equivalence ($E=mC^2$), 106,
        113–14, 164–65
mathematical physics, 24–25, 58, 83–
        84, 162–63
    Archimedes on, 24–25, 36–37, 41–
        42, 43, 44
    Descartes and, 33–34
    Galileo and, 23, 36, 42, 43
    Newton and, 58, 83–84, 97, 187
    Pierre Duhem and, 125–26, 162–63
mathematics, 24–25, 110–11. See also
        geometry
    Bacon and, 25–26, 28
    Bertrand Russell on, 111–12, 154–
        55, 236–37
    economics and, 89–90
    Jean d'Alembert and, 82, 87, 96–97
    and knowledge in the strong sense, 12, 33,
        73–74, 108
    Plato and, 33, 108, 112, 113
matrix mechanics, 126, 166, 172
matter
    Berkeley's critique of, 72–73, 93
    conservation of, 113–14
    theoretical conceptions about, 163
Maxwell, James Clerk. See electromagnetic
        wave theory of light
meaning, 135–36
measurement of scientific concepts, 147
Median stars, 38–39
medicine, 229. See also neurophysiology
    thought styles in, 189–90
Mendeleev, Dimitri, 175

Merton, Robert K., 187, 194
metaphysical stage (Comte), 121
metaphysics, 4, 140–41, 147. *See also*
    ontology; *specific topics*
    of Henri Bergson, 140–42, 144
    Pierre Duhem on, 125–27
methodological anarchy, 211–12
Meyerson, Émile, 146–47
Milky Way, 38, 101, 176–77
Mill, John Stuart, 119–21, 122
    induction and, 119
    Mill's Methods, 120
mind
    Bacon on, 26, 27
    Hume on, 92–93, 180
    nature of, 180
    study of, 92–93
mirroring the world, 155–57
modern science, 3
modernism, 218–19, 243–44
modernity, 219–20, 221–22
momentum (*mv*), 62–63, 170–71
    conservation of, 113–14, 163
Montesquieu, Baron de, 89–90
moral certainty, 71–72
motion. *See also under* heat
    absolute and relative, 54–55
    Newton's laws of, 53–54, 55–56
"muddle-headedness," 245
mv. See momentum
mv2, 62–63, 97. *See also* energy

natural ontological attitude, 241
natural phenomena, questions about, 21
natural philosophers, 22–23, 25
    Bacon and, 26, 27
    Cartesian, 59–60
    Descartes and, 34
    Galileo and, 36, 37, 42
    on heat, 103–4
    magical, 23–24
    Newton and, 58
    reason, reasoning, and, 23–24, 27
    science and, 24
    scientists and, 116
    terminology, 116
natural philosophy, 21–22, 31, 33, 78. *See also*
    under Bacon, Francis; Herschel, John
    arguments in, 31
    Berkeley and, 72–73
    Cartesian, 56, 64, 67, 69–70

emergence of, 21–22
    Galileo and, 42, 46
    goal of, 40, 46
    Hobbes and, 65, 67
    Leibniz and, 60–61, 68, 84–85
    Locke and, 70
    mathematics-based, 22
    Newton's, 51–52, 56, 57, 60 (*see also*
        Newton, Isaac)
    science and, 3
natural selection, 203–4, 217, 250–51
naturalistic epistemology, Quine's, 236–38
nature, 115–16
    defining, 4
Nemorarius, Jordanus, 22
Neurath, Otto, 153–54, 210
neurophysiology, 93–94, 181, 183, 237–38
    Charcot, hysteria, and, 182
    Hermann Helmholz and, 182–83
    hormones and, 181–82, 229
    Mach and, 126–28
    reasoning and, 17–18
neutrality. *See also* theory neutrality
    value, 86–87, 226–27, 256–57
neutrinos, 125, 169
neutrons, 169
Newton, Isaac, 58, 60
    alchemy and, 50–51, 52, 70
    Bacon and, 53, 55
    characterizations of, 50
    corpuscular theory of light, 57, 59, 97,
        101–2, 196, 200–1 (*see also* wave theory
        of light: vs. particle theory)
    Einstein and, 117–18, 206–7
    foundations of his mechanics, 54
    on God, 50–52, 60–61
    gravity theory, 50, 51–52, 55–56, 60–61,
        94–97, 116, 117–18, 131–32
    Huygens and, 54–55, 58–59, 70 (*see also*
        wave theory of light: vs. particle theory)
    induction and, 57–58
    Kepler and, 94, 116–18, 200
    laws of motion, 53–54, 55–56
    vs. Leibniz, 50, 54–55, 60–61, 62–63, 97
    mathematical physics and, 58, 83–84,
        97, 187
    method of Synthesis/method of
        Composition, 57, 58
    Opticks, 51–53, 56, 57, 60–61
    passion for knowing the Truth about
        Reality, 52

personality, 50
*The Principia: Mathematical Principles of
    Natural Philosophy*, 50, 51, 54, 55–56,
    57, 60, 70, 95, 187
    Robert Hooke and, 17, 50, 66
Nietzsche, Friedrich S., 134, 139, 140
    Foucault and, 221, 243–44
    perspectivalism, 139, 142, 144, 176, 185,
        186–87, 228, 247
    power and, 139, 221
    on science and truth, 139, 142
    on values, 139
    on will, 139
1960s and 1970s, anti-establishmentarianism
    of the, 198
1980s criticism of science, 243
Noether, Emmy, 169–70
nomological-deductive model, 235–37. *See
    also* covering law model
normal science, 197–98, 205, 206, 213–14
    defined, 200–1
    Kuhn and, 197–98, 200–2, 204–5, 206,
        214, 225
    nature of, 201–2, 205
    paradigms and, 200–2, 203, 204–5, 206
    tasks of, 197–98

Objectivism and Relativism (Bernstein), 19
Ockham, William of, 23
Oldenburg, Samuel, 55
Olympian Gods. *See* Giants and Gods; Gods
ontological attitude, natural, 241
ontology, 105–6. *See also* epistemology and
        ontology; metaphysics; *specific topics*
    Parmenidean-Rationalist, 253–54
    Pierre Duhem and, 126, 239, 240
    relativity theory and, 161, 207
operationalism, 147
Oresme, Nicolas, 22
Oughtred, William, 67
oxygen theory of combustion (Lavoisier), 17,
    102, 103–4, 200–1, 205

paradigm change, 201–2, 203, 205. *See also*
        paradigms
    Kuhn on, 196–97, 203, 204–6, 207
    new data and, 203
"paradigm," use of the term, 204, 225
paradigms, 201. *See also under* Kuhn, Thomas
    challenging, 202
    crises and anomalies as threats to, 201–2

examples, 199–201
    influence of, 202, 203
    nature of, 201–2
    normal science and, 200–2, 203, 204–5, 206
    thought styles and, 189–90, 199
Parmenidean-Rationalist ontology, 253–54
Parmenides, 7, 146–47, 151–52
particle physics, 125, 167, 168–71, 175, 233–
        34. *See also* atomic theory of matter;
        uncertainty principle; wave function
        collapse; wave–particle duality
particle theory of light. *See also* corpuscular
        theory of light; wave–particle duality
    Einstein and, 162, 164, 165–66
Pascal, Blaise, 67
Pauli, Wolfgang, 125
Peirce, Charles Sanders, 148, 151–52
    on abductive reasoning, 117, 148–49,
        168–69, 206
    Dewey and, 151–52
    on habits, 148, 153
    pragmatism and, 147–48, 149
    William James and, 147, 149, 151–52
    William Whewell and, 148–49, 168–69,
        194, 206
pendulum, period of a, 44
Persian Letters, The (Montesquieu), 89
perspectivalism, 140, 185, 240
    of Nietzsche, 139, 142, 144, 176, 185, 186–
        87, 228, 247
Petrarch, 80–100
Petty, William, 89–90
Phaedrus (Plato), 8, 14, 224
Phaenomena, 56, 57, 58
phenomenalism, Mach's, 154, 171, 211–12
philosophers. *See also* certainty: quest for;
        philosophes
    "carve nature at its joints," 13–14, 23, 86,
        98, 150
philosophes (intellectuals of 18th-century
        Enlightenment), 81, 84, 87–88
    reason and, 81, 84–85, 87–88
    and sensations as origin of knowledge, 87
    Voltaire and, 84–85
philosophy. *See also* natural philosophy
    analytic, 236–37
    Friedrich Schlegel on, 134
    linguistic turn in, 155–56
    purpose, 152
    vs. theology, 22–23
    Western, 64, 75–76

phlogiston, 17, 98, 102
photoeletric effect, Einstein on, 164, 165–66,
    168, 255
photons, 163, 171. *See also* wave–particle
    duality
photosynthesis, 98
physics, 164–65. *See also specific topics*
    mathematical (*see* mathematical physics)
    nature of, 110–11
physiocracy, 91–92
phytosynthesis, 98
Pickering, Andrew, 233–34
Planck, Max, 203
    atomic theory and, 128, 163–64, 165–66
    black-body radiation and, 106, 161–62,
        163–64
    Einstein and, 161–64, 165–66
    Ernst Mach and, 128
plate tectonics, 190, 207
Plato, 7, 8, 9–10, 12–13, 98–99. *See also*
    philosophers: "carve nature at its joints";
    Socrates
    Aristotle and, 10, 11–14, 22, 26
    Bacon and, 26
    Descartes and, 33
    as essentialist, 13–14
    Giants–Gods battle, 19, 63, 136 (*see also*
        Giants–Gods battle)
    Isocrates' criticism of, 152
    on knowledge, 7, 8, 9–10, 12–13, 26
    and knowledge in the strong sense, 9, 12,
        14, 108
    on logically necessary truth vs. opinion, 7
    mathematics and, 9, 33, 108, 112, 113
    Protagoras and, 12–13
    Socrates and, 13–14 (*see also* Socrates)
    sophists and, 7, 12–13, 222–23
Plato's dialogues, 7, 106, 179, 180. *See also*
    Socrates
    Phaedrus, 8, 14, 224
    Theaetetus, 8–9
pluralism, theoretical, 211–12
Poincaré, Henri, 145–46, 175
Polanyi, Michael, 185–86, 194, 201, 206, 209–
    10, 227–28
    Kuhn and, 194, 201, 209–10, 227
political agendas, science used to promote,
    1–2, 248, 250, 256, 258
Pope, Alexander, 50
Pope Urban VIII, 47
Popper, Karl, 215–17

background and life history, 215
empiricism and, 215, 217
falsification and, 201, 212, 215–17
Imre Lakatos and, 212
Marxism, psychoanalysis, and, 215–16
overview, 215
Paul Feyerabend and, 211–12, 215
philosophy of science, 201, 212, 215–
    16, 217
writings, 215, 217
positive stage (Comte), 121
positivism, 121. *See also* logical positivism
positrons, 170–71
postmodernism, 134–35, 244–45, 246,
    247, 249
    Alain Renaut and, 243–44
    Alan Sokal and, 247–48, 249–50
    criticism of, 243, 244, 248, 249–50
    Jean Lyotard on, 219
power, 139
    Foucault and, 219–20, 221, 232
    Nietzsche and, 139, 221
power relations/power relationships, 219–20,
    221, 225, 232
pragmatism, 141–42
    Charles Peirce and, 147–48, 149
    Dewey and, 151, 152
    empiricism and, 147, 151, 152
    goal of, 152
    vs. rationalism, 150
    William James and, 147, 149, 150
"prediction," connotations of the
    term, 235–36
predictive success, 5, 118–19
Priestly, Joseph
    Antoine Lavoisier and, 17, 102, 205
    experiments, 17, 102, 205
Principle of Sufficient Reason (PSR), 68–69,
    74–75, 83–84, 100, 146–47
probability theory, 90
Protagoras, 14
    Aristotle and, 12–13
    Giants and, 12–13, 14
    man–measure doctrine, 13, 14, 181
    Plato and, 12–13
psychoanalysis, 181–82
psychology, 89, 92–93, 127–28, 237. *See also*
    James, William; mind
    behavioral, 184
    clinical, 181–82
    Ernst Mach and, 126–28, 145–46

experimental, 122, 127, 182–83
    gestalt, 127–28, 184–86
Ptolemy (Claudius Ptolemaeus), 199–200
    Copernicus and, 17, 48, 200
Putnam, Hilary, 240–41
Pythagorean paradigm, 199–200

quantum electrodynamics (QED), 174–75
quantum field theory, 174
    energy and, 106, 171–72, 174
quantum foam, 172
quantum gravity, 247
quantum mechanics, 169–70, 173–74. See
    also uncertainty principle
quarks, 233
Quesnay, Francois, 91
Quetelet, Adolphe, 167
Quine, Willard van Orman, 210, 217, 235–
    37. See also Duhem–Quine thesis
    on analytic–synthetic distinction, 159
    Donald Campbell and, 237–38
    empiricism and, 124–25, 159, 238
    naturalistic epistemology, 236–38
    realism and, 237–38
    on truth, 237

radio waves, 161–62
radioactive decay, 125, 168, 169, 207, 216
random mutations, 168, 250–51
randomness, 167–68
"rational reconstruction" methodology, 213, 231
rationalism, 253–54
    Descartes and, 18–19, 35, 69–70, 148
    empiricism and, 45–46, 141, 142
    Gods and, 8, 12, 140–41, 207–8
    vs. pragmatism, 150
    time and, 140
rationalist definition of knowledge, 71
rationality, 212, 218–19
    Kant and, 130, 133
    knowledge and, 2, 4
    Kuhn and, 209, 214
    Paul Feyerabend and, 211–12
    science and, 209, 211, 212, 214–15,
        226, 228
Ravetz, Jerome, 226–28
    Dewey and, 256, 257–58
    know-how and, 227, 256, 257–58
    Kuhn and, 226–28
    on science, 226–27

realism, 15–16, 80, 85, 94, 99, 234, 235, 240–
    42, 255
    Donald Campbell and, 237–38
    Quine and, 237–38
    relativism and, 87 (see also relativism)
    social constructionism and, 242, 248, 249–
        50 (see also social constructionism)
    types of, 228–29, 240–41
reality
    defined, 8, 9–10
    nature of, 8, 10, 140, 142, 175, 207
    science and, 175, 207
reason, 84. See also Principle of
        Sufficient Reason
    defined, 9
    Dewey on, 152–53
    Galileo and, 40
    Jean d'Alembert on, 86, 87–88
    nature of, 142–43
    philosophes and, 81, 84–85, 87–88
    as subordinate to passions, 78, 136
reasoning, 142
    Aristotle on, 10–11, 12, 22–23, 28–29
    Bacon and, 25–27, 28, 67, 115–16, 117,
        180, 185–86, 187
    discursive, 136, 178, 180, 218–19, 221
    methods of (see abduction; deduction;
        induction)
Reichenbach, Hans, 154, 157–58, 241–42
relativism, 230, 243. See also cultural
        relativism
    epistemological, 211–12, 243
    vs. objectivism, 19
    opposition to, 192, 218, 240–41
    strong, 218, 230
    William James and, 150
relativity theories. See general theory of
        relativity; special theory of relativity
religion, 22–23, 251. See also Bible; Catholic
        Church; God
    Newton and, 50–51
    science education and, 1–2, 250–51
    William Whewell and, 119
religious criticism of science, 19–20
religious right, 20, 250–51
religious stage (Kierkegaard), 137
religious vs. scientific convictions, 2
Renaut, Alain, 243–44
representation of phenomena, 126
research laboratories. See laboratories

research programs, degenerative vs.
    progressive, 214–15
Richer, Jean, 95
Romantic movement, 81–82, 83, 132–33
Romantic philosophers/Romantics, 132,
    134–35, 137, 143–44. *See also specific
    philosophers*
Romanticism, 130, 133, 136–37
Rose, Alex, 4
Rousseau, Jean-Jacques, 82–83, 132–
    33, 136–37
Russell, Bertrand, 128–29, 140, 155–57
    atomism and, 155–56
    epistemology and, 1
    Frege and, 111–12, 154–55
    on logic and mathematics, 111–12, 154–
        55, 236–37
    logical positivism and, 154, 155, 156–57
    Wittgenstein and, 154–57
Russell Hanson, Norwood, 127–28, 185
Rutherford, Ernest, 128
    Niels Bohr and, 128, 168
    "solar system" model of the atom, 168

Salmon, Wesley, 241–42
Sapir, Edward, 180–81
Sapir–Whorf hypothesis, 180–81
Sarpi, Paolo, 38, 41
Saussure, Ferdinand de, 135–36, 221–
    22, 223–24
Schaffer, Simon, 232
Scheffler, Israel, 210–11, 235
    Kuhn and, 209–11
    Lakatos and, 212, 235
    Paul Feyerabend and, 211, 212, 218
    on science, 209–11, 235
    *The Structure of Scientific Revolutions
        (SSR)* and, 209, 211
Scheiner, Christopher, 39, 41
Schickard, Wilhelm, 67
Schlegel, Friedrich, 134–35
Schlick, Moritz, 153–54, 156–58
    logical positivism, 122, 153–54, 156–
        57, 210
Schopenhauer, Arthur, 138–39, 180–81
    Kierkegaard and, 138, 180–81
Schrödinger, Erwin, 163, 166
    wave mechanics, 157, 166, 172
Schrödinger equation, 164, 170, 171, 172–73
Schrödinger's cat, 173
Schweber, Sylvan, 234

science. *See also specific topics*
    criticism of (*see* academic left)
    goals of, 120, 126, 209–10
    nature of, 122, 175–76, 207, 209–10, 215–
        16, 228
    1980s criticism of, 243
science, history of
    Kuhn and, 194–95, 196–97, 200, 203–4,
        205, 211, 212–13, 214, 219–20, 231–32
    Pierre Duhem and, 124, 126–27, 213
    sequential revolutions in the, 203–4 (*see
        also* scientific revolution)
    social constructionism and, 136, 186–
        87, 239
    Whig, 196–97
    William Whewell and, 116, 119, 126, 213
science wars, 178, 218, 219, 225, 234,
        245, 246, 247, 248, 249–50. *See also
        specific topics*
    aftermath, 250, 258
    causes, 187, 207–8, 234–35, 253–54
    definition and nature of, 20
    fronts on which they were fought, 251–
        52, 253–54
    peak of, 247
    religion and, 251–52
"Science Wars" (special issue of *Social
        Text*), 248
scientific method, 63. *See also* Bacon, Francis:
        experimental method of
    Galileo and, 36–37
scientific revolution, 21. *See also*
        paradigm change
    Kuhn on, 203–4, 206 (*see also* Kuhn,
        Thomas)
scientific statements/scientific knowledge
        claims, 210
scientific theories, 101–2, 136, 210. *See also*
        theories; *specific topics*
    choosing between, 102–3
    explanatory, 105–6
    Pierre Duhem on, 126
    purpose of, 105–6
scientific thinking, 178–81
    macro level of, 190–91
self-evident truth, 9, 10–11, 12, 32–33,
        108–9, 113. *See also* analytic–synthetic
        distinction
semantic holism, 135–36
sensation, 87, 93–94, 127
    Henri Bergson on, 141–42

Hume on, 73, 74, 87, 183
Locke on, 70–71
philosophes on knowledge and, 87
primary vs. secondary, 32, 69–71, 72–73, 183
sensory intuitions, 77–78. *See also* intuition
Shapin, Steven, 232
skepticism, 69
    Catholic Church and, 18, 69
    defined, 64
    Descartes and, 30, 32
    empiricism and, 71–72, 73, 74–75
    Hume and, 73, 75–76, 136
    Kant and, 75–76, 133
    Locke and, 71, 73
skeptics, 25, 69, 74–75, 87–88, 136. *See also*
    specific philosophers
Smith, Adam, 92
social atomism, 179–80, 243–44
social construction of scientific knowledge
    (SCSK), 231–32, 238, 239, 243, 248
    books by advocates of, 186–87, 233–34 (*see*
        also Laboratory Life; Pickering, Andrew)
    David Bloor and, 232–33
    and the history of science, 136, 186–87, 239
    objectivity and, 227, 239, 245
    opponents of, 197–98, 238, 240–41, 245, 247
    and science as a belief system, 234–35
    "strong program," 231–32, 249–50
social construction of technologies
    (SCT), 252–53
Social Construction of What?, The
    (Hacking), 239
social constructionism, 186–87, 239, 243, 252–53
social physics/sociology, 89, 121, 167
social science, 84–85, 88–89, 94, 99, 121–22.
    See also specific social sciences
"socially constructed," origin of the term,
    186–87, 227
sociology. *See* social physics/sociology
Socrates, 14
    on knowledge, 8–9, 14, 179, 182
    on the mind, 8–9, 182
    nature and, 8–9, 13–14
    on speech vs. writing, 224–25
Sokal, Alan, 247–48
    postmodernism and, 247–48, 249–50
    writings, 248, 249

Sokal's hoax, 248–50
Sophist, The (Plato), 7
sophists, 7, 12–13
    Aristotle and, 12–13, 222–23
    Giants and, 7, 12–13, 207–8
    Plato and, 7, 12–13, 222–23
space
    defined, 54–55
    Descartes on matter and, 106
    nature of, 54–55, 106
space and time, 59–60, 142
    Kant and, 77–78
    Newton and, 54–55, 59–61, 164
    relativity theories and, 54–55, 142, 164–65
space-time, 164, 165, 172
special theory of relativity, Einstein's,
    106, 147, 157, 164–65, 170, 172–73, 174–75. *See also* mass–energy
    equivalence; space and time; speed
    of light
    Newtonian physics and, 54–55, 117–18, 164, 206–7
    ontology and, 161, 207
speed of light, 125, 161–62, 164. *See also*
    special theory of relativity
Spinoza, Baruch, 67, 68–69, 130, 179
    Descartes, Cartesianism, and, 33, 35, 67–68, 130, 179
    writings, 67, 68
Stallo, J. B., 128–29
stars, 29, 37, 38–39, 131–32
stellar parallax, 29
Stevens, Wallace, 183–84, 209
Stevin, Simon, 18–19, 41
strong sense of knowledge. *See* knowledge in
    the strong sense
Structure of Scientific Revolutions, The (SSR)
    (Kuhn), 61–62, 188, 194, 204, 206–8, 218, 242
    characterization of, 218
    description of, 194
    in historical context, 197–98, 199, 207–8, 218, 226–27, 235–37, 242, 248
    impact of, 199, 207–8, 235, 237–38, 242, 252
    Israel Scheffler on, 209, 211
    Paul Feyerabend and, 211, 212, 218
    Quine, Carl Hempel, and, 235–37
    responses to, 194, 199, 207–8, 209, 211, 212, 218
    theses, 194, 204, 207–8, 209, 219–20, 231

sufficient reason. *See* Principle of
    Sufficient Reason
sun as center of universe. *See* heliocentrism
sunlight, homogeneity vs. heterogeneity
    of, 53, 66
suppositions, 59–60
Swift, Jonathan, 82
Synthesis, Newton's method of, 57, 58
synthetic a priori truths, 76, 111, 118–19. *See
    also* analytic–synthetic distinction

tacit knowledge, 185–86, 206
Tarski, Albert, 210, 217
Tartaglia, 23
taxonomies, 99, 175, 204. *See also*
    classification schemes
telescopes, 39–40, 42–43, 94–95, 176–77
    Galileo and, 36, 38, 39, 40, 42–43, 47–48
    invention of, 36, 38
testability, 158–59
Theaetetus (Plato), 8–9
theological stage (Comte), 121
theology. *See also* religion
    vs. philosophy, 22–23
theoretical pluralism, 211–12
theories, 63. *See also* scientific theories
    mapping, theories, and the world, 238–39
theory change, 101–2, 118, 128, 175–76,
    194, 205, 214–15, 237, 256–57. *See also
    specific topics*
    Kuhn on, 194–95, 212, 213, 214–15
theory neutrality, 59–60, 157
    of observations, 157–58
thermodynamics, 105–7, 122, 167
    Fourier and, 105–7, 145 (*see also* Fourier,
        Joseph: theory of heat flow)
Thomson, Benjamin, 98, 104–5
thought collectives, 188, 189
    defined, 189
    Kuhn and, 194
    Ludwik Fleck's, 188, 189, 202, 216, 225, 230
thought styles, 188–93
    changes in, 189–91, 192–93
    defined, 189
    examples of, 189–90
    Kuhn and, 190–91, 192–93, 194
    Ludwik Fleck on, 188–91, 192–93, 194,
        199, 225, 230
    nature of, 189–90
    paradigms and, 189–90, 199
    thought collectives and, 188, 190–91, 225

three stages, law of, 121
time. *See also* duration; space and time
    definition and nature of, 54–55
    reality and, 140
tough–tender dichotomy, 149–50
Tracy, Antoine Destutt de, 93–94, 127
tree of knowledge, d'Alembert's, 86–87,
    98, 121–22
truth
    definitions, 9–10, 36, 210
    Dewey on, 153
Turgot, Anne Robert Jacques (Baron de
    l'Aulne), 84, 89, 90, 91

uncertainty principle, 170–72. *See also*
    matrix mechanics
unconscious of science, 219–21
underdetermination (of theory by data), 124,
    159, 236–37
universities, 21–23, 244
Urban VIII, Pope, 47

value neutrality, 86–87, 226–27, 256–57
van Fraassen, Bas, 228–29
verisimilitude, 217
versions (conceptual frameworks), 240–41
Vesalius, Andreas, 18–19, 23
Vienna Circle, 153–54, 155, 157–58
    logical positivism and, 155, 157–58, 188–
        89, 215
Volta, Alessandro, 107, 200
Voltaire, François-Marie Arouet, 84–85
vortex-based theory of gravity
    (Descartes), 51, 95

wave equation, 166, 170, 172–73
wave function, 172. *See also* Schrödinger
    equation
wave function collapse, 172–73, 224–25
wave mechanics, 166, 172, 173
wave–particle duality, 101–2, 162–63, 164,
    165–66, 171
wave theory of light, 59–60, 97, 101–2
    vs. particle theory, 59, 97, 101–2, 196,
        200–1, 255 (*see also* wave–particle
        duality)
weak sense of knowledge. *See* knowledge in
    the weak sense
Weber, Joseph, 232–33
Weber, Max, 186–87
Westfall, Richard S., 52

Wheeler, John, 172, 173–74
Whewell, William, 117–19, 122
  Bacon and, 116, 117, 119
  Charles Peirce and, 117, 148–49, 168–69,
    194, 206
  deduction and, 116, 117–19, 148–49,
    213, 236
  Fundamental Ideas and, 116, 117–19, 126,
    148–49, 213
  history of science and, 116, 119, 126, 213
  on induction, 116, 117–20, 148–49, 168–
    69, 206
  John Herschel and, 116, 119
  John Stuart Mill and, 119–20
  overview, 116
  Pierre Duhem and, 126, 213
  on predictive success, 118–19
  on reasoning, 116, 118
  religion and, 119
  writings, 116, 118, 119–20
Whig history of science, 196–97
Wigner, Eugene, 169–70
will, 42, 132–33, 139
William of Ockham, 23
Wittgenstein, Ludwig, 155–56, 178
  atomism and, 156
  Bertrand Russell and, 154–57
  logical positivism and, 154–55, 156–57
  Tractatus Logico-Philosophicus, 155–56, 178
Wolpoff, Milford, 202
Woolgar, Steve, 229, 230–31. *See also*
  Laboratory Life
world, defining, 4
Wundt, Wilhelm, 127, 182–83

zoology, 98, 99